MACHINE ANALYSIS WITH COMPUTER APPLICATIONS FOR MECHANICAL ENGINEERS

MACHINE ANALYSIS WITH COMPUTER APPLICATIONS FOR MECHANICAL ENGINEERS

James Doane

Frontier-Kemper Constructors, Indiana, USA

This edition first published 2016
© 2016 John Wiley & Sons Ltd

Registered office
John Wiley & Sons Ltd, The Atrium, Southern Gate, Chichester, West Sussex, PO19 8SQ, United Kingdom

For details of our global editorial offices, for customer services and for information about how to apply for permission to reuse the copyright material in this book please see our Web site at www.wiley.com.

Library of Congress Cataloging-in-Publication Data is available.

A catalog record for this book is available from the British Library.

ISBN: 978-1-118-54134-0

Set in 10/12 pt TimesLTStd-Roman by Thomson Digital, Noida, India

Printed in Singapore by C.O.S. Printers Pte Ltd

1   2016

Contents

Preface

The material presented in this book evolved from supplementary material prepared for teaching a course in machine analysis. One major thing I have learned from teaching Machine Analysis, and also from discussing with others who have taught the course, is that students do not understand the purpose of machine analysis and lack understanding of the big picture. This book focuses on learning the big picture with the use of computer methods. As an example, students typically learn kinematic analysis of linkage systems by calculating velocity and acceleration of a linkage mechanism in one "freeze frame" position. As a result, the students get answers in the form of vectors, which have no real meaning to them because it is not a complete solution. In order to have a complete understanding of the concept, students need to calculate velocity and acceleration for one complete cycle of motion of the mechanism. Although it is far too tedious to calculate by hand, students can develop computer code that will analyze the mechanism for its complete cycle of motion. From this broad analysis, students can now plot acceleration curves for the motion of the mechanisms and learn in a more visual manner. Because most students are visual learners, this text focuses on teaching the material in this more complete and visual fashion. Students will be taught how to build computer code throughout the book, and in the end they will be able to develop a tool to do a much more thorough analysis of mechanisms. This approach will make the learning process more effective, and it will also serve as a useful tool to use as practicing engineers. Some of the solution methods discussed in this book stem from graphical methods that may seem antiquated. Though some of these methods may be outdated, they present the material in a very visual way to help students understand the more complex analytical solution methods.

This book is primarily intended for use as an undergraduate text for a mechanical engineering course in Analysis of Machines. However, the book is presented in such a way that it would also be very beneficial to practicing engineers. Although Chapter 2 of the book reviews essential kinematic concepts, it is assumed that students possess the prerequisite materials from engineering statics and dynamics. Calculus will be used periodically throughout the text; therefore, students should also have a good working knowledge of derivatives and integrals. Basic knowledge of differential equations will also be beneficial for the topics of vibration. Some materials covered are more advanced and could be included in undergraduate courses if desired, but they are best suited for graduate studies.

Acknowledgments

This book would not have been possible without the support of my colleagues, students, friends, and family. The institutional support of the University of Evansville was crucial for its completion. The University's Department of Mechanical and Civil Engineering offered support by providing me time to work on the text as well as students to aid in the process. I am especially thankful to Dr. Phil Gerhart, Dr. Douglas Stamps, and Dr. Jim Allen for their advice and for sharing their experience of writing their own books.

Although many students aided in the process of writing this book, I would especially like to acknowledge Darwin Cordovilla for his assistance in the development of the solutions manual.

Finally, I would like to thank my family. I truly appreciate the understanding of my daughters, Cory and Rebecca, during the long and demanding process of writing a textbook. I am also extremely grateful for the support and help from my wife Alice, without whom I could not have completed this book.

About the companion website

This book is accompanied by a companion website:

www.wiley.com/go/doane0215

This website includes:

- Solution Manuals for every chapter (as PowerPoint slides)

1

Introductory Concepts

1.1 Introduction to Machines

1.1.1 Brief History of Machines

In our modern world, we are surrounded by machines, and they have become an integral part of our daily lives. We know that high-tech machines of today were not always in existence, but it is hard to imagine what it would be like without them because our lives would be drastically different. It is difficult to say when the first machine was developed. Knowledge of very early machines comes from archeology, but this work is difficult. The difficulty is partly due to the fact that it is rare to discover intact machines, but it is more common to discover early machine components. Over time machines developed from very crude to extremely elaborate, and often that development moved in parallel with the development of human culture.

In the very early years, machine development was difficult and slow. Sometimes advancements in technology were driven by military needs, and other times advancements were required for survival. Primitive man devised simple tools made of wood, stone, or bone that were essential for survival. Machines were developed to produce fire, and simple mechanisms were developed to trap animals for food. Numerous machines from different cultures were also developed to extract water. Archimedes (287–212 BC) developed a method for water extraction using a spiral screw, such as the one illustrated in Figure 1.1. Machines such as levers and inclined planes were used by the Egyptians to build numerous monuments such as the pyramids.

One important class of mechanisms developed through the ages is those used to measure time. Many machines were devised for measurement of the phases of the moon, but one particularly interesting device was discovered in a shipwreck in 1900. The Antikythera mechanism, which is schematically shown in Figure 1.2, is estimated to have been fabricated around 100 BC from a bronze alloy. This complex gear mechanism contains at least 30 gears and acts as an analog computer to calculate astronomical positions. The device likely was used to predict solar and lunar eclipses as well as display positions of the five known planets of the time.

A major contributor to machine inventions was Leonardo da Vinci (1452–1519). He recorded ideas and observations in thousands of pages of notebooks, mostly in the form of drawings. He was fascinated with nature and was way ahead of his time in understanding fluid flow and turbulence. In his study of human anatomy, he recognized mechanical function such as the joints acting as hinges. Leonardo used principles of engineering statics to analyze

Machine Analysis with Computer Applications for Mechanical Engineers, First Edition. James Doane.
© 2016 John Wiley & Sons, Ltd. Published 2016 by John Wiley & Sons, Ltd.
Companion Website: www.wiley.com/go/doane0215

Figure 1.1 Archimedes' screw. Source: Wikimedia [http://commons.wikimedia.org/wiki/File: Brockhaus_and_Efron_Encyclopedic_Dictionary_b3_020-4.jpg]

the mechanics of biomaterials such as bones and muscle. His knowledge of mechanics was also applied to machines, and Leonardo is believed to be responsible for dissecting machines into basic machine elements. Among his many machine designs was a water-powered milling machine that utilized primitive gears to transmit motion. He also developed concepts for converting rotary crank motion into reciprocating motion. He designed hoist systems to lift heavy loads using gears. He recognized the high level of friction in machines and designed multiple devices, such as bearings, to reduce friction. Leonardo's interest in anatomy and mechanics also led to his work to design a flying machine, as shown in Figure 1.3.

Galileo (1564–1642) investigated the behavior of pendulums, and he discovered that the period of pendulum is not affected by amplitude of motion. Christiaan Huygens (1629–1695)

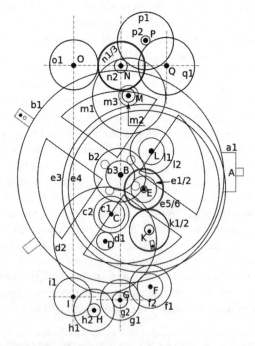

Figure 1.2 Schematic of the Antikythera mechanism. Source: Wikimedia [http://commons.wikimedia. org/wiki/File:Antikythera_mechanism_-_labelled.svg]

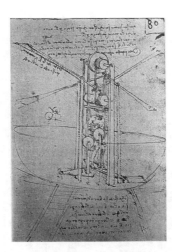

Figure 1.3 Sketch of flying machine by Leonardo da Vinci. Source: Wikimedia [http://commons.wikimedia.org/wiki/File:Leonardo_da_vinci,_Flying_machine.jpg]

was a Dutch scientist who worked in areas of mathematics, physics, astronomy, and horology (the science of measuring time). Huygens worked with clocks to make them more accurate, and he patented the first pendulum clock in 1656. Figure 1.4 shows a pendulum clock invented by Huygens and built around 1673.

Leonhard Euler (1707–1783) was a great mathematician of the eighteenth century. His contributions to mathematics and science were numerous and cover a breadth of topics. As an engineering student, you will see references to Euler in numerous classes. A very important contribution of Euler was the Euler–Bernoulli beam equation, which is extensively used to calculate deflection of beams. Euler introduced a rotating coordinate system critical for describing three-dimensional orientation of rigid bodies, which is vital to describing complex three-dimensional motion of mechanisms.

It is impossible to think of advancements in machines without discussing the development of steam engines. Steam engines, such as that shown in Figure 1.5, replaced the use of horses to generate power and allowed for operation of factories in cities.

James Watt (1736–1819), a Scottish engineer, experimented with steam and made improvements to the steam engine designed by Thomas Newcomen in the early 1700s. The Watt steam engine has several ingenious inventions that make it a vast improvement. Watt recognized that a great amount of energy was wasted in the Newcomen engine. While repairing a Newcomen engine, he realized that the cylinder (for the piston) was heated and cooled repeatedly. Watt thought that if the condensing step could be moved, the condenser could be kept cold at all times while the cylinder remained hot. One of Watt's inventions was to separate the condensing step to reduce wasted energy and it greatly improved efficiency. Watt developed a mechanism known as a straight line mechanism, which is a linkage mechanism that generates a straight line path to move pistons (see Section 1.5.5 for additional information on straight line mechanisms). Watt also utilized a governor mechanism as an early feedback system to regulate rotation speed. Lighter engines were of course later developed, such as Nikolaus Otto's four-stroke engine developed in the late 1800s. The sun gear mechanism used by Watt will be discussed in Chapter 8.

Gears are a vital part of many machines. Though crude gear designs allow for transfer of power, they are not well suited for higher speed operation. Robert Willis (1800–1875) made

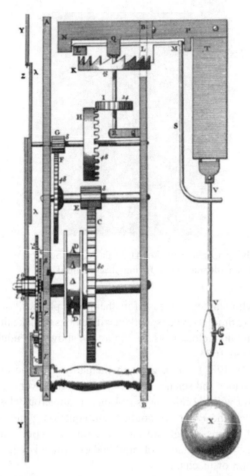

Figure 1.4 Huygens pendulum clock. Source: Wikimedia [http://commons.wikimedia.org/wiki/File: Huygens_clock.png]

significant contributions to the standardization and design of gears and gear teeth. Willis showed that involute curves (curves used for gear teeth – see Chapter 7 for details) allow for interaction of gears with different diameters without angular acceleration. Willis developed the use of a constant pressure angle to standardize gear manufacturing. A brief historical timeline of gear development is provided in Chapter 7.

German engineer Franz Reuleaux (1829–1905) is often noted as one of the greatest minds in machine theory of the nineteenth century and the father of kinematics. His extensive work in kinematics was published as a book in 1875, which was quickly translated into English as the title *The Kinematics of Machinery: Outlines of a Theory of Machines*. Both Willis and Reuleaux developed ideas that mechanisms are formed as kinematic chains, which can be analyzed by examining relative motion of element pairs. Reuleaux expanded on existing ideas of instant centers of rotation by calculation of centrodes, or paths of the instant center. Reuleaux developed ideas that mechanical motion was controlled by interactions and connections between the individual moving members of the machine.

Figure 1.5 Newcomen steam engine. Source: Wikimedia [http://commons.wikimedia.org/wiki/File: Newcomen_steam_engine_at_landgoed_groenedaal.jpg]

Ferdinand Freudenstein (1926–2006) is often referred to as the father of modern kinematics. He began making major contributions to machine analysis early in his career. The Freudenstein equation, which will be discussed and used in several sections relating to linkages, was actually developed in his Ph.D. dissertation. The equation is very useful in position analysis of linkage mechanisms as well as linkage design.

1.1.2 Why Study Machine Analysis?

It would be very rare to go through a day without the use of some type of machine. Today's machines come in many forms. Some machines are rather basic such as a bicycle or simple hand tools while others, such as cars and automated manufacturing equipment, can be very complex. Recent advancements in technology allow for machines to be automated and run at very high speeds. High-speed operations of machines offer many advantages but can add complications in design.

Mechanical engineers responsible for designing machines must have a strong understanding of machine kinematics and kinetics. Poor understanding of the kinematics and kinetics of machines can lead to unsatisfactory performance or even catastrophic failure of components. Acceleration analysis, as an example, must be performed for all portions of a machine's cycle to determine maximum values. Though this acceleration analysis is typically complicated, it is required to determine force values, which are then used to design machine elements based on allowable stress values or allowable deformations.

This text will examine the core subjects of kinematics and kinetics of machines. The primary focus will be to build a strong foundation of machine analysis; therefore, many advanced topics are outside the scope and will not be presented. Readers interested in exploring the more advanced topics, or more information about the core topics of this text, should review the bibliography sections at the end of each chapter for suggested resources.

1.1.3 Differences between Machine Analysis and Machine Design

It is fairly common in a 4-year mechanical engineering curriculum to take machine analysis and machine design as two separate courses. Both courses are important, but the content differs. It is somewhat common for students to confuse the two courses or be unclear why both courses are needed.

Machine analysis focuses on the kinematics and kinetics of mechanisms. Course material builds on concepts learned in engineering dynamics. The most common machine components covered in a machine analysis course include linkages, gears, and cams though others can be included. Machine analysis covers methods of designing the geometry of linkage mechanisms to perform specific tasks, as well as analyzing the kinematics of an existing linkage mechanism. Deflection of the machine members is often considered negligible, so they are commonly treated as rigid bodies. Analysis of gears focuses mostly on the interaction of teeth and behavior of gear trains. The focus on cams is developing the geometry of the cam to perform the desired motion. Though topics in machine analysis include forces, things such as deflection, stress, fatigue, and wear are not discussed.

The focus of machine design revolves more around designing machine elements for strength and rigidity. Much of the material covered in machine design will build on previous knowledge of mechanics of materials, such as combined loading conditions, failure criteria, curved beams, deflection of complex systems, and pressurized cylinders. Gears are studied to develop under-standing of contact stresses and bending stresses to avoid failure. Concepts of shaft design are covered, including stress concentrations, fatigue stress, and deflection. Other machine elements often discussed in machine design include bearings, clutches, brakes, fasteners, and springs.

Though the two topic areas are different, they work in parallel. An engineer's first focus will be the motion of the machine. The fundamental requirement is often focused around the idea of proper displacements. Once the displacement has been developed, the resulting acceleration can be determined. Using the accelerations, the study moves to kinetics to determine forces. Design work then moves to analysis of developed stresses and deformations. This design process is often iterative. Machine analysis is a phase of machine design. Therefore, one must often use knowledge of machine analysis and machine design through multiple iterations to develop the final design.

1.2 Units

1.2.1 Importance of Units

Engineering students get introduced and reintroduced to systems of units throughout their college lives. Nearly every engineering text, regardless of the subject, offers at least a short section devoted to units. However, most engineering students still get confused by the details of the different systems of units. That confusion, unfortunately, commonly continues past graduation and can cause serious (sometimes catastrophic) problems. In 1983, a Boeing 767 ran out of fuel at 41 000 feet because of an error when manually converting between kilograms and liters to determine the required amount of fuel. NASA lost a Mars orbiter due to a mismatch in unit systems. From these quick examples alone, you can determine that it is extremely important for engineering students to understand unit systems and unit conversions.

1.2.2 Unit Systems

Throughout this text, both the International System of units (SI from Systeme International) and the US customary unit system (inherited from the British Imperial System) will be used.

Table 1.1 Summary of unit systems

Quantity	SI unit (symbol)	US customary unit (symbol)
Time	Second (s)	Second (s)
Length	Meter (m)	Foot (ft)
Mass	Kilogram (kg)	Slug
Force	Newton (N)	Pound (lb)

Where applicable, figures will give dimensions in both sets of units. Example problems will include a sample from each system, and end-of-chapter problems will do the same. It is recommended to work problems from each category to become proficient at both. Regardless of individual preference, a mechanical engineering student needs to become fluent in both systems. Typically, a person will have better intuitive sense for one system compared with the other. As an example, a person raised using the US units will have a good sense for a distance in miles but may not even approximately determine a distance in kilometers. A general summary of the units is given in Table 1.1.

The base units for the SI system are time, mass, and length. Units for force are then defined using Newton's second law. A newton is the force required to give a one kilogram mass an acceleration of one meter per second squared.

$$1\,N = 1\,kg \cdot 1\,m/s^2 \tag{1.1}$$

Therefore, a newton will have units of $kg \cdot m/s^2$. In US customary units, the base units are length, time, and force. A slug (32.174 pounds mass) will accelerate at a rate of one foot per second squared if it has an applied force of one pound.

$$1\,lb = 1\,slug \cdot 1\,ft/s^2 \tag{1.2}$$

Manipulation of the equation will show that a slug will have units of $lb \cdot s^2/ft$. In some cases, you may see yet another unit for mass known as a blob. The blob is simply the inch version of a slug: $lb \cdot s^2/in$. Therefore, one blob is equal to 12 slugs. It is often convenient to express values in the SI system using prefixes. The common prefixes are given in Table 1.2.

Table 1.2 SI unit prefixes

Amount	Multiple	Prefix	Symbol
1 000 000 000 000	10^{12}	Tera	T
1 000 000 000	10^9	Giga	G
1 000 000	10^6	Mega	M
1 000	10^3	Kilo	k
100	10^2	Hecto	h
10	10	Deka	da
0.1	10^{-1}	Deci	d
0.01	10^{-2}	Centi	c
0.001	10^{-3}	Milli	m
0.000 001	10^{-6}	Micro	μ
0.000 000 001	10^{-9}	Nano	n
0.000 000 000 001	10^{-12}	Pico	p

1.2.3 Units of Angular Motion

For some students, a common source of confusion comes from units for angular dimensions. The common units of angular measure used in mathematics are the degree and the radian. A degree (°) is equal to 1/360 of a full revolution. In other words, there are 360° in a complete revolution. To define a radian, consider the concept of arc length. In a circle of radius r, if θ is expressed in radians, the arc length is defined by $s = r\theta$. Therefore, a radian is the central angle that will cause the arc length to equal the radius. There are 2π radians in a complete revolution.

$$1 \text{ revolution} = 2\pi \text{ radians} = 360° \tag{1.3}$$

It is also important to note that radians are a ratio. By simply rearranging the arc length equation, radians will be in terms of arc length over radius, both of which are length terms.

It is fairly common for students to have a better understanding of degrees, which leads to the tendency to use degrees as a default. However, a mistake of using the incorrect unit can cause very large error because a radian is considerably larger than a degree (1 radian is approximately 57.296 degrees). Much of the material in machine analysis will rely on a complete under-standing of when to use degrees versus when to use radians. In general, radians have geometrical meaning (such as circumference of a circle) and are far more convenient to use in mathematics. Degrees are rather artificial and do not directly tie into geometry. The idea to use 360 increments seems rather arbitrary, but it actually dates back to the Babylonians who used a sexagesimal system, which is a base 60 system passed down to the Babylonians by the ancient Sumerians. The sexagesimal system is also still used today to measure time (60 minutes in an hour and 60 seconds in a minute).

1.2.4 Force and Mass

Another common source of confusion comes from trying to distinguish between mass and force (or weight), especially in the English system of units. The terms mass and weight are often misused, but they are not the same. The basic relationship between mass and force was discovered by Isaac Newton and is known as Newton's second law given in Equation 1.4.

$$F = ma \tag{1.4}$$

Using SI units of kilogram for mass and meter per second squared for acceleration, the units for force will be $kg \cdot m/s^2$, which was previously defined as a newton. For US customary units, force would be defined as pounds (or more appropriately pound force that has the symbol lb_f) and the acceleration would have units of feet per second squared. This would cause the units of mass to be $lb_f \cdot s^2/ft$, which was previously defined as a slug.

The English Engineering system uses pound mass (lb_m) as the unit of mass. The relationship between pound force and pound mass is defined using

$$1 \text{ lb}_f = 32.17 \frac{lb_m \cdot ft}{s^2} \tag{1.5}$$

Using our definition of a slug, we can also write

$$1 \text{ slug} = 32.17 \text{ lb}_m \tag{1.6}$$

Weight is a force caused by gravity acting on a mass. Newton's second law can be transformed to give the relationship between weight and mass:

$$W = mg \tag{1.7}$$

where g is the acceleration due to gravity. In SI units, the acceleration due to gravity is 9.81 m/s^2 and in US customary units it is 32.17 ft/s^2.

It is sometimes convenient to use a proportionality constant g_c and rewrite Newton's second law as

$$F = \frac{ma}{g_c} \tag{1.8}$$

where

$$g_c = 1 \frac{\text{kg} \cdot \text{m/s}^2}{\text{N}} \quad \text{(SI units)} \tag{1.9a}$$

$$g_c = 1 \frac{\text{slug} \cdot \text{ft/s}^2}{\text{lb}_f} \quad \text{(US customary units)} \tag{1.9b}$$

$$g_c = 32.17 \frac{\text{lb}_m \cdot \text{ft/s}^2}{\text{lb}_f} \quad \text{(English Engineering units)} \tag{1.9c}$$

Note that it is not required to use the proportionality constant as long as you are thorough in keeping track of all units. The usefulness of the proportionality constant will be illustrated in a few quick examples.

Example Problem 1.1

What is the weight in pound force of an object that has a mass of 75 pounds mass?

Solution: For this problem, we will use the proportionality constant for English Engineering units. The weight is calculated using

$$W = \frac{ma}{g_c} = \frac{75 \text{ lb}_m \cdot 32.17 \frac{\text{ft}}{\text{s}^2}}{32.17 \frac{\text{lb}_m \cdot \text{ft/s}^2}{\text{lb}_f}}$$

Answer: $W = 75 \text{ lb}_f$

Example Problem 1.2

What is the weight in newtons of an object that has a mass of 65 kilograms?

Solution: For this problem, we will use the proportionality constant for SI units. The weight is calculated using

$$W = \frac{ma}{g_c} = \frac{65 \text{ kg} \cdot 9.81 \frac{\text{m}}{\text{s}^2}}{1 \frac{\text{kg} \cdot \text{m/s}^2}{\text{N}}}$$

Answer: $W = 637.65 \text{ N}$

Example Problem 1.3

What is the weight in pound force of an object that has a mass of 4 slugs?

Solution: For this problem, we will use the proportionality constant for US customary units. The weight is calculated using

$$W = \frac{ma}{g_c} = \frac{4\ \text{slug} \cdot 32.17\ \dfrac{\text{ft}}{\text{s}^2}}{1\ \dfrac{\text{slug} \cdot \text{ft}/\text{s}^2}{\text{lb}_f}}$$

Answer: $W = 128.68\ \text{lb}_f$

1.3   Machines and Mechanisms

1.3.1 Machine versus Mechanism

Before we jump into the concepts of machine analysis, we must first understand the idea of machines versus mechanisms. You may think that there is no real difference between a machine and a mechanism, and you may use the two terms interchangeably. In fact, it can be difficult to properly define a machine or mechanism because there is not really a clear division between the two. A general definition of a mechanism is that it is a fundamental device (or assembly of parts) to produce, transform, or control motion. For example, a mechanism can transform rotary motion to linear motion. Mechanisms will typically develop low forces. A machine can be thought of as a combination or assembly of mechanisms to do work, provide force, or transmit power. Machines have the primary purpose of completing work. A milling machine, as an example, is a manufacturing tool that uses a rotating cutter to remove material. The machine does work, and must provide large amounts of power to cut high-strength alloys. There are numerous mechanisms, or machine elements, within the milling machine. Lead screws, as an example, are machine elements used to transmit rotary motion to linear motion to move the table of the milling machine. The rotary cutter is powered by belts or gears, which allow for variations in operating speed.

1.3.2 Simple Machines

You most likely have been introduced to the simple machines: lever, wheel, pulley, inclined plane, and screw. The only reason to briefly discuss some of them here is to start the foundation of machines and mechanisms. A lot of more complex machines build off these fundamental simple machines. Probably, the most basic machine of all is the lever. A lever is simply a rigid member that rotates about a fulcrum (pivot point) to transmit force to another point. There are three classes of levers. First class levers, as shown in Figure 1.6a, have the fulcrum located between the applied force and the load. They are typically used to provide a mechanical advantage. Pulling a nail with a hammer is an example of first class levers. Second class levers, as shown in Figure 1.6b, have the load located between the fulcrum and the applied force. These levers are again used to provide a mechanical advantage. Every time you use a wheelbarrow, you are taking advantage of a second class lever. If the force is applied between the fulcrum and the load, as shown in Figure 1.6c, it is a third class lever. Third class levers actually lose the mechanical advantage, but they allow for large movement at the load. Third class levers occur frequently in the human body to provide large range of motion.

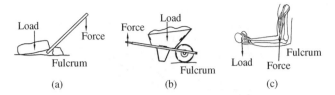

Figure 1.6   Three classes of levers: (a) first class; (b) second class; (c) third class

Though levers can give large mechanical advantage, compound levers can be used to generate the same mechanical advantage in a more compact design. The benefits of compound levers will be illustrated by comparing a simple lever arrangement in Example Problem 1.4 with the modified compound lever arrangement in Example Problem 1.5.

Another simple machine is the inclined plane, as shown in Figure 1.7a. The force required to push the body up the inclined plane will depend on the slope of the inclined plane. The portion of the total load multiplied by the ratio of rise to length of the slope will give the force required. As an example, if the slope length is five times the rise, the force needed will be one-fifth the load. One variation of an inclined plane is a wedge. Wedges do their job by moving, unlike stationary inclined planes. Chisels and hatchets are other common examples of wedges. It will be shown later in this chapter that cam mechanisms utilize the principles of inclined planes and wedges.

If we take the inclined plane and wrap it around a cylinder, we get the basic principle of a screw. There are a wide variety of uses for screw mechanisms. Archimedes used screws to raise water, and screw feeders are still commonly used in material handling. In plastics manufacturing, injection molding machines use a large screw to feed the plastic pellets to the dies. Screw mechanisms are used in tools such as clamps, drills, and presses. A common application in machine analysis is a worm gear, as illustrated in Figure 1.7b. The worm has an inclined plane wrapped in the form of a helix. As the worm spins about its central axis, the mating worm gear turns.

1.3.3 Static Machine Analysis

This text will obviously focus on machines in motion. However, before focusing on the kinematics and kinetics of machines in motion, let us look at how to analyze static forces in machines. The goal is to calculate output forces based on a given set of input forces. Static force analysis is presented here as a brief review and as a means of preparation for dynamic force analysis. It will be seen in Chapter 11 that dynamic force analysis builds off of the concepts of static force analysis. For complex machines, it is necessary to disassemble the machine and create multiple free body diagrams. Because this method is likely review from statics,

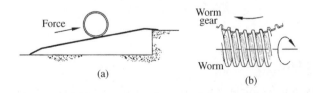

Figure 1.7   (a) Inclined plane. (b) Example of helical inclined plane

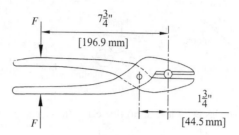

Figure 1.8

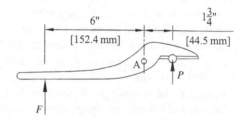

Figure 1.9

the process will be demonstrated in examples. Interested readers can reference engineering statics textbooks for more examples.

Example Problem 1.4

Force is applied to the handles of cutting shears in the location shown in Figure 1.8. Determine the magnitude of cutting force.

Solution: To determine the cutting force, we need to isolate one portion of the cutting shears and draw a free body diagram as shown in Figure 1.9.
Summation of moments about point A gives

$$\sum M_A = 0$$

$$F(6) - P(1.75) = 0$$

$$P = \frac{6}{1.75} F$$

Answer: $P = 3.43F$

Example Problem 1.5

The cutting shears from Example Problem 1.4 are modified using compound levers as shown in Figure 1.10. Determine the magnitude of the cutting force for the applied force shown.

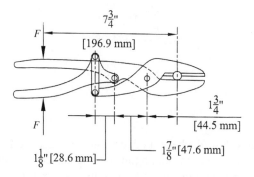

Figure 1.10

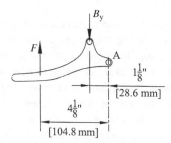

Figure 1.11

Solution: We again separate the mechanism and construct free body diagrams of individual components. Starting with the lower handle shown in Figure 1.11, the summation of moments gives

$$\sum M_A = 0$$

$$F(4.125) - B_y(1.125) = 0$$

$$B_y = \frac{4.125}{1.125} F = 3.67F$$

The force B_y is directed through the two-force member to the upper cutter. From the free body diagram shown in Figure 1.12,

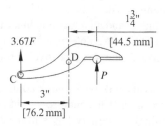

Figure 1.12

$$\sum M_\mathrm{D} = 0$$

$$3.67F(3) - P(1.75) = 0$$

$$P = \frac{3.67F(3)}{1.75}$$

Answer: $P = 6.29F$

The previous examples show the benefit of compound linkage mechanisms. Notice that the distance from the applied force to the cutting location remained the same for both examples. The modification to a compound linkage nearly doubled the cutting force.

1.3.4 Other Types of Machines

Obviously, the machines of today utilize more than the simple machines. Many different types of more complex machines exist, but this text will focus on only a few common machine elements. Sections to follow will briefly introduce the basic types of mechanisms covered in this text. Chapters to follow will provide further details needed for analysis and design of such mechanisms. The primary types of mechanisms covered are linkages, gears, and cams. While reading the sections to follow, try to think of actual examples of machines that use these types of mechanisms. The better you can understand the basic uses of these mechanisms, the better start you will have to being able to analyze them in future chapters.

1.4 Linkage Mechanisms

1.4.1 Introduction to Linkage Mechanisms

The first category of mechanism we will examine is linkage mechanisms. Linkage mechanisms will be introduced in this chapter, but Chapters 3–6 focus on the details of linkage mechanisms. In some forms, a linkage mechanism is a set of connected levers (or compound levers) used to provide a specific motion. A link is simply an individual rigid body, which is then interconnected in pairs to form a linkage mechanism. A joint is a point where pairs of links are connected. The complete assembly of links is known as a linkage mechanism or kinematic chain. This text will focus primarily on planar linkage mechanisms, which are mechanisms in which all links in the system move in parallel planes. Another classification would be spatial mechanisms, where the links are not all in parallel planes.

1.4.2 Types of Links

Links are numbered sequentially beginning with one for the stationary link, which typically represents the frame of the mechanism. The stationary link is commonly called the ground link. The driving link is numbered as link 2, and all remaining links are numbered in order. Points of rest are designated with the letter O. For example, in a four-bar linkage mechanism O_2 and O_4 are points of attachment for links 2 and 4, respectively. Link numbering is illustrated in Figure 1.13.

There are different ways to classify link types, but a common method is to classify by the number of connection points (or nodes) it contains. A link with two connection points is referred to as a binary link. Similarly, a ternary link will have three connection points and a quaternary link will have four. Figure 1.14 shows examples of each link type described.

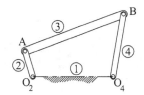

Figure 1.13   Numbering system for a linkage mechanism

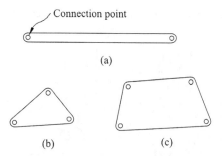

Figure 1.14   Link classification by number of nodes: (a) binary link; (b) ternary link; (c) quaternary link

Links can also be classified by their actual function. Examples of this type of classification will be presented in Section 1.5 during the discussion of common types of linkage mechanisms.

1.4.3 Types of Joints

Now that we have developed a general understanding of links, we can move on to types of joints, which are connections between links. Joints serve the function of controlling relative motion of connected links. Typically, joints, which are also referred to as kinematic pairs, are more confusing to students due to the layers of terminology. Classification of joints is also more confusing than that of links. Joints can be classified by the type of contact, the number of links connected, the method of maintaining joint contact, and the general motion of the joint. One thing to note is that there are several types of joints, but only a few of those are applicable in planar mechanisms. Since this text will focus extensively on planar mechanisms, this section will focus more on the joint types that relate to planar motion. This does not indicate that the others are not important. The other types will be only briefly discussed here (in an attempt to reduce confusion).

Joint types are separated into two major categories known as lower pairs and higher pairs. Lower pairs are joints with surface contact, and higher pairs are joints with point or line contact. Of the six lower pairs, only two apply to planar mechanisms. Figure 1.15 illustrates the two lower pairs significant for planar mechanisms. The first is known as a revolute joint, which is commonly designated by the symbol R. A revolute joint, as shown in Figure 1.15a, can be thought of as a basic hinge joint or pin joint. The second lower pair that applies to planar mechanisms is the prismatic pair, designated by the symbol P. A prismatic pair, as shown in Figure 1.15b, is a sliding joint constrained to move in one linear direction without rotation.

The remaining four lower pairs, which are illustrated in Figure 1.16, do not apply to planar mechanisms due to the fact that the resulting motion is three dimensional. The helical joint

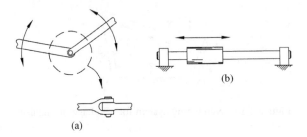

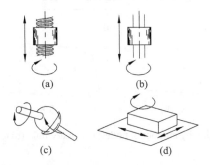

Figure 1.15   Lower pairs usable in planar mechanisms: (a) revolute joint; (b) prismatic joint

Figure 1.16   Lower pairs for spatial mechanisms: (a) helical joint; (b) cylindrical joint; (c) spherical joint; (d) planar joint

shown in Figure 1.16a is a linear screw allowing rotation and linear translation (yet the two motions are constrained by the pitch of the screw). A cylindrical joint, as shown in Figure 1.16b, is allowed to translate in a linear direction and rotate about its axis. Figure 1.16c shows a spherical joint, which is a ball and socket joint. A planar joint, as illustrated in Figure 1.16d, is like a block moving freely on a plane and allows motion in the Cartesian x–y plane and rotation about the z-axis.

The lower pairs for planar mechanisms shown in Figure 1.15 are both one-degree-of-freedom joints. Revolute joints only allow one angular rotation and prismatic joints only allow translation in one axial direction. Although the helical joint is not a joint used in planar mechanisms, it also is a one-degree-of-freedom joint because the angular motion and translation are constrained by the pitch of the helix. The cylindrical joint is a two-degree-of-freedom joint allowing independent rotation and translation. The spherical joint and planar joint are both three-degree-of-freedom joints.

1.5   Common Types of Linkage Mechanisms

The number of possible arrangements of links in a linkage mechanism is only limited by the imagination. However, many common applications can be achieved with basic four-bar linkage mechanism (the four bars are the fixed ground link and three moving links). More complicated linkage mechanisms are commonly built using a four-bar mechanism to drive others. Some mechanisms that have a physical form different from a typical four-bar mechanism can be modeled as an equivalent four-bar mechanism. Because of their frequent use and wide variety of applications, discussion of linkage mechanisms in this text will focus heavily on four-bar

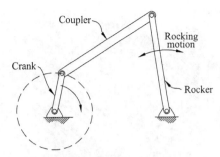

Figure 1.17   Crank–rocker mechanism

mechanisms. Some special configurations of four-bar mechanisms have been given names because they occur so frequently. Some of those common configurations will be briefly defined here, though more detail will be given in future chapters.

1.5.1   Crank–Rocker Mechanisms

The general four-bar mechanism can have many configurations. However, different names exist for the configurations based on the range of motion of links 2 and 4. Chapter 3 will further examine the different configurations. The configuration we will examine in this section is known as a crank–rocker mechanism, which is shown in Figure 1.17. The title of this mechanism comes from the fact that the driving link (link 2) is called a crank and the output link (link 4) is called a rocker. Link 3 is a floating link that connects the driver to the output and is commonly called the coupler. The term crank signifies that the driving link will complete a full revolution relative to the ground link. Typically, the crank will move in a continuous rotating motion at a constant rotational speed. The term rocker signifies that the output link oscillates in a rocking motion and is unable to complete a full revolution.

Crank–rocker mechanisms have many common applications. One very common application would be the mechanism used to move windshield wipers. The wipers are driven by a motor that causes the crank to continually rotate. The blade then moves with the output link in a rocking motion.

1.5.2   Slider–Crank Mechanisms

The next category of linkage mechanisms discussed is a slider–crank mechanism, which is shown in Figure 1.18. A slider–crank mechanism is a special case of a four-bar mechanism. The input link (link 2) is again a crank and moves in a continuous rotation. The output is now a sliding block, called a slider or piston, and is constrained to oscillate in a pure straight line

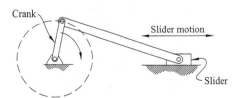

Figure 1.18   Slider–crank mechanism

Figure 1.19 Chevrolet V8 engine showing slider–crank mechanism. Reproduced from Mabie and Reinholtz, Mechanisms and Dynamics of Machinery, 4th edition, John Wiley & Sons. © 1987

motion. The link connecting the crank to the slider (link 3) is commonly known as the connecting rod but the name often changes depending on the application.

Common applications of slider–crank mechanisms include reciprocating engines and compressors. Figure 1.19 shows slider–crank mechanisms in a V8 engine.

1.5.3 Toggle Mechanisms

Toggle mechanisms generate large forces through a short distance and are commonly used in clamps and crushers. The toggle mechanism shown in Figure 1.20a is a multilink mechanism

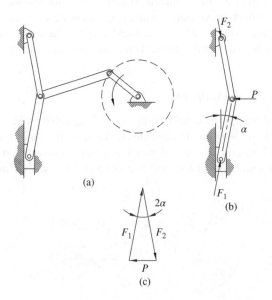

Figure 1.20 (a) Toggle mechanism. (b) Forces. (c) Force polygon

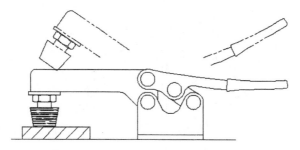

Figure 1.21   Toggle clamp

combining the four-bar crank–rocker mechanism and the slider–crank mechanism. The slider–crank portion is shown separated in Figure 1.20b to better illustrate the force magnification. A force P applied as shown will cause reaction forces F_1 and F_2. Figure 1.20c shows a force polygon.

From the force polygon, it can be seen that

$$\sin \alpha = \frac{P/2}{F_1}$$

$$F_1 = \frac{P}{2 \sin \alpha} \qquad (1.10)$$

From Equation 1.10, it can be seen that as α approaches zero the force F_1 approaches infinity. Therefore, toggle mechanisms can generate a large output force for a relatively small input force. A very common application of a toggle mechanism is a toggle clamp, such as the one shown in Figure 1.21.

1.5.4 Quick Return Mechanisms

Quick return mechanisms come in many forms, but all have the same fundamental goal. Commonly, mechanisms are developed to do work, and the cycle of motion can be separated into a working stroke and a return stroke. Mechanisms can be designed such that the time required for the working stroke equals that of the return stroke. Quick return mechanisms, as the name implies, are mechanisms that have a return stroke that is faster (takes less time) than the working stroke. The faster return occurs due to the geometry of the mechanism and occurs even with a constant input speed. The time ratio is the ratio of input displacement during the work stroke to the input displacement during the return stroke.

Several mechanism types can have a quick return. A slight adjustment to the slider–crank mechanism shown in Figure 1.18 such that the crank center of rotation is offset from the centerline of the slider (known as an offset slider–crank mechanism) will make a quick return mechanism. Offset slider–crank mechanisms and other types of quick return mechanisms will be explored further in Chapter 5.

1.5.5 Straight Line Mechanisms

The next class of linkage mechanism is in which a point on the mechanism generates a straight line motion, which sounds like a very simple task. A mechanism that converts rotary motion

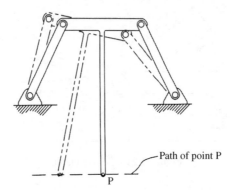

Figure 1.22 Straight line mechanism

to a straight line motion (over a limited interval) is known as a straight line mechanism. Developing a machine to generate a straight line is rather complex, and they developed throughout history. In early development of the steam engine, it was required for parts, such as piston rods, to move in a straight line. Many configurations have been developed and most only generate an approximate straight line due to the fact that exact straight line mechanisms are very complicated to design. Figure 1.22 shows one example of a straight line mechanism in which point P moves along the straight line path shown.

1.5.6 Scotch Yoke Mechanism

A scotch yoke mechanism, as shown in Figure 1.23, has the same fundamental objective of the slider–crank mechanism discussed in Section 1.5.2. The scotch yoke is a basic mechanism that converts rotary motion into reciprocating linear motion using a pin in a moving slot.

   An application for this mechanism is to open and close valves. The crank is connected to the slider by the pin, and as the crank turns the pin it moves the slider. Though the goal of the slider–crank and scotch yoke mechanisms is the same, their performance differs slightly. The scotch yoke mechanism will generate perfect simple harmonic motion (sine wave), whereas the slider–crank mechanism will produce a slightly distorted sine wave. Because the output is simple harmonic motion, the scotch yoke mechanism is commonly used in testing equipment to simulate simple harmonic vibration. A major disadvantage to a scotch yoke mechanism is the high contact pressure and wear.

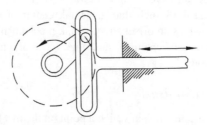

Figure 1.23 Scotch yoke mechanism

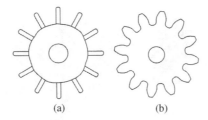

(a)   (b)

Figure 1.24   (a) Early gears with peg teeth. (b) Modern gears with involute teeth

1.6 Gears

1.6.1 Introduction to Gears

From its first discovery, very primitive man made many uses of the wheel. Eventually, wheels developed to more efficient forms including a solid wheel with a separate axle. Many other ideas, such as gears, sparked from the basic concepts of a wheel and axle. In very early development of gears, they were simply wheels with peg teeth, such as those shown in Figure 1.24a. Though the peg teeth served a useful purpose, these early gears did not operate smoothly due to nonuniform velocity transmission. Other tooth shapes were eventually developed to improve the performance of gears. Modern gears have a gear tooth shape of a complex involute curve, such as the one shown in Figure 1.24b. Chapter 7 will cover development of the involute curve as well as the interaction between mating gears.

The most basic definition of gears would be that they are toothed wheels used to transmit power. However, gears can accomplish much more. Gear sets have numerous applications, including automotive, aircraft, agricultural equipment, marine equipment, and manufacturing. Gear sets vary drastically in size. High-power applications, such as the gearbox shown in Figure 1.25, require very large gear sets. Smaller sets are used in hand tools, such as the one shown in Figure 1.26, to change speed.

Gears will be discussed in detail in Chapters 7 and 8. Chapter 7 will focus on gear geometry and terminology, while Chapter 8 will focus on combining gears into gear trains. There are

Figure 1.25   Example of a large power transmission gear unit. Reproduced from Michalec, Precision Gearing Theory and Practice, John Wiley & Sons, © 1966

Figure 1.26   Gear application. Reproduced from Michalec, Precision Gearing Theory and Practice, John Wiley & Sons, © 1966

several types of gears, but one major factor for choosing the appropriate gear would be the relative shaft positions. The two major shaft arrangements would be parallel and intersecting; though intersecting should be further divided into perpendicular and skewed.

1.6.2 Spur Gears

The first type of gear to be discussed will be spur gears. Spur gears are very common and have the simplest geometry. The easiest way to think of the function of two spur gears in contact is to think of two rolling cylinders in contact, as shown in Figure 1.27a. One cylinder would turn and drive the mating cylinder. If the cylinders have different diameters, there would be a speed difference between the two cylinders. One major disadvantage of having two rolling cylinders in contact would be the large potential of slipping between the two cylinders. Adding teeth to the cylinders will create spur gears, as shown in Figure 1.27b. The teeth eliminate this slip problem causing positive rolling contact. The driving gear, often called the pinion, will rotate in the opposite direction of the driven gear. The ratio of speeds of the two gears will depend on the number of teeth on the two gears, which will be discussed in detail in Chapter 7.

Spur gears have teeth that are parallel to the axis of rotation and they connect parallel shafts. Properly aligned spur gears will not produce end thrust. The pinion and gear will rotate in opposite directions if both have external teeth, or will rotate in the same direction if the gear has internal teeth.

Gears are standardized, which allows for interchangeability and less expensive manufacturing. Hobbing, which is illustrated in Figure 1.28, is a common process for producing spur gears. The cylindrical cutter (hob), as shown in Figure 1.28a, cuts the teeth in the gear blank. Figure 1.28b shows a hobbing operation, which is a special milling operation to progressively cut the gear teeth.

(a)   (b)

Figure 1.27   (a) Rolling cylinders. (b) Spur gears

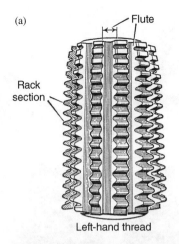

(a)

Figure 1.28 (a) Cutting hob. (b) Hobbing a spur gear. Reproduced from Mabie and Reinholtz, *Mechanisms and Dynamics of Machinery*, 4th edition, John Wiley & Sons, © 1987

1.6.3 Helical Gears

Though spur gears are very useful, they do have limitations. Spur gears will tend to be noisy due to the abrupt contact of the gear teeth. An improvement could be made by staggering the teeth, such as the layout shown in Figure 1.29b. A helical gear is developed as the individual spur gears within the staggered set approach zero thickness.

Helical gears are similar to spur gears in overall function. Unlike spur gears, the teeth on helical gears are cut at an angle to the gear's axis of rotation. Though the details will be presented in later chapters, the angled teeth allow for smoother operation and higher speeds. The trade-off for better performance is higher manufacturing costs.

Helical gears are extremely versatile and can be used in parallel or intersecting shaft arrangements. A parallel arrangement is shown in Figure 1.29a and an intersecting shaft arrangement is shown in Figure 1.29d. Helical gears tend to push apart, or exert a side thrust. One method of compensating for the side thrust is to combine two helical gears with opposite helical angles. Such a gear is known as a herringbone and is shown in Figure 1.29c. The hobbing process for a helical gear is shown in Figure 1.30.

(a)   (b)

(d)

(c)

Figure 1.29   Types of helical gears: (a) mounted on parallel shafts (most common type), gears have helices of opposite hand; (b) rotated spur gear laminations approach a helical gear as laminations approach zero thickness; (c) double helical or herringbone gears may or may not have a center space, depending on manufacturing method; (d) when mounted on nonparallel shafts, they are crossed helical gears, and usually have the same hand. ((a, d) Courtesy Boston Gear; (c) courtesy Horsburgh & Scott.) Reproduced from Juvinall and Marshek, Machine Component Design, 5th edition, John Wiley & Sons, © 2011

1.6.4 Bevel Gears

Bevel gears are an efficient means of transmitting power in intersecting shaft arrangements. Bevel gears connect shafts typically intersecting at 90° angles, although other angles are possible. The simplest way to visualize the interaction of bevel gears is to consider two cones in contact.

There are three types of bevel gears, as illustrated in Figure 1.31. The simplest type is a straight bevel, shown in Figure 1.31a. Spiral bevels, shown in Figure 1.31c, have curved teeth and allow for higher speeds. Spiral bevel gears have two or more teeth in contact at all times, which lowers tooth loading. The teeth engage more gradually than straight bevel gears, which cause quieter operation. Hypoid bevel gears, such as the set shown in Figure 1.31e, keep the benefits of spiral bevels but allow the drive shaft to be set lower.

Bevel gears can be used for speed reduction, but sometimes they are used to simply transmit power between intersecting shafts. Therefore, a 1 : 1 ratio is fairly common. Figure 1.31b shows a bevel gear set with no change in speed.

Figure 1.30 Hobbing a helical gear. Reproduced from Mabie and Reinholtz, Mechanisms and Dynamics of Machinery, 4th edition, John Wiley & Sons, © 1987

(a)

(b)

(c)

(d)

(e)

Figure 1.31 Types of bevel gears: (a) straight-tooth bevel gears; (b) straight-tooth bevel gears; special case of miter gears (1:1 ratio); (c) spiral bevel gears; (d) bevel gears mounted on nonperpendicular shafts; (e) hypoid gears. ((a, c, d, e) Courtesy Gleason Machine Division; (b) courtesy Horsburgh & Scott.) Reproduced from Juvinall and Marshek, Machine Component Design, 5th edition, John Wiley & Sons, © 2011

(a)

(b)

Figure 1.32   Worm gear sets: (a) single enveloping; (b) double enveloping. ((a) Courtesy Horsburgh & Scott; (b) courtesy Ex-Cell-O Corporation, Cone Drive Operations.) Reproduced from Juvinall and Marshek, Machine Component Design, 5th edition, John Wiley & Sons, © 2011

1.6.5 Worm Gears

Another common gearing for right angle shaft arrangements is worm gears. Worm gears allow for very high reduction ratios compared with other gearing systems. A worm gear set, such as those shown in Figure 1.32, consists of a worm meshing with a worm gear. The worm is essentially a threaded screw, and the worm gear can be thought of as a special helical gear.

Worm gear sets are commonly used in speed reducers, as shown in Figure 1.33, because they can have very large reductions in speed. A large concern with worm gear sets would be heat generation. The speed reducer shown also has fins to increase heat transfer.

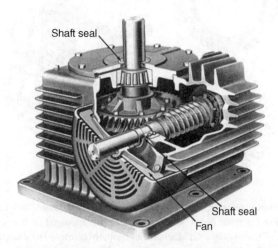

Shaft seal

Shaft seal

Fan

Figure 1.33   Worm gear speed reducer. (Courtesy Cleveland Gear Company.) Reproduced from Juvinall and Marshek, Machine Component Design, 5th edition, John Wiley & Sons, © 2011

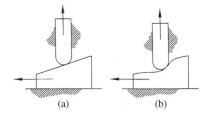

Figure 1.34   Basic concept of a cam: (a) simple wedge; (b) complex wedge

1.7   Cams

1.7.1   Introduction to Cams

Cam mechanisms are another major class of mechanisms and are used to develop nonuniform motion. In its basic form, a cam can be thought of as a specifically shaped wheel that causes a defined motion of a follower. Most commonly, cams are used to cause follower motion that is complex and accurately timed. As will be seen in later chapters, cam design often requires specific constraints on velocity and accelerations, which become more critical at high-speed operation.

In their most basic form, cams can be thought of as a specialized wedge. Figure 1.34a shows a simple wedge being moved horizontally along a flat surface. As the wedge moves to the left, it will push the rod above it upward. The surface of the wedge could be more complex, such as that shown in Figure 1.34b, so that the motion of the rod is in a very specific fashion. In the language of cam mechanisms, the wedge is the cam and the rod is a follower. The cam surface will typically describe a mathematical function so that the follower moves in the required fashion.

For practical purposes, cams will not be simple wedges. Cam mechanisms come in a wide variety of shapes, but a couple of the most common types are introduced below.

1.7.2   Disk Cams

Probably, the most common type of cam in use would be disk cams (sometimes called radial cams). To get a general understanding of a disk cam, we will continue the basic wedge concept from Figure 1.34. The wedge shown in Figure 1.35a has a similar concept but it now causes

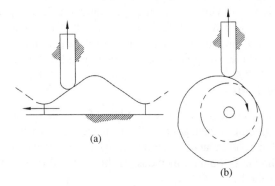

Figure 1.35   (a) Wedge for a rise and fall motion. (b) Wedge profile wrapped around a circle

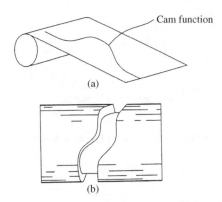

Figure 1.36 (a) Wrapping the cam function around the cylinder. (b) Drum cam

a very specific type of motion. If the wedge were moved to the left, it would cause the follower to rise and then fall back to its original position. We could have that motion continue by constantly moving the wedge to the left and right, though that is not very convenient, or by having several wedges lined up in a row (as shown). To make the continuous motion more practical, we take the top profile of the wedge and wrap it around a circle as shown in Figure 1.35b. That circular disk, known as a disk cam, can now rotate at a constant speed and cause the desired continuous rise and fall of the follower.

Obviously, the wedge profile can become far more complex to get nearly any desired motion of the follower. The profile is generally defined by a mathematical expression. The mathematical expression must be carefully picked to ensure proper follower motion as well as safe acceleration values. Chapter 9 will cover many common mathematical functions used in cam design and describe procedures for "wrapping" that function around the circle to create the disk cam geometry.

1.7.3 Cylindrical Cams

The basic structure of a cylindrical cam (also called a drum cam or barrel cam) uses a groove cut around the periphery of a cylinder. Figure 1.36a shows the idea of wrapping the mathematical cam function around a cylinder. The figure shows the cam function, which would be rolled around the cylinder. The result is a cylindrical cam such as the one shown in Figure 1.36b. The layout design process for a cylindrical cam is generally simple compared with disk cams because the cam function can theoretically be drawn on a paper template and rolled around a cylinder. The transfer of the cam function onto a disk cam is more complex and causes distortion. Figure 1.37 shows the process of cutting a cylindrical cam.

1.8 Solution Methods

Throughout the text several methods will be introduced for analyzing machine elements. Each method will provide specific advantages and disadvantages. Often a designer will utilize more than one method within a particular design process, so it is important to be familiar with several solution methods. The three common methods for this text will be graphical, analytical, and computer solution methods.

Figure 1.37 Cutting a cylindrical cam

1.8.1 Graphical Techniques

Graphical techniques often have the advantage of being quick and simple, especially for developing initial designs. As an example, it is often desired to design a linkage mechanism to perform a very specific task. Graphical techniques can be utilized easily to develop several possible configurations. Once an arrangement has been designed, it may be desirable to move to other solution methods to refine the mechanism.

Velocity and acceleration analysis can often be done graphically with the use of vector polygons. If drawn by hand, the polygons can provide a quick solution at a lower level of accuracy. Polygons can also be generated in CAD software at a higher level of accuracy.

Graphical techniques offer other advantages. One is that most machine analysis work will require creating drawings of the machine elements. Graphical techniques can often be done in conjunction with the drawing process. Another advantage is that graphical techniques are very visual, which can aid in understanding the solution.

1.8.2 Analytical Methods

Analytical methods require the development and solution of equations, making them the most familiar to engineering students. An obvious advantage of analytical methods over graphical methods would be improved accuracy. There are instances, however, when analytical methods are tedious or excessively complex. It can often be beneficial to use a combination of graphical and analytical methods.

1.8.3 Computer Solutions

The last solution method discussed includes the use of computer software. Machine analysis problems often become repetitive. Development of a simple computer code can greatly aid in the repetitive calculations. It is often desired to have solutions for several positions of a mechanism, and developing these solutions analytically can be very time consuming. Therefore, it is common to use computers to solve for velocity, acceleration, and forces in a mechanism through the complete range of motion for that mechanism.

Even though commercial software is available, it is fairly straightforward for an individual to develop solutions in spreadsheets or computer programs such as MathCAD or MATLAB®. Because of the major advantages of utilizing computers for mechanism analysis, this book will focus on not only the underlying theory but also how to develop computer code to solve for a wide variety of mechanisms. Chapter 6, as an example, will focus almost entirely on computer solution methods for linkage mechanisms.

It would be of limited use to develop a computer code to solve only one mechanism arrangement. For example, developing a computer code that only analyzes slider–crank linkage mechanisms would be less useful than a code to solve any arrangement of a four-bar linkage mechanism. To make computer code more versatile, the methodology for solving kinematics of mechanisms using computers is often quite different from the methods for solving kinematics analytically. Computer methods often take advantage of numerical optimization methods.

1.9   Methods of Problem Solving

1.9.1   Step 1: Carefully Read the Problem Statement

Most students do not like story problems. In essence, all engineering problems are just story problems. The critical first step is to carefully read the problem statement and completely understand the problem. Read the problem statement two or three times if necessary, but do not move on to step 2 until you completely understand the problem statement.

1.9.2   Step 2: Plan Your Solution

It is very important to think about the solution process required for the problem. Do not just start writing down equations aimlessly looking for one that may work. One of the biggest mistakes students make is that they memorize solution steps to certain styles of problems. Do not memorize solution procedures! Instead, understand how to plan the solution to any type of problem. Memorizing solution procedures may make you successful on homework problems, but it is most likely setting you up for failure on exams. Most professors will not write exam questions that are extremely similar to homework problems. You must be able to plan a solution to any style of problem to be really successful.

1.9.3   Step 3: Solve the Problem

Obviously, this is a critical step. However, it is important to understand that you need to solve the problem in a way to help ensure success. Solve the problem neatly and following your solution plan developed in step 2. Keep your work as organized as possible. Well-organized work will help you review problems later and will help your professor understand your thought process. Draw figures when necessary (as neatly as possible) because figures will definitely help you visualize the problem.

1.9.4   Step 4: Read the Problem Statement Again

This is a very important step that is often ignored. Once the problem is solved, read the problem statement again. The main purpose would be to verify that you solved everything asked for in the problem statement. This is also a good time to make sure that you included the proper units on you answers.

1.10   Review and Summary

This chapter served as a general introduction into the field of machine analysis and to illustrate the importance of machine analysis. Machines are continually becoming more complex requiring a deeper knowledge of machine theory. The remaining chapters will be discussing details of the different types of mechanisms introduced in this chapter. Therefore, a primary purpose of this chapter was to develop the common terms and basic types of mechanisms.

Problems

P1.1   Make a list of machines you use during the next week. Write a brief description of the machine.

P1.2   Human factors engineering, which focuses on aspects of the interface between man and machines, is an important aspect of machine design. Discuss issues such as safety, comfort, and efficiency that must be considered in machine design. Give specific examples of machines with good and poor human factors engineering.

P1.3   Research a topic of interest related to a machine failure. Discuss the cause(s) of the failure. How could the design have been changed to reduce the risk of the failure?

P1.4   For each of the linkage mechanisms shown in Figure 1.38, identify all missing link numbers and attachment points. Label each link by the number of connection points (binary, ternary, or quaternary).

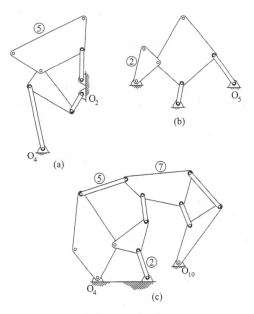

Figure 1.38

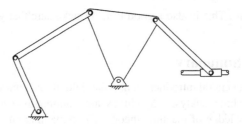

Figure 1.39

P1.5 For the linkage mechanism shown in Figure 1.39, identify all revolute and prismatic joints.

P1.6 Find four applications of a crank–rocker mechanism. Describe the application of each and draw a simplified diagram of each mechanism.

P1.7 Research slider–crank mechanisms and scotch yoke mechanisms. Discuss the similarities and differences between each.

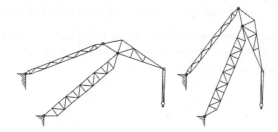

Figure 1.40

P1.8 Figure 1.40 shows a level luffing crane, which is an application of an approximate straight line four-bar mechanism. The crane allows for horizontal motion of the hook while it maintains an approximately constant elevation. Research the function of the level luffing crane and summarize your findings.

P1.9 Research different types of straight line mechanisms, such as Watt's mechanism and Robert's mechanism. Discuss the advantages and disadvantages of each type.

P1.10 Research applications of the different gear systems discussed in this chapter. Write a summary of your findings.

P1.11 Explain why helical gears provide smoother motion compared with spur gears.

P1.12 Research applications of the different cam systems discussed in this chapter. Write a summary of your findings.

P1.13 The two mechanisms shown in Figure 1.41 both convert rotary motion to linear oscillating motion of slider B. Figure 1.41a shows a traditional slider–crank

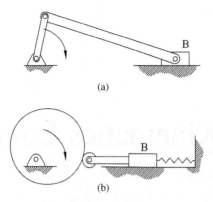

(a)

(b)

Figure 1.41

mechanism and Figure 1.41b uses an eccentric cam. Discuss similarities and differences of the motion of slider B for each arrangement.

Further Reading

Paz, E.B., Ceccarelli, M., Otero, J.E., and Sanz, J.M. (2010) *A Brief Illustrated History of Machines and Mechanisms*, Springer, New York.

2

Essential Kinematics Concepts

2.1 Introduction

Generally, engineering dynamics is prerequisite material for any student studying machine analysis. Therefore, the purpose of this chapter is not to provide a complete study of dynamics. Key kinematic topics are covered here only as a form of review. Students needing a more thorough coverage of the kinematics topics discussed in this chapter are encouraged to consult an engineering dynamics textbook (see bibliography section for some examples).

Standard dynamics textbooks clearly separate topics by analysis of particles and analysis of rigid bodies. In general terms, a particle is simply a body of negligible mass. Large bodies, such as a train or car, can be treated as a particle for certain applications. Therefore, a better description would be that a body can be treated as a particle when its dimensions have no relevance to its motion. Analysis of rigid bodies involves rotation of the body about its center of mass. Because of this rotation, the shape of the body, and the resulting moment of inertia, becomes very important.

Study of machine analysis focuses a lot on rotating bodies (cams, gears, linkages, etc.), and therefore focuses on rigid body analysis. For that reason, the emphasis of the chapter will be on reviewing rigid body kinematics. The majority of this book also focuses on planar motion, which is motion of a rigid body when all points on the body move in parallel planes. As a result, concepts of three-dimensional rigid body kinematics will not be reviewed in this chapter. The narrow focus on planar motion may seem limiting, but most machines operate in plane motion. More complex machines can typically be broken down into sets of planar motion mechanisms.

There are two main parts of dynamics. The first part is kinematics, which is a study of motion without references to the forces that caused the motion. In other words, kinematics deals with the geometric features of motion. Kinetics is the second part of dynamics and focuses on the motion and the forces acting on a body. It is essential to have a solid understanding of kinematics before working in the area of kinetics. The focus of this chapter will be areas of kinematics, and essential kinetics concepts will be reviewed in a later chapter.

Machine Analysis with Computer Applications for Mechanical Engineers, First Edition. James Doane.
© 2016 John Wiley & Sons, Ltd. Published 2016 by John Wiley & Sons, Ltd.
Companion Website: www.wiley.com/go/doane0215

2.2   Basic Concepts of Velocity and Acceleration

As a basic definition, velocity is the time rate of change of position. Instantaneous linear velocity (v) is defined as

$$v = \frac{ds}{dt} \tag{2.1}$$

and will commonly have units of ft/s or m/s. In Equation 2.1, s is position and t is time. Instantaneous angular velocity (ω) is defined as the time rate of change of angular displacement.

$$\omega = \frac{d\theta}{dt} \tag{2.2}$$

Angular velocity will have units of radians per second or revolutions per second. Velocity is a vector, so it has magnitude and direction. The velocity vector will always be tangent to the path of motion. The magnitude of velocity, which is a scalar quantity, is commonly referred to as speed.

Acceleration is the rate of change of velocity with respect to time. Therefore, acceleration is the first time derivative of velocity or the second time derivative of position. Linear acceleration (a) is given by

$$a = \frac{d^2 s}{dt^2} = \frac{dv}{dt} \tag{2.3}$$

and has units of ft/s^2 or m/s^2. Rotational acceleration (α) is the time derivative of rotational velocity and will have units of radians per second squared.

$$\alpha = \frac{d^2\theta}{dt^2} = \frac{d\omega}{dt} \tag{2.4}$$

Acceleration is also a vector, so it will have magnitude and direction. Unlike velocity, the acceleration vector will not generally be tangent to the path. A positive acceleration indicates increasing velocity and a negative acceleration indicates decreasing velocity (deceleration).

2.3   Translational Motion

Planar motion of a rigid body will fall into one of the three categories: translation, pure rotation, or general plane motion. In translational motion, any line segment on the rigid body will remain in the same orientation throughout the entire motion. In other words, rotational velocity and acceleration are both zero. It is common to think of translation as motion with a straight line trajectory, which is not always the case. In fact, translation is further subdivided into rectilinear translation and curvilinear translation. In rectilinear translation, all points on the rigid body move along straight lines. In curvilinear translation, all points on the rigid body move along congruent curved paths. Figure 2.1 shows a pallet moving along a monorail conveyor system to illustrate both types of translation. The pallet will move along different portions of the monorail conveyor system, but it will always remain parallel to the ground. Figure 2.1a shows the pallet moving along a straight line. The trajectory of two points shown will be straight lines, which is

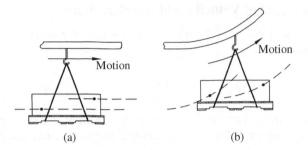

(a)   (b)

Figure 2.1   (a) Rectilinear translation. (b) Curvilinear translation

rectilinear translation. In Figure 2.1b, the pallet is moving along a curved portion of the monorail system, but the pallet remains in a level position. The trajectory of the two points will now be curved paths while the pallet remains parallel to the floor, which is curvilinear translation.

2.4   Rotation about a Fixed Axis

2.4.1 Velocity

The next type of motion to consider is rotation about a fixed axis, which is commonly referred to as pure rotation. A point on a body in pure rotation will have a circular trajectory. Multiple points on a body in pure rotation will have trajectories making concentric circles centered about the fixed rotation point. Referring back to a crank–rocker mechanism discussed in Chapter 1 as an example, both the crank and rocker will move in pure rotation.

Kinematic analysis of fixed axis rotation is relatively simple. Our first step will be velocity analysis. Consider a link in pure rotation about point O as shown in Figure 2.2. We wish to find the velocity of point A on the link located by vector $\vec{r}$.

The instantaneous linear velocity of a point on a rotating body will be proportional to the distance of that point from the center of rotation. The velocity of point A is defined by the cross product

$$\vec{v} = \vec{\omega} \times \vec{r} \tag{2.5}$$

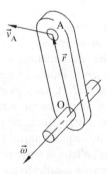

Figure 2.2   Velocity for fixed axis rotation

The velocity of point A ($\vec{v}_A$) is the absolute velocity because it is relative to a fixed pivot point. For planar motion, the velocity magnitude can be represented as the scalar

$$v = r\omega \qquad\qquad (2.6)$$

Example Problem 2.1

For the gear system shown in Figure 2.3, gear A has a radius of 191 mm and rotates counterclockwise at 10 rad/s. Gear B has a radius of 96 mm. Gear C is on the same shaft as gear B and has a radius of 252 mm. Gears B and C rotate together as one unit. Gear D has a radius of 161 mm. Determine the rotation speed and direction of gear D.

Solution: The velocity of point p (the point of contact between gears A and B as shown in Figure 2.4) can be determined from the rotation of gear A.

$$v_p = r_A \omega_A = 191\text{ mm}\left(10\frac{\text{rad}}{\text{s}}\right) = 1910\frac{\text{mm}}{\text{s}} \downarrow$$

If no slipping occurs between gears A and B, the rotational speed of gear B will be (note that rotation direction of B is opposite of A)

$$\omega_B = \frac{v_p}{r_B} = \frac{1910\frac{\text{mm}}{\text{s}}}{96\text{ mm}} = 19.896\frac{\text{rad}}{\text{s}}$$

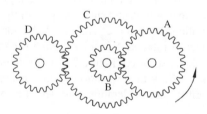

Figure 2.3

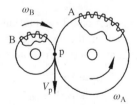

Figure 2.4

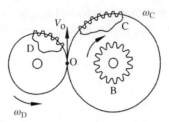

Figure 2.5

Because gears B and C move as one unit, the rotation speed and direction of gear C will be the same as gear B. Next we can look at the intersection between gears C and D as shown in Figure 2.5.

$$v_O = r_C \omega_C = 252 \text{ mm} \left(19.896 \frac{\text{rad}}{\text{s}} \right) = 5013.79 \frac{\text{mm}}{\text{s}} \uparrow$$

If no slipping occurs between gears C and D, the rotational speed of gear D will be (note change in rotation direction)

$$\omega_D = \frac{v_O}{r_D} = \frac{5013.79 \frac{\text{mm}}{\text{s}}}{161 \text{ mm}}$$

Answer: $\omega_D = 31.1 \dfrac{\text{rad}}{\text{s}}$ (ccw)

2.4.2 Acceleration

The absolute accelerations for pure rotation can be determined from differentiation of the velocity terms. The direction of the acceleration vector is more complicated than that of velocity. Acceleration is typically separated into two components: tangential and normal. The tangential acceleration component will be tangent to the path of the point and will only exist if the body has angular acceleration. Normal acceleration will be normal to the path of the point and directed toward the center of rotation. Consider again a rigid link in pure rotation. Both accelerations for the rigid link are shown in Figure 2.6. The rotational speed of the body is $\vec{\omega}$ and the rotational acceleration is $\vec{\alpha}$. With the position vector $\vec{r}$ being the vector from point O to point A (not shown for clarity), the accelerations are determined from the following cross products:

$$\vec{a}_t = \vec{\alpha} \times \vec{r}$$
$$\vec{a}_n = \vec{\omega} \times (\vec{\omega} \times \vec{r})$$

(2.7)

For planar motion, the accelerations can also be represented in scalar form. The rotational velocity and acceleration will always be in the $\pm \hat{k}$ direction, counterclockwise rotation being positive. The magnitude of the tangential and normal accelerations will be

$$a_t = r\alpha$$
$$a_n = r\omega^2 = v\omega$$

(2.8)

and their directions can be determined by inspection.

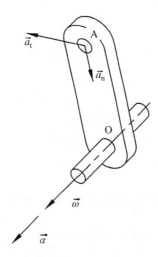

Figure 2.6   Acceleration for fixed axis rotation

Example Problem 2.2

The right triangular plate shown in Figure 2.7 is in pure rotation about point O and is decelerating at a rate of 3 rad/s. At the instant shown, it has a rotation speed of 5 rad/s clockwise. Determine the tangential and normal acceleration vectors for points A and B.

Solution: The rotational speed and acceleration can both be expressed as vectors. The rotation speed is clockwise, which is a negative vector. The deceleration indicates the acceleration direction is counterclockwise, which is positive.

$$\vec{\omega} = -5\hat{k} \,\text{rad/s} \qquad \vec{\alpha} = 3\hat{k} \,\text{rad/s}^2$$

We will start by finding the acceleration of point A. The position vector shown in Figure 2.8 will be

$$\vec{r}_A = 4\cos(160)\hat{i} + 4\sin(160)\hat{j}$$

$$\vec{r}_A = -3.7588\hat{i} + 1.3681\hat{j}$$

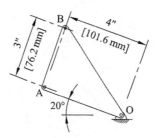

Figure 2.7

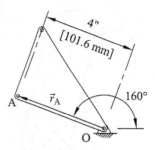

Figure 2.8

The tangential and normal acceleration vectors of point A are determined from the cross products.

$$\left(\vec{a}_A\right)_t = \vec{\alpha} \times \vec{r} = 3\hat{k} \times \left(-3.7588\hat{i} + 1.3681\hat{j}\right)$$

$$\left(\vec{a}_A\right)_n = \vec{\omega} \times \left(\vec{\omega} \times \vec{r}\right) = -5\hat{k} \times \left(-5\hat{k} \times \left(-3.7588\hat{i} + 1.3681\hat{j}\right)\right)$$

$$\left(\vec{a}_A\right)_n = -5\hat{k} \times \left(6.8405\hat{i} + 18.7940\hat{j}\right)$$

Answer: $\left(\vec{a}_A\right)_t = -4.1\hat{i} - 11.3\hat{j} \quad (\text{in.}/\text{s}^2)$
$\quad\quad\quad \left(\vec{a}_A\right)_n = 94.0\hat{i} - 34.2\hat{j} \quad (\text{in.}/\text{s}^2)$

Next we move to point B. The angles in Figure 2.9 need to be determined in order to get the position vector for point B.

$$\tan\theta = \frac{3}{4} \Rightarrow \theta = 36.87°$$

$$\phi = 160° - \theta = 160° - 36.87° = 123.13°$$

Length OB is 5″ because the triangle is a 3-4-5 triangle. Position vector for point B is

$$\vec{r}_B = 5\cos(123.13)\hat{i} + 5\sin(123.13)\hat{j}$$

$$\vec{r}_B = -2.7327\hat{i} + 4.1872\hat{j}$$

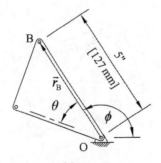

Figure 2.9

Cross products are used to get acceleration values for point B.

$$\left(\vec{a}_B\right)_t = \vec{\alpha} \times \vec{r} = 3\hat{k} \times \left(-2.7327\hat{i} + 4.1872\hat{j}\right)$$

$$\left(\vec{a}_B\right)_n = \vec{\omega} \times \left(\vec{\omega} \times \vec{r}\right) = -5\hat{k} \times \left(-5\hat{k} \times \left(-2.7327\hat{i} + 4.1872\hat{j}\right)\right)$$

$$\left(\vec{a}_B\right)_n = -5\hat{k} \times \left(20.9360\hat{i} + 13.6635\hat{j}\right)$$

Answer: $\left(\vec{a}_B\right)_t = -12.6\hat{i} - 8.2\hat{j}$ $(\text{in.}/\text{s}^2)$
$\left(\vec{a}_B\right)_n = 68.3\hat{i} - 104.7\hat{j}$ $(\text{in.}/\text{s}^2)$

2.5 General Plane Motion

2.5.1 Introduction

Planar motion that is a combination of translation and pure rotation is known as general plane motion. Consider the slider–crank mechanism shown in Figure 2.10a. The crank will move in pure rotation and the slider will move in linear translation. The coupler (link 3) will move in general plane motion. The general plane motion can be broken down into translation and rotation, as shown in Figure 2.10b.

2.5.2 Velocity Difference and Relative Velocity

Motion can be described in different ways, but in machine analysis it is often described with respect to the frame of the machine. The velocity of point A in Figure 2.2, for example, is relative to a fixed pivot location that is likely attached to a frame. Absolute motion is motion described with respect to the frame, even if the frame is in motion. Velocity analysis becomes more complex when the pivot location is not stationary. Figure 2.11a shows a link pivoting

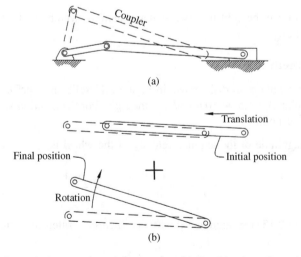

(a)

(b)

Figure 2.10   (a) General plane motion of the coupler. (b) Combination of translation and rotation

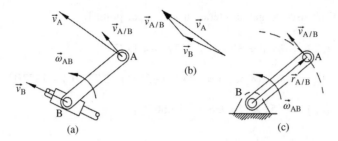

Figure 2.11 (a) Velocity difference. (b) Vector polygon. (c) Representation of the velocity of point A with respect to point B

about point B, which is moving on a slide. When motion is described relative to another moving link, it is called relative motion. Relative motion of two points can be determined by the vector difference of the absolute motion of each point.

As represented by the vector polygon shown in Figure 2.11b, the absolute velocity of point A ($\vec{v}_A$) now is determined using velocity difference.

$$\vec{v}_A = \vec{v}_B + \vec{v}_{A/B} \tag{2.9}$$

As shown in Figure 2.11c, the velocity of point A with respect to point B ($\vec{v}_{A/B}$) is determined by treating point B as a fixed pivot and determining the velocity of point A as if it were in pure rotation.

$$\vec{v}_{A/B} = \vec{\omega}_{AB} \times \vec{r}_{A/B} \tag{2.10}$$

Substituting Equation 2.10 into Equation 2.9 gives

$$\vec{v}_A = \vec{v}_B + \vec{\omega}_{AB} \times \vec{r}_{A/B} \tag{2.11}$$

The vector equation can be split into two scalar equations, which can be solved for the two unknowns $\vec{v}_A$ and $\vec{\omega}_{AB}$.

Example Problem 2.3

The 5 inch (127 mm) diameter disk shown in Figure 2.12 rolls on a surface without slipping. The velocity of point O is 1.5 ft/s (0.46 m/s) to the right. For the position shown, calculate the absolute velocity of point P.

Solution: The magnitude of the angular velocity of the wheel is

$$\omega = \frac{v_O}{r} = \frac{1.5 \, \text{ft/s}}{2.5 \, \text{in.} (1/12)} = 7.2 \frac{\text{rad}}{\text{s}}$$

Referring to Figure 2.13, the angular velocity can then be written in vector form as

$$\vec{\omega} = -7.2 \hat{k}$$

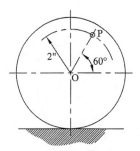

Figure 2.12

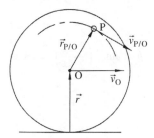

Figure 2.13

The velocity of point P with respect to point O is

$$\vec{v}_{P/O} = \vec{\omega} \times \vec{r} = -7.2\hat{k} \times \frac{2}{12}\left(\cos 60\hat{i} + \sin 60\hat{j}\right)$$

$$\vec{v}_{P/O} = \vec{\omega} \times \vec{r} = -7.2\hat{k} \times \left(0.0833\hat{i} + 0.1443\hat{j}\right)$$

$$\vec{v}_{P/O} = 1.0390\hat{i} - 0.5998\hat{j}$$

The absolute velocity of point P is calculated from Equation 2.9.

$$\vec{v}_P = \vec{v}_O + \vec{v}_{P/O}$$

$$\vec{v}_P = 1.5\hat{i} + \left(1.0390\hat{i} - 0.5998\hat{j}\right)$$

Answer: $\vec{v}_P = 2.54\hat{i} - 0.60\hat{j}$ (ft/s)

Figure 2.14 shows the velocity vector and the horizontal and vertical components. The velocity can be represented as magnitude

$$|\vec{v}_P| = \sqrt{2.54^2 + (-0.60)^2} = 2.61 \text{ ft/s}$$

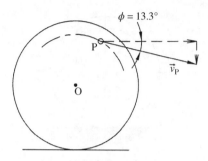

Figure 2.14

and direction

$$\phi = \tan^{-1}\left(\frac{0.60}{2.54}\right) = 13.3°$$

Answer: $v_P = 2.61 \text{ ft/s} \angle -13.3°$

Example Problem 2.4

In the linkage position shown in Figure 2.15, link 2 rotates counterclockwise at a constant rate of 65 rpm. The distance AP is 5 inches (127 mm). Determine the rotation speed of link 3 and the velocity of point P.

Solution: The link numbers are determined as presented in Chapter 1. Link O_2A will be link 2, the ternary link ABP will be link 3, and link O_4B is link 4.

First, the rotational speed of the input link must be converted to radians per second.

$$\omega_2 = 65 \frac{\text{rev}}{\text{min}}\left(2\pi \frac{\text{rad}}{\text{rev}}\right)\left(\frac{1 \text{ min}}{60 \text{ s}}\right) = 6.81 \frac{\text{rad}}{\text{s}}$$

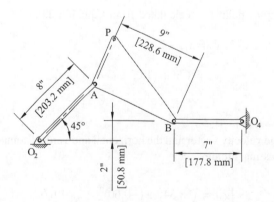

Figure 2.15

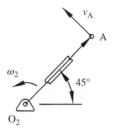

Figure 2.16

Link 2 is in pure rotation, as shown in Figure 2.16. From Equation 2.5,

$$\vec{v}_A = \vec{\omega}_2 \times \vec{r}_2$$

$$\vec{v}_A = 6.81\hat{k} \times (8\cos 45\hat{i} + 8\sin 45\hat{j})$$

$$\vec{v}_A = -38.52\hat{i} + 38.52\hat{j}$$

We can now move to link 3. To find the angular orientation, we can use right triangles as shown in Figure 2.17. From triangle O₂aA,

$$Aa = 8\sin 45 = 5.66 \text{ in.}$$

Moving to triangle AbB, and knowing that distance from A to b will be $5.66 - 2 = 3.66$,

$$\sin \alpha = \frac{3.66}{9}$$
$$\alpha = 24°$$

To find the velocity of point B, we can use the relative velocity equation. Referring to Figure 2.18a,

$$\vec{v}_B = \vec{v}_A + \vec{v}_{B/A} = \vec{v}_A + \vec{\omega}_3 \times \vec{r}_{B/A}$$

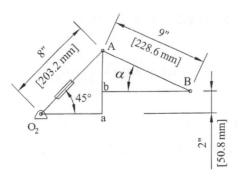

Figure 2.17

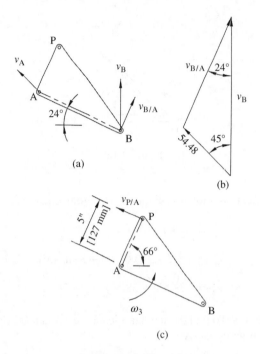

Figure 2.18

The velocity vector at point B must act vertically because link 4 is horizontal and in pure rotation. The magnitude of the velocity of point A is

$$v_A = \sqrt{(-38.52)^2 + 38.52^2} = 54.48 \frac{\text{in.}}{\text{s}}$$

The relative velocity of point B with respect to point A can be determined using law of sines. From Figure 2.18b,

$$\frac{v_{B/A}}{\sin 45} = \frac{54.48}{\sin 24} \Rightarrow v_{B/A} = 94.71 \frac{\text{in.}}{\text{s}}$$

The rotation speed of link 3 can then be determined as

$$\omega_3 = \frac{v_{B/A}}{AB} = \frac{94.71 \frac{\text{in.}}{\text{s}}}{9 \text{ in.}}$$

Answer: $\omega_3 = 10.52 \frac{\text{rad}}{\text{s}}$

The velocity of point P can now be determined.

$$\vec{v}_P = \vec{v}_A + \vec{v}_{P/A} = \vec{v}_A + \vec{\omega}_3 \times \vec{r}_{P/A}$$

From Figure 2.18c, the vector $\vec{r}_{P/A}$ will be

$$\vec{v}_A = 5 \cos 66\hat{i} + 5 \sin 66\hat{j} = 2.03\hat{i} + 4.57\hat{j}$$

The velocity of point P is

$$\vec{v}_P = -38.52\hat{i} + 38.52\hat{j} + 10.52\hat{k}(2.03\hat{i} + 4.57\hat{j})$$

$$\vec{v}_P = -38.52\hat{i} + 38.52\hat{j} + 21.36\hat{j} - 48.08\hat{i}$$

Answer: $\vec{v}_P = -86.6\hat{i} + 59.9\hat{j}$ (in./s)

2.5.3 Relative Acceleration

Differentiation of the relative velocity equation gives the relative acceleration equation.

$$\vec{a}_A = \vec{a}_B + \vec{a}_{A/B} \tag{2.12}$$

Unlike velocity analysis, the relative acceleration term will have two components. The term $\vec{a}_{A/B}$ will be rotational, so it will have normal and tangential components, which were given earlier for rotation about a fixed point. Substituting the normal and tangential acceleration terms gives

$$\vec{a}_A = \vec{a}_B + \vec{a}_t + \vec{a}_n$$

$$\vec{a}_A = \vec{a}_B + \vec{\alpha}_{AB} \times \vec{r}_{A/B} + \vec{\omega}_{AB} \times \left(\vec{\omega}_{AB} \times \vec{r}_{A/B}\right) \tag{2.13}$$

Consider the system shown in Figure 2.19a. The acceleration of point B can be determined using Equation 2.13, which is represented in Figure 2.19b as a vector polygon. The normal and tangential vectors are shown in Figure 2.19c and are determined by assuming pure rotation about point B.

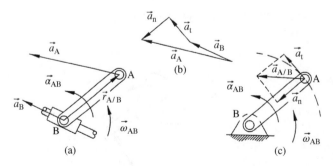

Figure 2.19 (a) Acceleration of the system. (b) Vector polygon. (c) Representation of the acceleration of point A with respect to point B

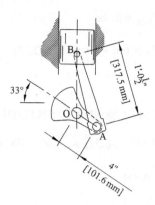

Figure 2.20

Example Problem 2.5

Figure 2.20 shows a slider–crank mechanism. The crank (link OA) rotates at a constant speed of 1200 rpm clockwise. For the position shown, determine the velocity and acceleration of the piston (point B).

Solution: The relative acceleration equation will be used for this problem. The acceleration of point B can be expressed as

$$\vec{a}_B = \vec{a}_A + \vec{a}_{B/A}$$

We will start with the crank (link 2), which is shown in Figure 2.21. The crank is in pure rotation at a constant rotational speed.

$$\omega_2 = 1200 \frac{\text{rev}}{\text{min}} \left(\frac{2\pi \text{ rad}}{\text{rev}}\right)\left(\frac{1 \text{ min}}{60 \text{ s}}\right) = 125.7 \frac{\text{rad}}{\text{s}}$$

From Equation 2.5,

$$\vec{v}_A = \vec{\omega}_2 \times \vec{r}_A = -125.7\hat{k} \times \left(4\cos(-33)\hat{i} + 4\sin(-33)\hat{j}\right)$$

$$\vec{v}_A = -273.85\hat{i} - 421.69\hat{j} \quad (\text{in./s})$$

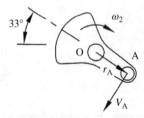

Figure 2.21

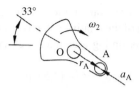

Figure 2.22

Referring to Equation 2.7, the tangential component of acceleration will be zero due to the rotational acceleration being zero. Therefore, the acceleration of A will only be equal to the normal component. Referring to Figure 2.22,

$$\vec{a}_A = \vec{\omega}_2 \times (\vec{\omega}_2 \times \vec{r}_A)$$

$$\vec{a}_A = -125.7\hat{k} \times \left(-125.7\hat{k} \times \left(4\cos(-33)\hat{i} + 4\sin(-33)\hat{j}\right)\right)$$

$$\vec{a}_A = -125.7\hat{k} \times \left(-125.7\hat{k} \times \left(3.3547\hat{i} + -2.1786\hat{j}\right)\right)$$

$$\vec{a}_A = -125.7\hat{k} \times \left(-273.85\hat{i} - 421.6858\hat{j}\right)$$

$$\vec{a}_A = -53006\hat{i} + 34423\hat{j} \quad (\text{in.}/\text{s}^2)$$

Next we move on to the connecting rod (link AB). The angular position of the connecting rod can be determined using law of sines. From Figure 2.23a,

$$\frac{12.5}{\sin 123} = \frac{4}{\sin \alpha} \Rightarrow \alpha = 15.57°$$

The velocity of point B is determined using the relative velocity equation.

$$\vec{v}_B = \vec{v}_A + \vec{v}_{B/A} = \vec{v}_A + \vec{\omega}_3 \times \vec{r}_{B/A}$$

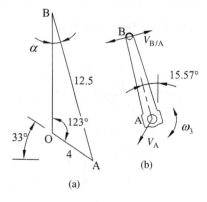

(a)   (b)

Figure 2.23

As shown in Figure 2.23b, the direction of the rotational velocity of link 3 is currently unknown. Therefore, the direction of the relative velocity vector is unknown (though it lies on a line perpendicular to line AB). Assuming a counterclockwise rotational velocity,

$$\vec{v}_B = -273.85\hat{i} - 421.69\hat{j} + \omega_3\hat{k} \times \left(-12.5\sin(15.57)\hat{i} + 12.5\cos(15.57)\hat{j}\right)$$

$$\vec{v}_B = -273.85\hat{i} - 421.69\hat{j} - 12.04\omega_3\hat{i} - 3.36\omega_3\hat{j}$$

The vector equation can be separated into two scalar equations. Taking the x-direction equation (note that the velocity of B is vertical, so its x-component will be zero),

$$0 = -273.85 - 12.04\omega_3 \Rightarrow \omega_3 = -22.74 \, \text{rad/s}$$

The negative sign indicates that the assumed counterclockwise rotation direction was wrong and the rotation direction is clockwise. Taking the y-direction equation,

$$v_B = -421.69 - 3.36(-22.74)$$

Answer: $v_B = 345.2 \, \text{in./s} \downarrow$

The acceleration of point B is determined using relative acceleration. Referring to Figure 2.24,

$$\vec{a}_B = \vec{a}_A + \vec{\alpha}_3 \times \vec{r}_{B/A} + \vec{\omega}_3 \times \left(\vec{\omega}_3 \times \vec{r}_{B/A}\right)$$

The direction of the rotational acceleration is again unknown, so we will assume a positive counterclockwise direction.

$$\vec{a}_B = -53006\hat{i} + 34423\hat{j} + \alpha_3\hat{k} \times \left(-12.5\sin(15.57)\hat{i} + 12.5\cos(15.57)\hat{j}\right)$$
$$- 22.74\hat{k} \times \left(-22.74\hat{k} \times \left(-12.5\sin(15.57)\hat{i} + 12.5\cos(15.57)\hat{j}\right)\right)$$

$$\vec{a}_B = -53006\hat{i} + 34423\hat{j} + \alpha_3\hat{k} \times \left(-3.36\hat{i} + 12.04\hat{j}\right)$$
$$-22.74\hat{k} \times \left(-22.74\hat{k} \times \left(-3.36\hat{i} + 12.04\hat{j}\right)\right)$$

$$\vec{a}_B = -53006\hat{i} + 34423\hat{j} - 12.04\alpha_3\hat{i} - 3.36\alpha_3\hat{j} + 1737.47\hat{i} - 6225.98\hat{j}$$

$$\vec{a}_B = -51268\hat{i} + 28197\hat{j} - 12.04\alpha_3\hat{i} - 3.36\alpha_3\hat{j}$$

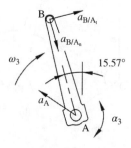

Figure 2.24

The vector equation can be separated into two scalar equations. We will again start with the x-direction equation because acceleration of point B will be purely vertical.

$$0 = -51268 - 12.04\alpha_3 \Rightarrow \alpha_3 = -4134 \quad (\text{rad}/\text{s}^2)$$

The y-direction equation gives

$$a_B = 28197 - 3.36(-4134)$$

Answer: $a_B = 3507 \, \text{ft}/\text{s}^2\uparrow$

2.5.4 Instant Center of Rotation

A rigid body in plane motion can always be thought of as being in pure rotation about some fixed instantaneous center of rotation. The location of the instant center will change for a body in general plane motion, and note that its position will not necessarily be located on the actual rigid body. The velocity of the instant center will be zero; however, the acceleration will generally not be zero. Therefore, the instant center of rotation is only used in velocity analysis.

The instant center of rotation can be found on any rigid body in general plane motion if the velocity directions of two points are known. Consider the rigid body in Figure 2.25a, where the direction of the velocities at points A and B is known. The velocity vector is always perpendicular to the radial line to the center of rotation. Therefore, if a line is drawn perpendicular to each known velocity vector, the intersection of those two lines will be the instant center of rotation (labeled IC on the figure). The direction of velocity vectors at any other point on the rigid body can then be determined. Figure 2.25b shows lines drawn from the instant center to points C and D. The velocity at each point must be perpendicular to those lines. The directions of the velocity must be consistent with the rotational direction of the rigid body and the magnitudes can be determined based on the distance of that point to the instant center.

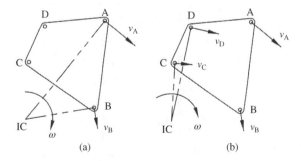

Figure 2.25 (a) Locating the instant center of rotation. (b) Determining velocity of points using the instant center

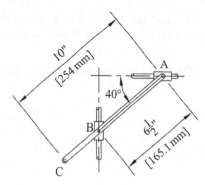

Figure 2.26

Example Problem 2.6

Link ABC shown in Figure 2.26 is connected to two blocks. At the instant shown, block B has a downward velocity of 1.2 m/s. Determine the angular velocity of the rod and the velocity magnitude of point C.

Solution: The velocity of point B is downward and the velocity of point A will be constrained to a horizontal direction. The intersection of lines drawn perpendicular to the velocities will be the instant center, as shown in Figure 2.27. The distance from B to the instant center is required to determine the angular velocity of the rod.

$$\cos 40 = \frac{d_B}{0.1651 \text{ m}} \Rightarrow d_B = 0.126 \text{ m}$$

Because the rod can be considered in pure rotation about the instant center, the angular velocity of the bar can now be determined from the known velocity and radial distance from the instant center.

$$\omega = \frac{v_B}{d_B} = \frac{1.2 \frac{m}{s}}{0.126 \text{ m}}$$

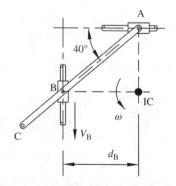

Figure 2.27

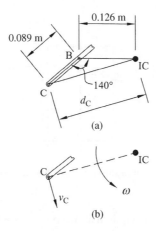

Figure 2.28

Answer: $\omega = 9.5\,\dfrac{\text{rad}}{\text{s}}$ (ccw)

The distance from point C to the instant center is required to find the velocity of point C. From Figure 2.28a and using law of cosines,

$$d_C^2 = 0.089^2 + 0.126^2 - 2(0.089)(0.126)\cos 140$$

$$d_C = 0.202\text{ m}$$

The velocity of C will be perpendicular to the line from C to the instant center as shown in Figure 2.28b. The direction of the velocity will be in the direction shown due to the counterclockwise rotation. The magnitude of the velocity can be determined using equations for pure rotation.

$$v_C = d_C\omega = 0.202\text{ m}\left(9.5\,\frac{\text{rad}}{\text{s}}\right)$$

Answer: $v_C = 1.9\,\dfrac{\text{m}}{\text{s}}$

2.6 Computer Methods

2.6.1 Numerical Differentiation

As mentioned in Chapter 1, computer methods will be utilized throughout the text. As an introduction to using computer methods, numerical differentiation will be discussed here as a tool for estimating velocity and acceleration. Numerical differentiation is a very useful tool in the analysis of machines. In complicated linkage mechanisms, as an example, acceleration calculations can be very tedious and are often required for several positions along a complete cycle of motion. A simpler approach would be to utilize numerical differentiation.

Different methods exist for doing numerical differentiation, but the central difference method is a very common method. Figure 2.29 shows the plot of a function $f(x)$. The slope is required at point p, the function value at x, on that function and will be estimated using central

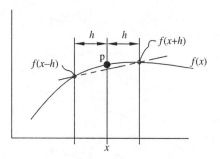

Figure 2.29   Central difference method for estimating slope

difference. The function values are required at two points, one before and one after the point p. Each point will be a distance h from point p as shown.

The slope of the line passing through the points $f(x+h)$ and $f(x-h)$ will be the approximation of the slope at point p. Equation 2.14 is the slope of the line (rise over run) and represents the central difference approximation for the first derivative.

$$f'(x) \approx \frac{f(x+h)-f(x-h)}{2h} \qquad (2.14)$$

The second derivative can be found in a similar fashion by finding the slope of the first derivative plot between half interval points. Equation 2.15 is the central difference equation for the second derivative.

$$f''(x) \approx \frac{f(x+h)-2f(x)+f(x-h)}{h^2} \qquad (2.15)$$

2.6.2 Illustrative Example

To illustrate the application of central difference for velocity and acceleration, let us consider an example. Figure 2.30 shows a 60 inch (1524 mm) long ladder leaning against a 30 inch (726 mm) tall wall.

The ladder starts in a vertical position against the wall. The bottom corner of the ladder is then pulled at a constant rate $v = 2$ in./s. The x and y coordinates of the ladder's center of gravity are recorded at 1 second intervals and shown in Table 2.1. For this problem, we will use the central difference method of numerical differentiation to determine the velocity and acceleration of the center of gravity at 1 second intervals for 15 seconds of motion and plot the results.

Though this problem can be solved analytically, we will use the available data to solve for velocity and acceleration. The pertinent information from the table will be the time and the coordinates of the center of gravity. The velocity in the x direction will be calculated here for a time of 3 seconds to illustrate the process.

$$f'(t) = \frac{f(t+\Delta t)-f(t-\Delta t)}{2\Delta t} = \frac{0.270 \text{ in.} - 0.035 \text{ in.}}{2(1 \text{ s})} = 0.1175 \frac{\text{in.}}{\text{s}} \qquad (2.16)$$

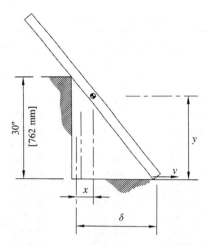

Figure 2.30 Illustrative example

Table 2.1 Data for illustrative problem

Time (s)	δ (in.)	Center of gravity position	
		x (in.)	y (in.)
0	0	0.000	30.000
1	2	0.004	29.934
2	4	0.035	29.737
3	6	0.117	29.417
4	8	0.270	28.987
5	10	0.513	28.461
6	12	0.858	27.854
7	14	1.131	27.186
8	16	1.882	26.471
9	18	2.565	25.725
10	20	3.359	24.962
11	22	4.259	24.192
12	24	5.259	23.426
13	26	6.352	22.671
14	28	7.531	21.937
15	30	8.787	21.213

The acceleration in the x direction at 3 seconds is calculated as

$$f''(t) = \frac{f(t + \Delta t) - 2f(t) + f(t - \Delta t)}{\Delta t^2}$$

$$= \frac{0.270 \text{ in.} - 2(0.117 \text{ in.}) + 0.035 \text{ in.}}{1 \text{ s}^2} = 0.072 \frac{\text{in.}}{\text{s}^2}$$

(2.17)

The process is continued for all times in both the x and y directions. Table 2.2 gives the results for all velocity and acceleration values. Notice that velocity and acceleration cannot be

Table 2.2   Velocity and acceleration of center of gravity

Time (s)	Velocity (in./s)		Acceleration (in./s^2)	
	x	Y	x	y
0	—	—	—	—
1	0.0175	−0.1316	0.0263	−0.1304
2	0.0560	−0.2581	0.0507	−0.1226
3	0.1175	−0.3749	0.0722	−0.1110
4	0.1983	−0.4784	0.0895	−0.0961
5	0.2941	−0.5663	0.1020	−0.0797
6	0.4001	−0.6375	0.1100	−0.0626
7	0.5120	−0.6918	0.1139	−0.0461
8	0.6258	−0.7303	0.1137	−0.0309
9	0.7383	−0.7545	0.1112	−0.0175
10	0.8470	−0.7663	0.1062	−0.0061
11	0.9500	−0.7677	0.0999	0.0034
12	1.0465	−0.7607	0.0930	0.0106
13	1.1357	−0.7472	0.0854	0.0164
14	1.2173	−0.7287	0.0779	0.0205
15	—	—	—	—

calculated for the first term because $f(t - \Delta t)$ does not exist. Similarly, the last point cannot be determined because $f(t + \Delta t)$ does not exist.

Figure 2.31 shows a plot of the velocities and Figure 2.32 shows the accelerations.

To demonstrate the accuracy of the numerical method, we can compare the estimated values obtained using central difference with the actual analytical solution. The distance the bottom corner of the ladder has moved is determined by

$$\delta(t) = v \cdot t = 2t \tag{2.18}$$

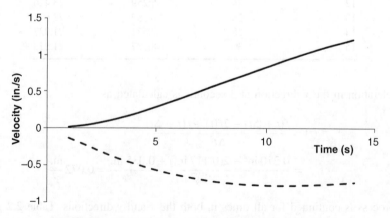

Figure 2.31   Velocity plots using central difference method

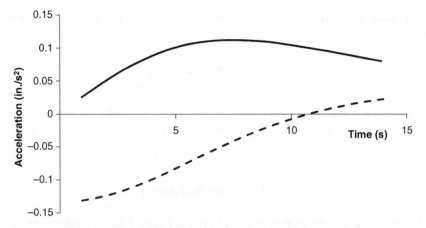

Figure 2.32 Acceleration plots using central difference method

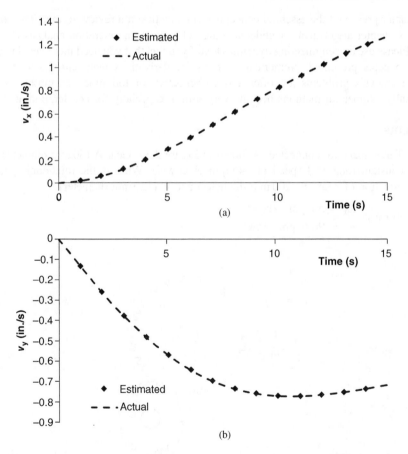

(a)

(b)

Figure 2.33 Comparison of actual velocity and estimated velocity: (a) velocity in the *x* direction; (b) velocity in the *y* direction

The angle between the floor and the lower corner of the ladder can be determined from the right triangle.

$$\theta(t) = \tan^{-1}\left(\frac{h_{\text{wall}}}{\delta(t)}\right) = \tan^{-1}\left(\frac{30}{2t}\right) \tag{2.19}$$

Without proof, the analytical equations for the center of gravity velocity in the x and y directions are given below, where v is the constant velocity of the lower corner of the ladder.

$$v_x(t) = v\left(1 - \sin^3 \theta(t)\right) \tag{2.20}$$

$$v_y(t) = -v\left(\sin^2 \theta(t)\cos \theta(t)\right) \tag{2.21}$$

The analytical solutions are plotted as dashed lines in Figure 2.33. The values obtained from central difference are shown as diamonds.

2.7 Review and Summary

This chapter presented the essential concepts of kinematics in a review format. The concepts from this chapter are critical for understanding velocity and acceleration analysis of mechanisms. Please use any engineering dynamics text if you feel that you need more detailed review of the concepts presented. Numerical methods for derivatives were introduced. Complex machine analysis problems are often too complicated (or too time consuming) to solve analytically. Numerical methods become very useful, especially for acceleration analysis.

Problems

P2.1　Three gears are connected as shown in Figure 2.34. Gear A (80 mm diameter) has a constant rotational speed of 145 rpm clockwise. What is the rotational speed and direction of gears B (200 mm diameter) and C (120 mm diameter)?

Answer: $\quad \omega_B = 58$ rpm　(ccw)
$\qquad\quad \omega_C = 96.6$ rpm　(cw)

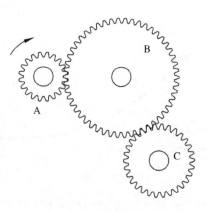

Figure 2.34

P2.2 The disk shown in Figure 2.35 has a radius of 4 inches (101.6 mm) and rotates about the fixed point O at a constant rate of 75 rpm clockwise. Determine the magnitude and direction of the velocities of points A and B.

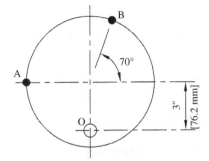

Figure 2.35

P2.3 Determine the magnitude and direction of the normal and tangential accelerations of point A for the disk in Problem P2.2.

P2.4 At the instant shown, the disk in Figure 2.35 has an angular velocity of 10 rad/s clockwise and is accelerating at a rate of 1.4 rad/s². Determine the velocity and acceleration of point B.

$$\vec{v}_B = 67.6\hat{i} - 13.7\hat{j}$$
Answer:
$$\vec{a}_B = -127.3\hat{i} - 677.8\hat{j}$$

P2.5 At the instant shown in Figure 2.36, the 5 inch radius gear A rotates counterclockwise at a rate of 8 rad/s and is decelerating at a rate of 2.3 rad/s². Gear A drives the 8.4 inch radius gear B. The 5.5 inch radius pulley C is on the same shaft as gear B. A belt around pulley C drives pulley D, which has a 9 inch radius. What is the rotational velocity and acceleration of pulley D?

$$\omega_D = 2.91 \text{ rad/s} \quad \text{(cw)}$$
Answer:
$$\alpha_D = 1.37 \text{ rad/s}^2 \quad \text{(ccw)}$$

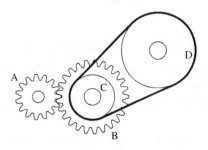

Figure 2.36

P2.6 The hexagon shown in Figure 2.37 is rotating about point O at a speed of 5.6 rad/s clockwise and accelerating at a rate of 2.7 rad/s^2. Determine the point along the perimeter that will have the highest velocity magnitude. Calculate the velocity and acceleration at that point.

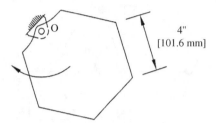

4"
[101.6 mm]

Figure 2.37

P2.7 In the mechanism shown in Figure 2.38, the 8 inch diameter gear A turns at a constant speed of 1100 rpm clockwise and drives the 20 inch diameter gear B. Attached to gear B is a 15 inch diameter disk. Determine the angular velocity of the connecting rod CD and the velocity of slider D for the instant shown.

Answer: $\omega_{CD} = 5.88 \, \text{rad/s}$ (cw)
$v_D = 412 \, \text{in./s} \leftarrow$

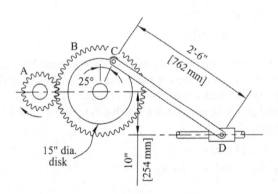

2'-6"
[762 mm]

B
A
25°
15" dia.
disk
10"
[254 mm]
D

Figure 2.38

P2.8 Continue Example Problem 2.4 to determine the rotation speed of link 4 (BO$_4$).

P2.9 The slider–crank mechanism shown in Figure 2.39 has a constant crank (link OA) rotation speed of 850 rpm clockwise. For the position shown when the crank is horizontal determine the velocity and acceleration of the piston (point B) at the instant shown.

P2.10 The slider–crank mechanism in Figure 2.39 is decelerating. At the instant shown, the crank is rotating counterclockwise at a speed of 640 rpm. It is decelerating at a constant

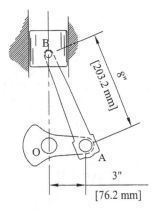

Figure 2.39

rate and comes to a complete stop in 1.2 minutes. Determine the velocity and acceleration of the piston at the instant shown.

Answer: $v_B = 234 \text{ in.}/\text{s} \uparrow$
$a_B = 4411 \text{ in.}/\text{s}^2 \uparrow$

P2.11 The spool shown in Figure 2.40 is unwinding such that point O moves straight down. At the instant shown, the spool has an angular velocity of 2 rad/s and an angular acceleration of 3.5 rad/s² (both counterclockwise). Find the acceleration of point O and point A.

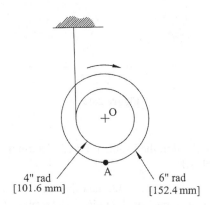

4" rad
[101.6 mm]

6" rad
[152.4 mm]

Figure 2.40

P2.12 The disk shown in Figure 2.41 rotates about point O at a constant speed of 45 rpm. Connecting rod AB is attached to a rack that turns the 6 inch radius gear. What is the rotational speed and rotational acceleration of the gear at the instant shown?

Answer: $\omega = 5.095 \text{ rad}/\text{s}$
$\alpha = 24.72 \text{ rad}/\text{s}^2$

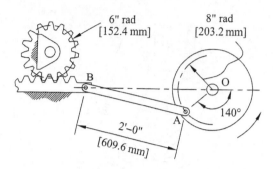

6" rad
[152.4 mm]

8" rad
[203.2 mm]

B

O

140°

A

2'-0"
[609.6 mm]

Figure 2.41

P2.13   The linkage mechanism shown in Figure 2.42 is driven by link O_2A, which rotates at a constant speed of 90 rpm clockwise. Determine the rotational velocity and acceleration of link O_4B in the position shown when the drive link is vertical.

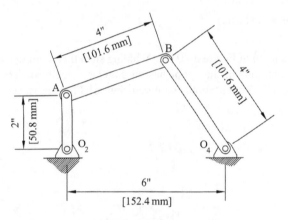

4"
[101.6 mm]

B

4"
[101.6 mm]

A

2"
[50.8 mm]

O_2

O_4

6"
[152.4 mm]

Figure 2.42

P2.14   For the linkage described in the Problem P2.13, determine the velocity magnitude of the midpoint of link AB.

P2.15   For the linkage shown in Figure 2.42, the drive link O_2A rotates at 5 rad/s clockwise and is accelerating at a rate of 1.2 rad/s^2. Determine the rotational velocity and acceleration of link O_4B in the position shown.

Answer: $\omega = 2.44\ \text{rad/s}$ (cw)
$\alpha = 16.73\ \text{rad/s}^2$ (ccw)

P2.16   The linkage mechanism shown in Figure 2.43 is driven by link O_2A. At the instant shown, the driving link is rotating at 8 rad/s clockwise and accelerating at 2.5 rad/s^2. Determine the rotational velocity and acceleration of links AB and O_4B in the position shown.

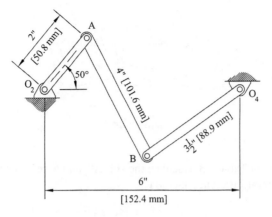

Figure 2.43

P2.17 Velocity measurements were taken experimentally for the linkage mechanism shown in Figure 2.44. At the instant shown, the velocity at point A was measured to be 9.4 in./s and the velocity at point B was measured to be 14.4 in./s (both in the directions shown). Do those velocity terms seem valid? Explain your answer.

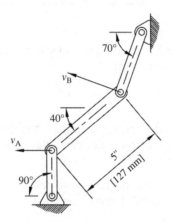

Figure 2.44

P2.18 The crank in the slider–crank mechanism shown in Figure 2.45 oscillates by $\theta(t) = (\pi/6)\sin(\pi t)$. What is the velocity and acceleration of the slider at 1.25 seconds?

$$\theta = -0.37 \text{ rad}$$
Answer: $\omega = -1.163 \text{ rad/s}$
$$\alpha = 3.654 \text{ rad/s}^2$$

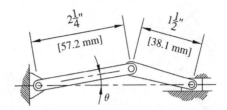

Figure 2.45

P2.19 From the data in Table 2.3, calculate the velocity and acceleration at time 2.5 seconds using the central difference method.

Table 2.3   Position data for Problems P2.19–P2.23

Time (s)	x (in.)	Time (s)	x (in.)	Time (s)	x (in.)
0	−0.92	3.5	2.27	7.0	3.38
0.5	2.40	4.0	2.69	7.5	2.48
1.0	3.09	4.5	0.42	8.0	−0.43
1.5	1.42	5.0	−1.54	8.5	−4.14
2.0	−3.20	5.5	−1.58	9.0	−2.44
2.5	−3.76	6.0	−2.33	9.5	1.82
3.0	0.29	6.5	0.40	10	2.48

P2.20 From the data in Table 2.3, calculate the velocity and acceleration at time 4.0 seconds using the central difference method.

Answer: $v = -1.85$ in./s
$a = 0$

P2.21 From inspection of the data provided, estimate the time when velocity is maximum. What is the magnitude of the velocity at that time? What is the magnitude of acceleration at that time?

P2.22 From the data in Table 2.3, calculate the velocity and acceleration from 5.0 through 6.0 seconds using the central difference method.

P2.23 From the data in Table 2.3, develop a spreadsheet to calculate the velocity and acceleration for all times using the central difference method. Plot the results.

P2.24 Figure 2.46 shows an eccentric cam that moves a swinging arm follower. The angular position of the arm is given in the table for 20° increments of cam rotation. Develop a spreadsheet to calculate the angular velocity and acceleration of the swinging arm using central difference method. The cam rotation speed is 80 rpm. Plot the results.

P2.25 For the slider–crank mechanism shown in Figure 2.39, develop a table of slider displacement values for crank rotations at 20° increments for one complete cycle. The crank rotates counterclockwise at a constant rate of 300 rpm. Using central difference, calculate the slider velocity and acceleration and plot the results for the complete cycle.

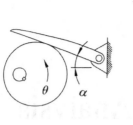

θ	α	θ	α
0	14.98	180	9.73
20	19.89	200	6.44
40	23.26	220	3.58
60	24.67	240	1.40
80	24.30	260	0.17
100	22.57	280	0.18
120	19.92	300	1.72
140	16.71	320	4.92
160	13.23	340	9.59

Figure 2.46

Further Reading

Beer, F.P. and Johnston, E.R. (2007) *Vector Mechanics for Engineers: Dynamics*, McGraw-Hill, New York.
Gray, G.L., Costanzo, F., and Plesha, M.E. (2010) *Engineering Mechanics: Dynamics*, McGraw-Hill, New York.
Greenwood, D.T. (1988) *Principles of Dynamics*, Prentice Hall, New Jersey.
Hibbeler, R.C. (2010) *Engineering Mechanics: Dynamics*, Prentice Hall, New Jersey.
Ramous, A.J. (1972) *Applied Kinematics*, Prentice Hall, New Jersey.

3

Linkage Position Analysis

3.1 Introduction

Analysis of linkage mechanisms begins with concepts concerning the overall mechanism arrangement. Much can be determined about a linkage mechanism simply by examining the length of links and how those links are assembled. Simply changing the arrangement of links in a mechanism will affect the mechanism's motion.

Though numerous types of linkage mechanisms exist, a large portion of this chapter will focus on position analysis of four-bar mechanisms. Four-bar linkage mechanisms are the simplest movable closed linkage possible, yet they are used in a wide variety of machines. Four-bar linkages are commonly found in tools, automotive components, construction equipment, exercise equipment, and even in the ligament arrangement in the human knee joint.

There are many methods that can be used for linkage position analysis. A basic positional analysis method is graphical position analysis, which simply requires drawing the linkage mechanism to scale. The scaled drawing can be done by hand or with the use of CAD software. An obvious advantage of CAD software is the increase in accuracy over hand-drawn methods. Regardless of how the scaled drawing is produced, the method of graphical position analysis is the same. The first step requires drawing each link to scale and in the desired arrangement. Once the scaled drawing is complete, all required positional information can simply be measured off the drawing. A major drawback to graphical position analysis is the fact that it is a time-consuming process and, therefore, is not feasible for position measurements in multiple locations of the linkage.

Though graphical position analysis is simple and straightforward, more accurate results can often be obtained by analytical position analysis (especially over results obtained from hand-drawn graphical methods). Analytical position analysis requires the development of geometric equations that will give positional outputs based on inputs of link lengths and angular position of driving link. A major advantage of analytical position analysis is the ability to automate the calculations to allow for position analysis for multiple arrangements of the linkage mechanism, because sometimes it is important to trace the path of a point on a linkage for one complete cycle. Several analytical position analysis methods exist, but this chapter will focus on methods utilizing geometrical analysis and position vectors.

Machine Analysis with Computer Applications for Mechanical Engineers, First Edition. James Doane.
© 2016 John Wiley & Sons, Ltd. Published 2016 by John Wiley & Sons, Ltd.
Companion Website: www.wiley.com/go/doane0215

Another important aspect of position analysis deals with the functionality of linkage mechanism. Some measures of functionality include mobility, transmission angle, toggle positions, and link rotation. Functionality measures are a critical part of linkage mechanism analysis because they help the designer determine how effective and efficient a particular linkage mechanism will be for a certain task.

Regardless of method used for position analysis, results from position analysis can serve many purposes. Some mechanisms may only require position analysis to determine functionality of the mechanism. For other mechanisms, typically those with higher velocity or acceleration, results from position analysis are an essential first step toward kinematic and kinetic analysis.

3.2 Mobility

3.2.1 Rigid Body Degrees of Freedom

We will begin our study of linkage analysis with a measure of functionality known as mobility, which is the system's number of degrees of freedom. The number of independent coordinates necessary to completely describe the position of an object is known as degrees of freedom (DOF). Consider a rigid body moving in a two-dimensional plane as shown in Figure 3.1. To determine the mobility of the rigid body, we must determine how many distinct ways it can move in the plane. For example, we could move the body horizontally along the x-axis or vertically along the y-axis. The rigid body could also be rotated about the z-axis. Therefore, it can be determined that a rigid body in plane motion will have two translational degrees of freedom (Δx and Δy) and one rotational degree of freedom ($\Delta \theta$).

3.2.2 Joint Mobility

In a linkage mechanism, the links are considered rigid bodies, and the individual links are connected by joints. Each joint serves as a kinematic constraint, which removes degrees of freedom. Different joints in a linkage mechanism allow different degrees of freedom between the links joined. Joints in planar mechanisms will allow either one degree of freedom or two degrees of freedom. Joints allowing one degree of freedom between connecting members are known as full joints. Recalling the six lower pairs described in Section 1.4.3, one rotating joint (revolute pair) and one translational joint (prismatic pair) are full joints. As an example,

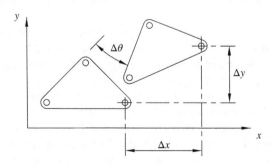

Figure 3.1 Degrees of freedom for plane motion of a rigid body

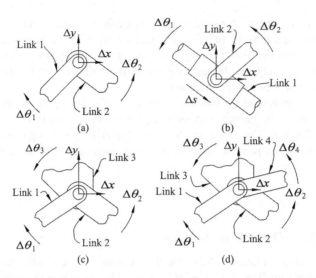

Figure 3.2   Degrees of freedom for (a) a full joint, (b) a half joint, (c) a double joint, and (d) a triple joint

Figure 3.2a shows two links connected by a revolute pair (full joint). If separated, each link would have three degrees of freedom giving a total of six degrees of freedom for the system. With the two links connected by a full joint, the system now has a total of four degrees of freedom (Δx, Δy, $\Delta\theta_1$, and $\Delta\theta_2$) because the connection forces link 1 and link 2 to move together at the joint. Therefore, the full joint reduces the system degrees of freedom by two.

Joints allowing two degrees of freedom are known as half joints. Higher pairs, which are planar joints with point contact, are half joints because they allow two degrees of freedom (rotation and translation). Figure 3.2b shows two links connected by a half joint. Again, the total degrees of freedom of the system would be six if the links were not connected. The system has five degrees of freedom (Δx, Δy, Δs, $\Delta\theta_1$, and $\Delta\theta_2$) once the connection is made. Therefore, the half joint reduces the degrees of freedom of the system by one.

Some mechanisms contain joints, called multiple joints, connecting k links at a single connection point. As an example, the double joint shown in Figure 3.2c connects a link to two other links with one joint. Because there are three links involved, the total degrees of freedom would be nine (three links times three degrees of freedom per link) if the links were not connected. Once the connection is made, there are a total of five degrees of freedom (Δx, Δy, $\Delta\theta_1$, $\Delta\theta_2$, and $\Delta\theta_3$) for the system, which gives a reduction of four degrees of freedom for a double joint.

Figure 3.2d shows an example of a triple joint, which is a joint connecting four links together. Unconnected, the four links would have 12 degrees of freedom. As seen in the figure, the triple joint has six degrees of freedom (Δx, Δy, $\Delta\theta_1$, $\Delta\theta_2$, $\Delta\theta_3$, and $\Delta\theta_4$) resulting in a reduction of six degrees of freedom from a triple joint.

In general, the order of a multiple joint is one less than the number of links connected. A double joint is second order and connects three links, meaning that the joint acts as two full joints. Each of the two full joints reduces the degrees of freedom by two, which gives a total reduction of four. Similarly, a triple joint is third order because it connects four links. A triple joint acts as three full joints, each removing two degrees of freedom giving a reduction of six degrees of freedom. This concept can be expanded to multiple joints of any order.

3.2.3 Determining Mobility of a Planar Linkage Mechanism

With an understanding of joint mobility, we can now examine degrees of freedom of complete linkage mechanisms. Consider N unconnected rigid links in a plane with one of the links acting as the fixed frame or ground link (see Section 1.4.2 for discussion of ground link). The remaining $N - 1$ moving links will each have three degrees of freedom making a total of $3(N - 1)$ degrees of freedom for the system. Connections must now be made between links, and each connection reduces the total mobility of the system. Every full joint removes two degrees of freedom and every half joint removes one degree of freedom. The total mobility of a linkage system can then be determined from Kutzbach's equation.

$$M = 3(N - 1) - 2j_1 - j_2 \tag{3.1}$$

where M is mobility (degrees of freedom), N is number of links, j_1 is number of full joints, and j_2 is number of half joints.

All mechanisms will have a positive, nonzero mobility. A system with mobility of zero, known as a structure, will not be capable of motion. Negative mobility results in a preloaded structure, which is not capable of motion and requires links to have some internal stress in order to be connected.

It is very important to note that Kutzbach's equation is general and exceptions do exist. However, there is no rule to predict when Kutzbach's equation will give invalid results. Therefore, it can be beneficial to have an alternative method of mobility analysis. Another method, which is based on concepts of graph theory, is called loop mobility criteria. Burton Paul discusses this method in detail.

Example Problem 3.1

For each linkage mechanism shown in Figure 3.3, (a) determine the number of links, (b) label all full and half joints, and (c) determine mobility.

Solution: The first linkage mechanism shown is a basic four-bar mechanism. The first step is to draw the kinematic diagram as shown in Figure 3.4.

a. The link numbers are shown on the kinematic diagram in circles. Do not forget to count the ground link.

$$N = 4$$

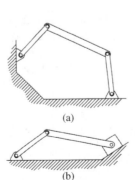

(a)

(b)

Figure 3.3

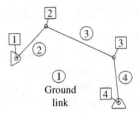

Figure 3.4

b. All joints are revolute joints, which are full joints. Each joint number is shown on the kinematic diagram in squares.

$$j_1 = 4 \qquad j_2 = 0$$

c. The mobility is determined from Kutzbach's equation (Equation 3.1).

$$M = 3(N - 1) - 2j_2 - j_1$$
$$M = 3(4 - 1) - 2(4) - 0$$

$$M = 1$$

The second linkage mechanism shown is a four-bar slider–crank mechanism. The first step is to draw the kinematic diagram, which is shown in Figure 3.5.

a. The link numbers are shown on the kinematic diagram in circles. The slider counts as link 4.

$$N = 4$$

b. Joints (shown in squares) 1–3 are full revolute joints. Joint 4 is a sliding full joint, which only has one DOF of translation.

$$j_1 = 4 \qquad j_2 = 0$$

c. The mobility is determined from Kutzbach's equation (Equation 3.1).

$$M = 3(N - 1) - 2j_2 - j_1$$
$$M = 3(4 - 1) - 2(4) - 0$$

$$M = 1$$

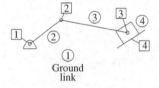

Figure 3.5

Example Problem 3.2

For the linkage mechanism shown in Figure 3.6, (a) determine the number of links, (b) label all full and half joints, and (c) determine mobility.

Solution: The first step is to draw the kinematic diagram as shown in Figure 3.7.

a. The link numbers are shown on the kinematic diagram in circles.

$$N = 6$$

b. Because of the multiple joint (double joint) connecting links 2, 3, and 4, there are five revolute joints and one sliding full joint (joint 6).

$$j_1 = 6$$

Point contact between link 5 and the ground counts as one half joint.

$$j_2 = 1$$

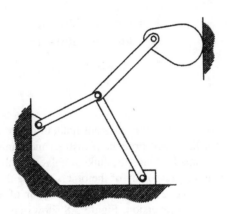

Figure 3.6

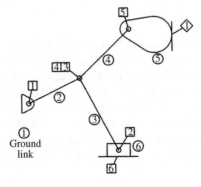

Figure 3.7

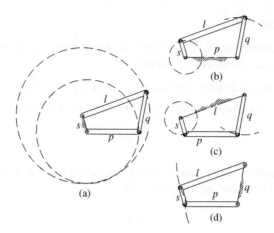

Figure 3.8 Grashof linkage inversions

c. The mobility is determined from Kutzbach's equation (Equation 3.1).

$$M = 3(N - 1) - 2j_2 - j_1$$
$$M = 3(6 - 1) - 2(6) - 1$$

$$M = 2$$

3.3 Inversion

Kinematic inversion is a process of fixing different links to create different mechanisms. A closed chain consisting of n links will produce n distinct mechanisms simply by choosing different links as the frame (link 1). Since the links involved in all inversions are identical, relative motion of a link with respect to any of the others will be the same for all inversions. Even though kinematic inversion does not affect relative motion of links, the absolute motion of links will be different for each inversion. Figure 3.8 shows an example of the kinematic inversions of a four-bar mechanism.

3.4 Grashof's Criterion

3.4.1 Introduction

In some four-bar linkage mechanisms, such as a crank–rocker mechanism, it is critical that at least one link (the crank) is capable of a full 360° of rotation. A relationship developed by Grashof (1883) aids in the analysis of the movability of the linkage mechanism. Grashof's criterion identifies the ability of at least one link in a four-bar linkage to complete 360° of rotation.

Consider a four-bar planar mechanism with all revolute joints. Let s be the shortest link length, l be the longest link length, and p and q be the lengths of the two remaining links. Grashof's condition states that for at least one link to rotate 360° the sum of the lengths of the longest and shortest links must be less than or equal to the sum of the other two links.

$$s + l \leq p + q \tag{3.2}$$

Using the Grashof condition, we can predict the rotation ability of links within a four-bar linkage based solely on the length of all links and without regard to order of attachment.

3.4.2 Grashof Linkage

A four-bar linkage mechanism that satisfies the Grashof condition with the inequality $(s + l < p + q)$ is known as a class I kinematic chain, or Grashof linkage. Motion of a Grashof linkage mechanism depends on the arrangement of the four links. Double crank mechanisms are created by grounding the short link. Figure 3.8a shows a Grashof linkage with the short link grounded.

Both links connected to the ground link make full rotations causing both of them to be cranks. When the short link serves as the input link (link 2), you get a crank–rocker mechanism. Figure 3.8b and c shows two inversions of the same Grashof linkage, but now a link adjacent to the short link is grounded. The short link, known as the crank, will make a full rotation. The rocker, the link opposite to the short link, will oscillate in a rocking motion. A mechanism, as shown in Figure 3.8d, is known as a Grashof double rocker. A double rocker is created when the coupler is the shortest link.

3.4.3 Non-Grashof Linkage

Failure to satisfy the Grashof condition $(s + l > p + q)$ leads to a class II kinematic chain, or non-Grashof linkage. None of the links of a non-Grashof linkage will be capable of completing a full revolution, which causes all inversions to be triple rocker mechanisms.

3.4.4 Special Case Grashof Linkage

For cases when the Grashof condition is satisfied by the equality $(s + l = p + q)$, the linkage is known as a class III kinematic chain, or special case Grashof linkage. Full rotation of a link will be possible; however, special case Grashof linkages have collinear arrangements, called change points, which occur at two locations during one complete cycle. One collinear position is shown in Figure 3.9a. Motion of the linkages after change points will have two possible outcomes, as shown in Figure 3.9b and c. Because of the uncertainty, change points should be avoided or controlled.

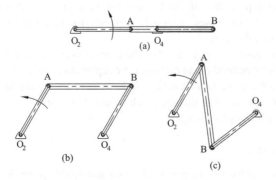

Figure 3.9 Change points of a class III kinematic chain

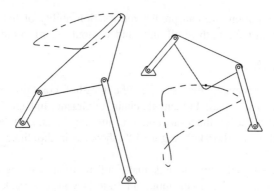

Figure 3.10   Coupler curve examples

3.5   Coupler Curves

3.5.1   Basic Concepts of Coupler Curves

In a four-bar mechanism, the two links adjacent to the ground link (typically labeled as links 2 and 4) move in pure rotation. However, motion of the coupler link (typically labeled as link 3) can be very complex. A curve, known as a coupler curve, can be created that represents the path of motion of a coupler point (a particular point on the coupler) during one cycle of motion. The number of coupler curves that can be created is infinite, due to the unlimited choices of coupler points. Figure 3.10 shows two different four-bar linkage mechanisms and coupler curves for a specific coupler point. It is interesting to note that the first mechanism is a Grashof linkage (crank–rocker) while the second mechanism is a non-Grashof mechanism. Therefore, closed coupler curves will be generated regardless of Grashof condition.

There are several methods for generating coupler curves. A time-consuming yet simple method is to draw the mechanism to scale in several different positions, marking the location of the coupler point for each. Once several positions have been determined, the points can be connected with a spline to create the coupler curve. CAD software expedites the process, and a coupler curve can be created relatively easily using this method.

Another method for generating coupler curves is to analytically determine coordinate values for the location of the coupler point for several positions. The coordinates for each location can then be plotted to create the coupler curve. Because this method requires calculating coordinates for several positions of the linkage mechanism, this method is best suited for computers. Section 3.11 discusses computer methods for position analysis that could be utilized to generate coupler curves.

There are also methods to directly calculate the coupler curve as a mathematical expression. Because of the coupler's general plane motion, the coupler curve shape is often very difficult to express mathematically. In general, four-bar linkage coupler curves are of the sixth order. Certain configurations of four-bar linkage mechanisms, however, will generate less complex fourth- or second-order coupler curves.

3.5.2   Double Points

Any point on a curve that has two tangents is known as a double point. It is fairly common for coupler curves to contain double points in the form of cusps or crunodes. To illustrate a cusp,

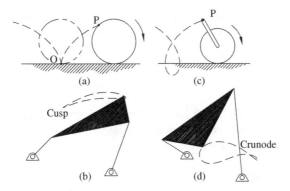

Figure 3.11   Double points: (a) cusp; (b) coupler curve with a cusp; (c) crunode; (d) coupler curve with a crunode

consider the path traced by a point located on the surface of a rolling wheel as shown in Figure 3.11a. The curve created by point P is known as a cycloid. As the wheel rotates into the position represented by the phantom line, point P follows the cycloid to point O. Point O is the instantaneous center of rotation in that position and therefore has zero velocity. As the wheel continues to roll, point P moves away from point O in the opposite direction. Figure 3.11b shows the coupler curve with a cusp. At a cusp, velocity is zero yet motion is continuous.

A crunode is a point where a curve crosses itself. Figure 3.11c shows a rolling wheel and the trace of the path of point P, which is now located at a radius larger than that of the wheel. An example of a coupler curve containing the crunode can be seen in Figure 3.11d.

3.5.3   Hrones and Nelson Atlas

As stated earlier, an infinite number of coupler curves exist for any four-bar linkage mechanism. Sometimes in design, or synthesis, a particular coupler curve shape is desired (see Chapter 5 for more discussion). The desired coupler curve shape can be obtained with the proper choice of coupler point location in proportions of links. The trial and error process required in finding the proper combination of coupler point location and linkage proportions would be very tedious.

An atlas entitled *Analysis of the Four-Bar Linkage: Its Applications to the Synthesis of Mechanisms* was developed by John A. Hrones and George L. Nelson (1951) and serves as a useful tool to aid in the design of a linkage mechanism to generate a specific coupler curve. The atlas contains large-scale (11 inch by 17 inch) plots of over 7000 coupler curves. Coupler curves were generated using several linkage proportions, which give the length of links (coupler, rocker, and frame) relative to a unit length of the crank. For each linkage proportion given, multiple coupler curves are plotted from a matrix of coupler point locations.

A designer can look through the Hrones and Nelson atlas to find a coupler curve shape appropriate for a particular design. The linkage arrangement that produces the desired coupler curve can then be scaled up or down without affecting the general shape of the coupler curve. Another piece of valuable information can be determined from the drawings in the Hrones and Nelson atlas. The coupler curves in the atlas are shown as dashed lines. Closer inspection shows that the length of the dashes is not consistent. From the start of one dash (point a) to the start of the successive dash (point b), it represents five degrees of rotation of the input link.

Longer dashes, therefore, represent faster velocity compared with short dashes. By examining dash lengths, average velocity can be determined from the Hrones and Nelson atlas by the following equation:

$$v_{ab} = \frac{(ab)\omega}{5} \tag{3.3}$$

where v_{ab} is average velocity of point moving along trajectory from point a to point b (inches per second), ab is length between a and b along trajectory (inches), and ω is the angular speed of drive crank (degrees per second).

3.6 Cognate Linkages

3.6.1 The Roberts–Chebyshev Theorem

Samuel Roberts (1827–1913) and Pafnuty Chebyshev (1821–1894), nineteenth-century mathematicians and members of the Royal Society, both independently studied four-bar linkage coupler curves. Though the theorem title includes both names, the discoveries were made at different times (Roberts in 1875 and Chebyshev in 1878). The theorem claims that three different planar four-bar linkages will trace identical coupler curves. It is important to note that the three different four-bar linkages are geometrically different and will most likely have different velocity and acceleration characteristics. The only commonality is the trace of the coupler curve at a particular point.

The following presentation of how to develop cognate linkages differs from the original works of Roberts and Chebyshev. We will use Cayley diagrams, named after the nineteenth-century British mathematician Arthur Cayley, to construct cognates in a very graphical and less mathematical way. The Cayley diagram (discussed in the following section) is a very quick and easy method to find the dimensions of the links and shape of the coupler for each cognate.

3.6.2 Steps for Determining Cognates

We wish to find the two cognates for the four-bar linkage shown in Figure 3.12a. The process begins by temporarily decoupling the fixed points O_2 and O_4. While holding the position

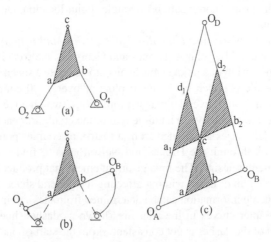

Figure 3.12 Construction of a Cayley diagram

of link 3 fixed, rotate links 2 and 4 so that they are collinear with link 3 (Figure 3.12b) creating two new points O_A and O_B.

The Cayley diagram, shown in Figure 3.12c, is then constructed using the new configuration of the linkage system using the following steps:

Step 1: A line parallel to line ac is drawn passing through point O_A.

Step 2: A line parallel to line bc is drawn passing through point O_B.

Step 3: The intersection of the two new parallel lines locates point O_D.

Step 4: A line parallel to line $O_A O_B$ is drawn through point c to create line $a_1 c b_2$.

Step 5: Line ac is extended to the intersection of line $O_B O_D$ to locate point d_2.

Step 6: Line bc is extended to the intersection of line $O_A O_D$ to locate point d_1.

The Cayley diagram gives important dimensional information for the two cognates. One cognate will consist of link 3, defined by triangle $a_1 c d_1$, and links 2 and 4 with lengths defined by $O_A a_1$ and $O_D d_1$, respectively. The second cognate will consist of link 3, defined by triangle $b_2 c d_2$, and links 2 and 4 with lengths defined by $O_B b_2$ and $O_D d_2$, respectively. Final arrangements of the two cognates cannot be defined until the length of each ground link is determined.

The next step is to rotate links 2 and 4 of the original linkage back to the initial fixed pivot points O_2 and O_4 while maintaining all the connections between other links. Figure 3.13a shows the result of this rotation known as the Roberts diagram, which shows the final arrangement of the cognates. Separating the linkage mechanisms gives the original linkage and the two cognates as shown in Figure 3.13b.

It should be noted that the Roberts diagram can be created without the use of the Cayley diagram. Referring again to Figure 3.13a, line $a_1 c$ is parallel to line $O_2 a$ and line $O_2 a_1$ is parallel to line ac. In other words, lines $a_1 c$ and $O_2 a_1$ can be determined by completing the parallelogram $O_2 a_1 c a$. Similarly, lines $c b_2$ and $O_4 b_2$ can be determined by completing the parallelogram $O_4 b c b_2$. The remaining sides of the coupler links for the two cognates can be constructed by noting that the three coupler links form similar triangles. Angles $a_1 d_1 c$ and $c d_2 b_2$ will be equal to angle acb of the original coupler. Similarly, angles $a_1 c d_1$ and $c b_2 d_2$ will be equal to angle abc of the original coupler. The final lines, $O_D d_1$ and $O_D d_2$, are constructed by completing the parallelogram $O_D d_1 c d_2$.

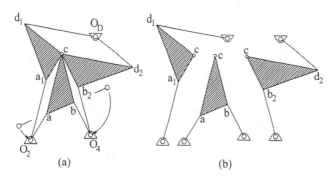

Figure 3.13 (a) Roberts diagram. (b) Final cognate linkages

The method described above is not limited to four-bar linkage mechanisms with a ternary coupler link. The Roberts–Chebyshev theorem works for four-bar linkage mechanisms with the coupler point located on a binary coupler, and it can be expanded to generate a cognate for slider–crank mechanisms. First let us consider a case where the coupler point is located on a binary coupler, as shown in Figure 3.14a.

3.6.3 Cognates for Binary Link

Figure 3.14a shows a four-bar linkage mechanism and the coupler curve for point P located on the binary coupler.

The cognates are constructed using the following steps (steps 1–6 are illustrated in Figure 3.14b and steps 7–9 are illustrated in Figure 3.14c):

Step 1: Draw a construction line parallel to O_2A through point P.

Step 2: Draw a construction line parallel to AB through point O_2.

Step 3: Point A_1 will be located at the intersection of the two construction lines.

Step 4: Draw a construction line parallel to O_4B through point P.

Step 5: Draw a construction line parallel to AB through point O_4.

Step 6: Point B_1 will be located at the intersection of the two construction lines.

Step 7: Point O_C is located on line O_2O_4 such that $AP/PB = O_2O_C/O_CO_4$.

Step 8: Point C_1 is located such that $AP/PB = A_1C_1/C_1P$.

Step 9: Point C_2 is located such that $AP/PB = PC_2/C_2B_1$.

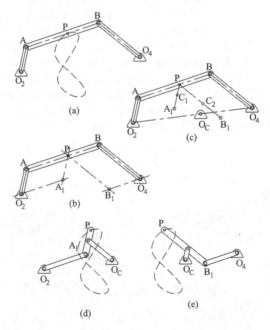

Figure 3.14 Construction of cognates for a four-bar mechanism with a binary coupler

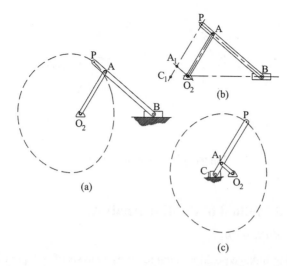

Figure 3.15   Construction of cognates for a slider–crank mechanism

Once all points have been located, the two cognates can be drawn. The original linkage mechanism and both cognates are shown in Figure 3.14 along with coupler curves for each. Figure 3.14a shows the original linkage and Figure 3.14d and e shows the two cognates. It can be seen that all three coupler curves are identical.

3.6.4   Cognates for Slider–Crank Mechanism

For slider–crank mechanisms, the Roberts–Chebyshev theorem is modified to state that two different slider–crank mechanisms will trace identical coupler curves. Figure 3.15a shows a four-bar slider–crank mechanism and the coupler curve for point P. The process of finding cognates is similar to that for the binary link described in the previous section. However, point O_4 is located at infinity causing only one cognate to be possible. A line parallel to line O_2A is drawn through point P, and a line parallel to AB is drawn through point O_2. The intersection defines point A_1. Point C_1 is located based on ratios developed for PAB being equal to ratios developed for C_1A_1P. Steps for locating points A_1 and C_1 are shown in Figure 3.15b. The final cognate is shown in Figure 3.15c along with the coupler curve.

3.7   Transmission Angle

It is useful to have some term that would give a good indication of the effectiveness of transferring motion to the output link. Linkage mechanisms that transfer this motion effectively will have smoother operation and will have the highest mechanical advantage. The transmission angle (μ), which is the acute angle between the coupler and the output link, is commonly used to indicate the effectiveness of motion transfer. Figure 3.16 shows the transmission angle for a typical four-bar linkage mechanism. A transmission angle of 90° is the optimal value. For small values of μ, less than 45°, force transferred to the output link tangentially becomes small and more of the force is exerted along the length (radially) of the output link.

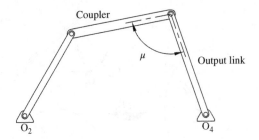

Figure 3.16   Transmission angle

3.8   Geometrical Method of Position Analysis

3.8.1   Essential Mathematics

Geometrically solving linkage position often requires analysis of oblique triangles, which are triangles that do not contain a right angle. Typically, the length of all sides of the triangle will be known, and the analysis involves finding unknown angles. Two essential trigonometric tools for solving oblique triangles are the law of sines and the law of cosines.

The law of sines states that the length of any side of the triangle is proportional to the sine of the opposite angle. Therefore, the law of sines for the oblique triangle shown in Figure 3.17 gives

$$\frac{a}{\sin \alpha} = \frac{b}{\sin \beta} = \frac{c}{\sin \gamma} \qquad (3.4)$$

For the same oblique triangle, the law of cosines gives the following set of equations:

$$a^2 = b^2 + c^2 - 2bc \cos \alpha$$
$$b^2 = a^2 + c^2 - 2ac \cos \beta \qquad (3.5)$$
$$c^2 = a^2 + b^2 - 2ab \cos \gamma$$

3.8.2   Common Approaches for Four-Bar Mechanisms

It is often desirable to define positions of all links of a single degree-of-freedom mechanism, based only on the angular position of the input link. First, we will analyze a four-bar linkage mechanism in the open arrangement, as shown in Figure 3.18a. Equations can be developed to give transmission angle (μ) and output angle (θ_4) in terms of the input angle (θ_2) by drawing a line connecting point A to point O_4, which divides the mechanism into two triangles. These

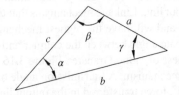

Figure 3.17   Oblique triangle for law of sines and law of cosines

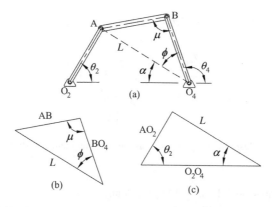

Figure 3.18 Position analysis for an open arrangement four-bar mechanism

two triangles, upper triangle ABO_4 and lower triangle O_2AO_4, can be analyzed individually using basic geometry.

The law of cosines for the upper triangle, shown in Figure 3.18b, gives the following:

$$L^2 = (AB)^2 + (BO_4)^2 - 2(AB)(BO_4)\cos \mu \tag{3.6}$$

Similarly, the law of cosines for the lower triangle, shown in Figure 3.18c, yields the following:

$$L^2 = (AO_2)^2 + (O_2O_4)^2 - 2(AO_2)(O_2O_4)\cos \theta_2 \tag{3.7}$$

The length L must be the same for the upper and lower triangles. Therefore, the right-hand sides of Equations 3.6 and 3.7 must be equal. Equating the upper and lower triangle equations gives

$$(AB)^2 + (BO_4)^2 - 2(AB)(BO_4)\cos \mu = (AO_2)^2 + (O_2O_4)^2 - 2(AO_2)(O_2O_4)\cos \theta_2 \tag{3.8}$$

The angle μ shown in Figure 3.18a is the transmission angle. As stated in Section 3.7, the transmission angle is a very useful term. Rearranging Equation 3.8 will give transmission angle in terms of angular position of input link.

$$\mu = \cos^{-1}\left[\frac{(AB)^2 + (BO_4)^2 - (AO_2)^2 - (O_2O_4)^2 + 2(AO_2)(O_2O_4)\cos \theta_2}{2(AB)(BO_4)}\right] \tag{3.9}$$

Another useful measure is the angle of the output link. Figure 3.18a shows the angle of the output link as θ_4. The output angle can be calculated from

$$\theta_4 = 180° - (\phi + \alpha) \tag{3.10}$$

where ϕ and α are determined using law of cosines for the upper and lower triangles.

$$(AB)^2 = L^2 + (BO_4)^2 - 2L(BO_4)\cos \phi$$
$$(AO_2)^2 = L^2 + (O_2O_4)^2 - 2L(O_2O_4)\cos \alpha \tag{3.11}$$

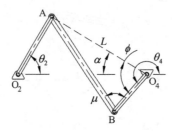

Figure 3.19 Position analysis for a crossed arrangement four-bar mechanism

We can rearrange Equation 3.11 to get angles ϕ and α.

$$\phi = \cos^{-1}\left[\frac{L^2 + (BO_4)^2 - (AB)^2}{2L(BO_4)}\right]$$

$$\alpha = \cos^{-1}\left[\frac{L^2 + (O_2O_4)^2 - (AO_2)^2}{2L(O_2O_4)}\right]$$

(3.12)

Equation 3.12 can be substituted into Equation 3.10 to give the final expression for output angle.

A similar analysis can be done for a four-bar mechanism in the crossed arrangement, such as the one shown in Figure 3.19. As with the analysis of open arrangement, a line connecting point A to point O_4 is drawn to divide the mechanism into two triangles (triangle ABO_4 and triangle O_2AO_4). Using the same procedure as before of analyzing each triangle individually, it can be determined that the equation for transmission angle is identical to that obtained for the open arrangement. Therefore, the transmission angle for the crossed arrangement is calculated using Equation 3.9. The equation for the output angle, however, differs slightly from that obtained for the open arrangement. Referring to Figure 3.19, the output angle is defined by the following, where angles ϕ and α are calculated using Equation 3.12 derived earlier for open arrangement.

$$\theta_4 = 180° + (\phi - \alpha)$$

(3.13)

Example Problem 3.3

For the four-bar linkage mechanism shown in Figure 3.20, determine the transmission angle and the value of θ_4.

Solution: The transmission angle can be found using Equation 3.9.

$$\mu = \cos^{-1}\left[\frac{(AB)^2 + (BO_4)^2 - (AO_2)^2 - (O_2O_4)^2 + 2(AO_2)(O_2O_4)\cos\theta_2}{2(AB)(BO_4)}\right]$$

$$\mu = \cos^{-1}\left[\frac{12^2 + 7.5^2 - 7^2 - 19^2 + 2(7)(19)\cos(35)}{2(12)(7.5)}\right]$$

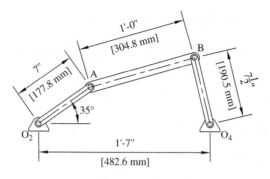

Figure 3.20

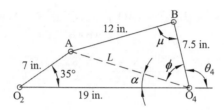

Figure 3.21

Answer: $\mu = 87.4°$

The next step is to divide the mechanism into two oblique triangles as shown in Figure 3.21. The dividing line length is determined from Equation 3.7.

$$L^2 = (AO_2)^2 + (O_2O_4)^2 - 2(AO_2)(O_2O_4)\cos\theta_2$$
$$L^2 = 7^2 + 19^2 - 2(7)(19)\cos(35)$$
$$L = 13.86''$$

To determine θ_4, we must first find the values of α and ϕ, which can be determined by using Equation 3.12.

$$\phi = \cos^{-1}\left[\frac{L^2 + (BO_4)^2 - (AB)^2}{2L(BO_4)}\right]$$
$$\phi = \cos^{-1}\left[\frac{13.86^2 + 7.5^2 - 12^2}{2(13.86)(7.5)}\right]$$
$$\phi = 59.87°$$

$$\alpha = \cos^{-1}\left[\frac{L^2 + (O_2O_4)^2 - (AO_2)^2}{2L(O_2O_4)}\right]$$
$$\alpha = \cos^{-1}\left[\frac{13.86^2 + 19^2 - 7^2}{2(13.86)(19)}\right]$$
$$\alpha = 16.84°$$

The value of θ_4 can now be calculated using Equation 3.10.

$$\theta_4 = 180° - (\phi + \alpha) = 180° - (59.87° + 16.84°)$$

Answer: $\theta_4 = 103.29°$

Example Problem 3.4

For the four-bar linkage mechanism shown in Figure 3.22, determine the transmission angle and the value of θ_4.

Solution: The transmission angle can be found using Equation 3.9.

$$\mu = \cos^{-1}\left[\frac{(AB)^2 + (BO_4)^2 - (AO_2)^2 - (O_2O_4)^2 + 2(AO_2)(O_2O_4)\cos\theta_2}{2(AB)(BO_4)}\right]$$

$$\mu = \cos^{-1}\left[\frac{4.5^2 + 4^2 - 3.5^2 - 8^2 + 2(3.5)(8)\cos(45)}{2(4.5)(4)}\right]$$

Answer: $\mu = 90.6°$

The next step is to divide the mechanism into two oblique triangles as shown in Figure 3.23. The dividing line length is determined from Equation 3.7.

$$L^2 = (AO_2)^2 + (O_2O_4)^2 - 2(AO_2)(O_2O_4)\cos\theta_2$$

$$L^2 = 3.5^2 + 8^2 - 2(3.5)(8)\cos(45)$$

$$L = 6.05''$$

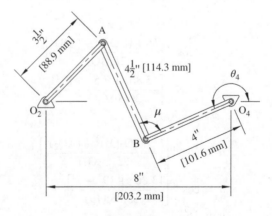

Figure 3.22

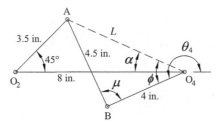

Figure 3.23

To determine θ_4, we must first find the values of α and ϕ, which can be determined using Equation 3.12.

$$\phi = \cos^{-1}\left[\frac{L^2 + (BO_4)^2 - (AB)^2}{2L(BO_4)}\right]$$

$$\phi = \cos^{-1}\left[\frac{6.05^2 + 4^2 - 4.5^2}{2(6.05)(4)}\right]$$

$$\phi = 48.05°$$

$$\alpha = \cos^{-1}\left[\frac{L^2 + (O_2O_4)^2 - (AO_2)^2}{2L(O_2O_4)}\right]$$

$$\alpha = \cos^{-1}\left[\frac{6.05^2 + 8^2 - 3.5^2}{2(6.05)(8)}\right]$$

$$\alpha = 24.11°$$

The value of θ_4 can now be calculated using Equation 3.13.

$$\theta_4 = 180° + (\phi - \alpha) = 180° + (48.05° - 24.11°)$$

Answer: $\theta_4 = 203.94°$

Now that we have developed expressions for angular positions of links, another useful positional measure would be coordinate positions of specific points on the mechanism. For example, it may be useful to have Cartesian coordinates of points A and B (two end connection points for the coupler) for the mechanism shown in Figure 3.24 in terms of only the angular position of the input link (θ_2). Referring to Figure 3.24, the coordinate location of point A can easily be expressed in terms of input angle.

$$a_x = (AO_2)\cos\theta_2$$
$$a_y = (AO_2)\sin\theta_2$$
(3.14)

Representing the coordinate location of point B in terms of the input angle is not as direct and requires the use of equations of circles. The equation of a circle with radius r and center point located at (h, k) is defined as $r^2 = (x - h)^2 + (y - k)^2$. Point B (Figure 3.24) is constrained

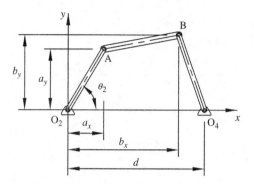

Figure 3.24   Coordinate locations of points A and B

to move on a circular arc about point A creating a circle of radius AB and center located at the point (a_x, a_y). Writing an equation of a circle that defines that path of point B gives the following:

$$(AB)^2 = (b_x - a_x)^2 + (b_y - a_y)^2 \qquad (3.15)$$

Point B is also constrained to move on a circular arc about point O_4. The equation of that circle, having radius BO_4 and center point $(d, 0)$, is defined by the following:

$$(BO_4)^2 = (b_x - d)^2 + b_y^2 \qquad (3.16)$$

Equations 3.15 and 3.16 give two nonlinear equations with the two unknown values of b_x and b_y. Though the two equations can be solved analytically, the derivation is tedious. Therefore, a more graphical approach will be used.

Referring to Figure 3.25a, an intermediate angle λ can be calculated from known values of a_x and a_y using trigonometric equations for right triangle AO_AO_4.

Values of a_x and a_y are determined using Equation 3.14 in terms of the input angle.

$$\lambda = \tan^{-1}\left(\frac{d - a_x}{a_y}\right) \qquad (3.17)$$

Length AO_4 can be determined using the Pythagorean theorem for the same right triangle.

$$AO_4 = \sqrt{a_y^2 + (d - a_x)^2} \qquad (3.18)$$

Angle ϕ can be determined using law of sines for the oblique triangle ABO_4.

$$\phi = \sin^{-1}\left(\frac{BO_4 \sin \mu}{AO_4}\right) \qquad (3.19)$$

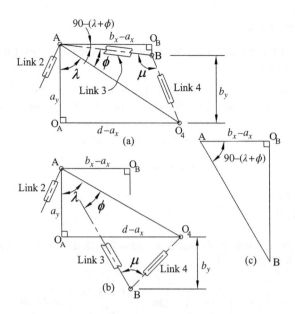

Figure 3.25 Graphical approach for determining coordinate location of point B

Now using right triangle AO_BB

$$\cos[90 - (\lambda + \phi)] = \frac{b_x - a_x}{AB} \tag{3.20}$$

Rearranging Equation 3.20, noting that $\cos[90 - (\lambda + \phi)] = \sin(\lambda + \phi)$, gives the x-coordinate of point B.

$$b_x = a_x + AB \sin(\lambda + \phi) \tag{3.21}$$

The y-coordinate of point B can be determined by rearranging Equation 3.16.

$$b_y = \sqrt{(BO_4)^2 - (b_x - d)^2} \tag{3.22}$$

A minor difference occurs when analyzing the coordinates of point B for a crossed arrangement as shown in Figure 3.25b. The method for determining angles λ and ϕ is identical to the procedure presented above for open arrangement. The right triangle AO_BB, shown in Figure 3.25c for clarity, involves a different angle as given below.

$$\cos[90 - (\lambda - \phi)] = \frac{b_x - a_x}{AB} \tag{3.23}$$

Rearranging Equation 3.23 gives the x-coordinate of point B for a crossed arrangement.

$$b_x = a_x + AB \sin(\lambda - \phi) \tag{3.24}$$

The y-coordinate of point B for the crossed arrangement can be found using Equation 3.22, just as the case for open arrangement.

Example Problem 3.5

For the four-bar linkage mechanism from Example Problem 3.3, determine the coordinate position of points A and B.

Solution: The transmission angle was calculated in Example Problem 3.3 to be $\mu = 87.4°$. The coordinate position of point A can be determined using Equation 3.14.

$$a_x = (AO_2)\cos \theta_2 = 7 \cos 35$$

$$a_y = (AO_2)\sin \theta_2 = 7 \sin 35$$

Answer: $\begin{aligned} a_x &= 5.734 \text{ in.} \\ a_y &= 4.015 \text{ in.} \end{aligned}$

To determine coordinate position of point B, we must first find the values of λ and ϕ. The angle λ is calculated from Equation 3.17.

$$\lambda = \tan^{-1}\left(\frac{d - a_x}{a_y}\right) = \tan^{-1}\left(\frac{19 - 5.734}{4.015}\right) = 73.16°$$

Distance AO_4 is calculated from Equation 3.18.

$$AO_4 = \sqrt{a_y^2 + (d - a_x)^2} = \sqrt{4.015^2 + (19 - 5.734)^2} = 13.86 \text{ in.}$$

The angle ϕ is calculated from Equation 3.19.

$$\phi = \sin^{-1}\left(\frac{BO_4 \sin \mu}{AO_4}\right) = \sin^{-1}\left(\frac{7.5 \sin 87.4}{13.86}\right) = 32.72°$$

The x-coordinate of point B can then be determined from Equation 3.21.

$$b_x = a_x + AB \sin(\lambda + \phi) = 5.734 + 12 \sin(73.16 + 32.72)$$

Equation 3.22 gives the y-coordinate of point B.

$$b_y = \sqrt{(BO_4)^2 - (b_x - d)^2} = \sqrt{7.5^2 - (17.276 - 19)^2}$$

Answer: $\begin{aligned} b_x &= 17.276 \text{ in.} \\ b_y &= 7.299 \text{ in.} \end{aligned}$

Example Problem 3.6

For the four-bar linkage mechanism from Example Problem 3.4, determine the coordinate position of points A and B.

Solution: The transmission angle was calculated in Example Problem 3.4 to be $\mu = 90.6°$. The coordinate position of point A can be determined using Equation 3.14.

$$a_x = (AO_2)\cos\theta_2 = 3.5\cos 45$$

$$a_y = (AO_2)\sin\theta_2 = 3.5\sin 45$$

Answer: $\quad a_x = 2.475 \text{ in.}$
$\qquad\quad a_y = 2.475 \text{ in.}$

To determine coordinate position of point B, we must first find the values of λ and ϕ. The angle λ is calculated from Equation 3.17.

$$\lambda = \tan^{-1}\left(\frac{d - a_x}{a_y}\right) = \tan^{-1}\left(\frac{8 - 2.475}{2.475}\right) = 65.87°$$

Distance AO_4 is calculated from Equation 3.18.

$$AO_4 = \sqrt{a_y^2 + (d - a_x)^2} = \sqrt{2.475^2 + (8 - 2.475)^2} = 6.054 \text{ in.}$$

The angle ϕ is calculated from Equation 3.19.

$$\phi = \sin^{-1}\left(\frac{BO_4 \sin\mu}{AO_4}\right) = \sin^{-1}\left(\frac{4\sin 90.6}{6.054}\right) = 41.35°$$

The x-coordinate of point B can then be determined from Equation 3.24.

$$b_x = a_x + AB\sin(\lambda - \phi) = 2.475 + 4.5\sin(65.87 - 41.35)$$

Equation 3.22 gives the y-coordinate of point B.

$$b_y = \sqrt{(BO_4)^2 - (b_x - d)^2} = \sqrt{4^2 - (4.342 - 8)^2}$$

Answer: $\quad b_x = 4.342 \text{ in.}$
$\qquad\quad b_y = 1.618 \text{ in.}$

Next we will develop an expression to locate the coordinate position of the slider in a slider–crank mechanism. Referring to the offset slider–crank mechanism in Figure 3.26, an expression for the location of the slider with respect to θ_2 and θ_3 can easily be determined.

$$x_B = O_2A\cos\theta_2 + AB\cos\theta_3 \qquad\qquad (3.25)$$

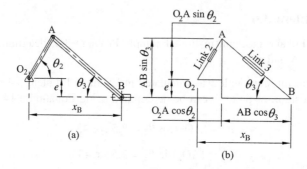

Figure 3.26 Position analysis for an offset slider–crank mechanism

However, we want to be able to express the position only in terms of the input angle θ_2. From Figure 3.26b, the vertical dimensions between point A and point B give the following:

$$AB \sin \theta_3 = e + O_2A \sin \theta_2 \tag{3.26}$$

Rearranging gives

$$\sin \theta_3 = \frac{e + O_2A \sin \theta_2}{AB} \tag{3.27}$$

Equation 3.27 gives an expression for the sine of θ_3 in terms of the input angle; however, we need an expression for the cosine of θ_3 in terms of the input angle to substitute into Equation 3.25. Perhaps the most straightforward way to develop the expression for cosine of an angle when the sine of the angle is known would be to represent sine and cosine graphically. The general procedure for determining cosine of an angle based on knowing the sine of that angle is discussed in Supplementary Concept 3.1.

Supplementary Concept 3.1

For right triangles such as the one shown in Figure 3.27, the sine of an angle is defined as the length of the side opposite to the angle divided by the length of the hypotenuse.

$$\sin \alpha = \frac{\text{opp}}{\text{hyp}}$$

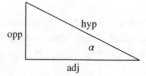

Figure 3.27

Treating the lengths of the opposite side and hypotenuse as known, the length of the adjacent side can be determined by the Pythagorean theorem to be

$$adj = \sqrt{hyp^2 - opp^2}$$

The cosine of α is defined as the length of the adjacent side divided by the length of the hypotenuse.

$$\cos \alpha = \frac{adj}{hyp}$$

Plugging in the expression for the adjacent side determined by the Pythagorean theorem gives

$$\cos \alpha = \frac{\sqrt{hyp^2 - opp^2}}{hyp} \tag{3.28}$$

Using this graphical approach to find the cosine of θ_3, the lengths of the opposite side and hypotenuse can be determined from Equation 3.27. Knowing that $\sin \alpha = opp/hyp$, the numerator of Equation 3.27 gives the length of the opposite side to be $e + O_2A \sin \theta_2$. Similarly, the denominator of Equation 3.27 gives AB as the length of the hypotenuse. Using the determined values of lengths of opposite side and hypotenuse in Equation 3.28 gives the following:

$$\cos \theta_3 = \frac{\sqrt{(AB)^2 - (e + O_2A \sin \theta_2)^2}}{AB} \tag{3.29}$$

Equation 3.29 can then be plugged into Equation 3.25 to give the final expression for the position of the slider as a function of the input angle.

$$x_B = O_2A \cos \theta_2 + \sqrt{(AB)^2 - (e + O_2A \sin \theta_2)^2} \tag{3.30}$$

Example Problem 3.7

For the four-bar slider–crank mechanism shown in Figure 3.28, determine (a) the horizontal position x_B of the slider for $\theta_2 = 32°$ and (b) the value of θ_2 where x_B is maximum.

Solution:

a. The horizontal position of the slider is given by Equation 3.30.

$$x_B = O_2A \cos \theta_2 + \sqrt{(AB)^2 - (e + O_2A \sin \theta_2)^2}$$
$$x_B = 6 \cos 32 + \sqrt{10^2 - (1.75 + 6 \sin 32)^2}$$

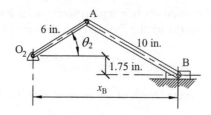

Figure 3.28

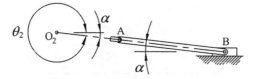

Figure 3.29

Answer: $x_B = 13.789$ in.

The arrangement for maximum x_B occurs when link 2 and link 3 are collinear, as shown in Figure 3.29. Then angle α is determined as

$$\sin \alpha = \frac{e}{O_2A + AB}$$

$$\alpha = \sin^{-1}\left(\frac{1.75}{6 + 10}\right) = 6.28°$$

b. The angle θ_2 for the maximum horizontal slider location can then be calculated.

$$\theta_2 = 360° - \alpha = 360 - 6.28$$

Answer: $\theta_2 = 353.72°$

3.9 Analytical Position Analysis

3.9.1 *Loop Closure Equation*

Another method for solving position involves the use of position vectors to represent the links, as shown in Figure 3.30. For a closed linkage system, summation of the position vectors must result in a value of zero. The summation, known as the loop closure equation, is given below.

$$\vec{R}_2 + \vec{R}_3 - \vec{R}_4 - \vec{R}_1 = 0 \tag{3.31}$$

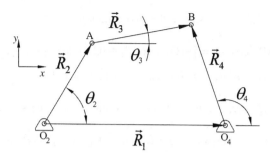

Figure 3.30 Loop closure

Each of the position vectors can be represented in terms of unit vectors $\hat{i}$ and $\hat{j}$. For convenience, the x-axis is aligned with link 1.

$$\vec{R}_1 = (O_2O_4)\hat{i}$$
$$\vec{R}_2 = (O_2A)\cos\theta_2\hat{i} + (O_2A)\sin\theta_2\hat{j}$$
$$\vec{R}_3 = (AB)\cos\theta_3\hat{i} + (AB)\sin\theta_3\hat{j} \tag{3.32}$$
$$\vec{R}_4 = (O_4B)\cos\theta_4\hat{i} + (O_4B)\sin\theta_4\hat{j}$$

The loop closure equation can then be expressed using Cartesian coordinates by substituting Equation 3.32 into Equation 3.31.

$$\left[(O_2A)\cos\theta_2\hat{i} + (O_2A)\sin\theta_2\hat{j}\right] + \left[(AB)\cos\theta_3\hat{i} + (AB)\sin\theta_3\hat{j}\right]$$
$$- \left[(O_4B)\cos\theta_4\hat{i} + (O_4B)\sin\theta_4\hat{j}\right] - (O_2O_4)\hat{i} = 0 \tag{3.33}$$

Collecting $\hat{i}$ terms gives

$$(O_2A)\cos\theta_2 + (AB)\cos\theta_3 - (O_4B)\cos\theta_4 - (O_2O_4) = 0 \tag{3.34}$$

and collecting $\hat{j}$ terms gives the following:

$$(O_2A)\sin\theta_2 + (AB)\sin\theta_3 - (O_4B)\sin\theta_4 = 0 \tag{3.35}$$

It is typical that all link lengths are known. Treating the input angle θ_2 as a known value, Equations 3.34 and 3.35 give two simultaneous nonlinear equations with two unknowns (θ_3 and θ_4).

Rearranging Equations 3.34 and 3.35 to isolate θ_3 gives the following:

$$(AB)\cos\theta_3 = -(O_2A)\cos\theta_2 + (O_4B)\cos\theta_4 + (O_2O_4) \tag{3.36}$$

$$(AB)\sin\theta_3 = -(O_2A)\sin\theta_2 + (O_4B)\sin\theta_4 \tag{3.37}$$

In order to eliminate θ_3, we can square the two equations

$$(AB)^2\cos^2\theta_3 = [-(O_2A)\cos\theta_2 + (O_4B)\cos\theta_4 + (O_2O_4)]^2$$

$$(AB)^2\sin^2\theta_3 = [-(O_2A)\sin\theta_2 + (O_4B)\sin\theta_4]^2 \tag{3.38}$$

and then add them together. Noting that $\cos^2\theta_3 + \sin^2\theta_3 = 1$, the summation gives

$$(AB)^2 = [-(O_2A)\cos\theta_2 + (O_4B)\cos\theta_4 + (O_2O_4)]^2 + [-(O_2A)\sin\theta_2 + (O_4B)\sin\theta_4]^2 \tag{3.39}$$

By expanding the right-hand side of Equation 3.39 and collecting like terms, we get the following:

$$\begin{aligned}(AB)^2 = {} & (O_2A)^2 + (O_4B)^2 + (O_2O_4)^2 - 2(O_2A)(O_4B)\cos\theta_2\cos\theta_4 \\ & + 2(O_2O_4)(O_4B)\cos\theta_4 - 2(O_2A)(O_2O_4)\cos\theta_2 - 2(O_2A)(O_4B)\sin\theta_2\sin\theta_4\end{aligned} \tag{3.40}$$

Next we can rearrange terms

$$\begin{aligned}(O_2A)^2 & - (AB)^2 + (O_4B)^2 + (O_2O_4)^2 + 2(O_2O_4)(O_4B)\cos\theta_4 - 2(O_2A)(O_2O_4)\cos\theta_2 \\ & = 2(O_2A)(O_4B)\cos\theta_2\cos\theta_4 + 2(O_2A)(O_4B)\sin\theta_2\sin\theta_4\end{aligned} \tag{3.41}$$

and divide by $2(O_2A)(O_4B)$.

$$\begin{aligned}\frac{(O_2A)^2 - (AB)^2 + (O_4B)^2 + (O_2O_4)^2}{2(O_2A)(O_4B)} & + \frac{(O_2O_4)}{(O_2A)}\cos\theta_4 - \frac{(O_2O_4)}{(O_4B)}\cos\theta_2 \\ & = \cos\theta_2\cos\theta_4 + \sin\theta_2\sin\theta_4\end{aligned} \tag{3.42}$$

Utilizing the trigonometric identity $\cos(\theta_2 - \theta_4) = \cos\theta_2\cos\theta_4 + \sin\theta_2\sin\theta_4$ gives the equation developed by Ferdinand Freudenstein in the 1950s.

$$K_1\cos\theta_4 - K_2\cos\theta_2 + K_3 = \cos(\theta_2 - \theta_4) \tag{3.43}$$

where

$$K_1 = \frac{O_2O_4}{O_2A}$$

$$K_2 = \frac{O_2O_4}{O_4B}$$

$$K_3 = \frac{(O_2A)^2 - (AB)^2 + (O_4B)^2 + (O_2O_4)^2}{2(O_2A)(O_4B)}$$

Freudenstein's equation allows for the output angle θ_4 to be calculated based on known link lengths and input angle θ_2. Freudenstein's equation is also very important in four-bar linkage synthesis to get a desired value (or multiple values) of θ_4 for a given value of θ_2. Use of Freudenstein's equation for linkage synthesis is discussed in Chapter 5.

Using Freudenstein's equation, it is also possible to show that for any given input angle, there are in fact two possible output angles (such as open and crossed positions). Obtaining the two output angles requires the use of tangent half-angle formulas, which converts an equation in sine and cosine to a quadratic polynomial.

$$\sin\theta = \frac{2\tan(\theta/2)}{1+\tan^2(\theta/2)} \qquad \cos\theta = \frac{1-\tan^2(\theta/2)}{1+\tan^2(\theta/2)} \tag{3.44}$$

Using the tangent half-angle formula to replace the terms $\cos(\theta_4)$ and $\sin(\theta_4)$ in Equation 3.42 along with utilizing expressions for K_1, K_2, and K_3 from Equation 3.43 gives

$$K_3 + K_1\left(\frac{1-\tan^2 x}{1+\tan^2 x}\right) - K_2\cos\theta_2 = \cos\theta_2\left(\frac{1-\tan^2 x}{1+\tan^2 x}\right) + \sin\theta_2\left(\frac{2\tan x}{1+\tan^2 x}\right) \tag{3.45}$$

where $x = \theta_4/2$. Collecting terms and multiplying by $1+\tan^2 x$ gives

$$K_3(1+\tan^2 x) + (K_1 - \cos\theta_2)(1-\tan^2 x) - K_2\cos\theta_2(1+\tan^2 x) = 2\sin\theta_2\tan x \tag{3.46}$$

Expanding Equation 3.46

$$\begin{aligned} K_3 + K_3\tan^2 x + K_1 - K_1\tan^2 x - \cos\theta_2 + \cos\theta_2\tan^2 x \\ - K_2\cos\theta_2 - K_2\cos\theta_2\tan^2 x = 2\sin\theta_2\tan x \end{aligned} \tag{3.47}$$

and collecting terms gives

$$K_3 + (K_3 - K_1 + \cos\theta_2 - K_2\cos\theta_2)\tan^2 x + K_1 - \cos\theta_2 - K_2\cos\theta_2 = 2\sin\theta_2\tan x \tag{3.48}$$

Recalling that $x = \theta_4/2$, Equation 3.48 can be written in a simpler form.

$$a\tan^2\left(\frac{\theta_4}{2}\right) + b\tan\left(\frac{\theta_4}{2}\right) + c = 0 \tag{3.49}$$

where

$$\begin{aligned} a &= K_3 - K_1 + \cos\theta_2 - K_2\cos\theta_2 \\ b &= -2\sin\theta_2 \\ c &= K_1 - (1+K_2)\cos\theta_2 + K_3 \end{aligned}$$

Equation 3.49 is a quadratic equation in the form $ay^2 + by + c = 0$, where $y = \tan(\theta_4/2)$. Solution to the quadratic equation will be in the form

$$\tan\left(\frac{\theta_4}{2}\right) = \frac{-b \pm \sqrt{b^2 - 4ac}}{2a} \tag{3.50}$$

Table 3.1 Possible outcomes for output angle from Equation 3.5 1

Real and equal	Only one possible position is possible due to geometry
Real and unequal	Two values of output angle correspond to two possible orientations of output link
Complex conjugate	Links will not connect for the given link lengths and input angle
	• Non-Grashof linkage beyond toggle position
	• Links will not connect in any position

The final solution for both values of θ_4 comes from rearranging Equation 3.50 to get the following:

$$\theta_4 = 2 \tan^{-1} \left(\frac{-b \pm \sqrt{b^2 - 4ac}}{2a} \right) \tag{3.51}$$

Because Equation 3.51 has the $\pm$ condition, it will give two answers relating to the two arrangements of the linkage mechanism. Depending on the values of a, b, and c, the two solutions can be any of the three types described in Table 3.1.

Example Problem 3.8

A four-bar mechanism has the dimensions $O_2O_4 = 12.5$, $O_2A = 4.75$, $AB = 6$, and $O_4B = 3.75$ (all dimensions in inches). Determine any possible output angle for an input angle of $25°$ and draw the linkage arrangement.

Solution: Before output angles can be calculated, the individual K_i terms need to be determined.

$$K_1 = \frac{O_2O_4}{O_2A} = \frac{12.5}{4.75} = 2.6316$$

$$K_2 = \frac{O_2O_4}{O_4B} = \frac{12.5}{3.75} = 3.3333$$

$$K_3 = \frac{(O_2A)^2 - (AB)^2 + (O_4B)^2 + (O_2O_4)^2}{2(O_2A)(O_4B)}$$

$$K_3 = \frac{(4.75)^2 - (6)^2 + (3.75)^2 + (12.5)^2}{2(4.75)(3.75)} = 4.4035$$

Individual terms of a, b, and c can now be determined.

$$a = \cos \theta_2 - K_1 - K_2 \cos \theta_2 + K_3$$

$$a = \cos 25 - 2.6316 - 3.3333 \cos 25 + 4.4035 = -0.3428$$

$$b = -2 \sin \theta_2 = -2 \sin 25 = -0.8452$$

$$c = K_1 - (K_2 + 1)\cos \theta_2 + K_3$$

$$c = 2.6316 - (3.3333 + 1)\cos 25 + 4.4035 = 3.1078$$

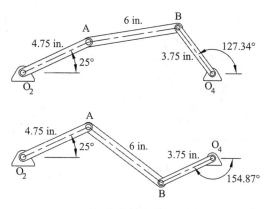

Figure 3.31

Output angles are calculated using Equation 3.51.

$$\theta_4 = 2\tan^{-1}\left(\frac{-b \pm \sqrt{b^2 - 4ac}}{2a}\right)$$

$$\theta_4 = 2\tan^{-1}\left(\frac{0.8452 \pm \sqrt{(-0.8452)^2 - 4(-0.3428)(3.1078)}}{2(-0.3428)}\right)$$

Answer: $\begin{aligned}(\theta_4)_1 &= 127.34° \\ (\theta_4)_2 &= -154.87°\end{aligned}$

Both possible output arrangements are shown in Figure 3.31.

Example Problem 3.9

A four-bar mechanism has the dimensions $O_2O_4 = 6$, $O_2A = 12$, $AB = 5$, and $O_4B = 3.25$ (all dimensions in inches). Determine any possible output angle for an input angle of $38°$ and draw the linkage arrangement.

Solution: Before output angles can be calculated, the individual K_i terms need to be determined.

$$K_1 = \frac{O_2O_4}{O_2A} = \frac{6}{12} = 0.5$$

$$K_2 = \frac{O_2O_4}{O_4B} = \frac{6}{3.25} = 1.8462$$

$$K_3 = \frac{(O_2A)^2 - (AB)^2 + (O_4B)^2 + (O_2O_4)^2}{2(O_2A)(O_4B)}$$

$$K_3 = \frac{(12)^2 - (5)^2 + (3.25)^2 + (6)^2}{2(12)(3.25)} = 2.1226$$

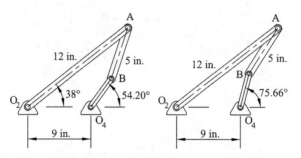

Figure 3.32

Individual terms of a, b, and c can now be determined.

$$a = \cos \theta_2 - K_1 - K_2 \cos \theta_2 + K_3$$
$$a = \cos 38 - 0.5 - 1.8462 \cos 38 + 2.1226 = 0.9558$$

$$b = -2 \sin \theta_2 = -2 \sin 38 = -1.2313$$

$$c = K_1 - (K_2 + 1)\cos \theta_2 + K_3$$
$$c = 0.5 - (1.8462 + 1)\cos 38 + 2.1226 = 0.3798$$

Output angles are calculated using Equation 3.51.

$$\theta_4 = 2 \tan^{-1} \left(\frac{-b \pm \sqrt{b^2 - 4ac}}{2a} \right)$$

$$\theta_4 = 2 \tan^{-1} \left(\frac{1.2313 \pm \sqrt{(-1.2313)^2 - 4(0.9558)(0.3798)}}{2(0.9558)} \right)$$

Answer: $\begin{aligned} (\theta_4)_1 &= 54.20° \\ (\theta_4)_2 &= 75.66° \end{aligned}$

Both possible output arrangements are shown in Figure 3.32.

3.9.2 Complex Number Notation

Alternatively, the vectors in the loop closure equation can be represented using complex number notation.

$$\vec{R}_1 = (O_2O_4)e^{j\theta_1}$$
$$\vec{R}_2 = (O_2A)e^{j\theta_2}$$
$$\vec{R}_3 = (AB)e^{j\theta_3}$$
$$\vec{R}_4 = (O_4B)e^{j\theta_4}$$

$$(3.52)$$

Aligning the ground link with the real axis causes the value of θ_4 to be equal to zero. Therefore, the vector $\vec{R}_4 = (O_4B)e^{j(0)} = (O_4B)$. Substituting Equation 3.52 into Equation 3.31 gives the following:

$$(O_2A)e^{j\theta_2} + (AB)e^{j\theta_3} - (O_4B)e^{j\theta_4} - O_4B = 0 \tag{3.53}$$

Supplementary Concept 3.2

Euler's identity relates complex exponentials to circular trigonometric functions. The power series expansion of the exponent function is given as

$$e^x = 1 + \frac{x}{1!} + \frac{x^2}{2!} + \frac{x^3}{3!} + \cdots \qquad -\infty < x < \infty$$

Taking the series using jx gives

$$e^{jx} = \sum_{n=0}^{\infty} \frac{(jx)^n}{n!} = \left(1 - \frac{x^2}{2} + \frac{x^4}{24} \cdots\right) + j\left(x - \frac{x^3}{6} + \frac{x^5}{120} \cdots\right)$$

The two series developed are the series for cosine and sine, which gives the Euler's identity

$$e^{\pm j\theta} = \cos\theta \pm j\sin\theta \tag{3.54}$$

where $j = \sqrt{-1}$.

Using the Euler's identity given in Equation 3.54, Equation 3.53 can be written as the following:

$$(O_2A)[\cos\theta_2 + j\sin\theta_2] + (AB)[\cos\theta_3 + j\sin\theta_3] - (O_4B)\left[\cos\theta_4 + j\sin\theta_4\hat{j}\right] - O_2O_4 = 0 \tag{3.55}$$

Equation 3.55 can now be separated into real components

$$(O_2A)\cos\theta_2 + (AB)\cos\theta_3 - (O_4B)\cos\theta_4 - (O_2O_4) = 0 \tag{3.56}$$

and imaginary components

$$j(O_2A)\sin\theta_2 + j(AB)\sin\theta_3 - j(O_4B)\sin\theta_4 = 0 \tag{3.57}$$

Equation 3.57 can be divided by j to give the following:

$$(O_2A)\sin\theta_2 + (AB)\sin\theta_3 - (O_4B)\sin\theta_4 = 0 \tag{3.58}$$

Equations 3.56 and 3.58 obtained by using complex notation are identical to Equations 3.34 and 3.35 obtained by using Cartesian coordinates.

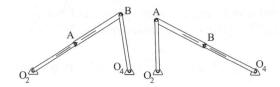

Figure 3.33   Toggle positions of a four-bar mechanism

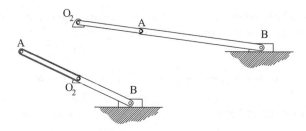

Figure 3.34   Toggle positions of a slider–crank mechanism

3.10   Toggle Positions

In a four-bar mechanism, positions exist when the coupler becomes collinear with an adjacent link. These collinear positions mark extreme limit positions, or toggle positions, of the mechanism. Figure 3.33 shows the two extreme positions of a four-bar mechanism and Figure 3.34 shows the two extreme positions of a slider–crank mechanism.

3.11   Computer Methods for Position Analysis

3.11.1   Position Analysis Using Spreadsheets

One approach for utilizing computers to perform position analysis would be analytical solutions with spreadsheet software. Analytical methods discussed in this chapter are very time consuming, which makes it difficult to analyze multiple positions of a linkage mechanism. Spreadsheets, however, can make calculations for multiple positions very efficiently. As an example, we will consider position analysis of a slider–crank mechanism as discussed in Section 3.8.2.

Inputs are needed that give information about the length of each link. In the example shown in Figure 3.35, the input link is 3 inches long, the coupler is 5.5 inches long, and there is also an offset of 2.25 inches. An input for rotation speed of the coupler is also shown as 200 rpm, though that value will not be used in the calculations of position. It is convenient to have a figure posted within the spreadsheet to define the input terms.

Once all input values have been defined, position calculations are performed for multiple arrangements of the mechanism. Figure 3.36 shows calculations for position (x_B) based on input values of θ_2 (in 0.1° increments). Values for position are determined using Equation 3.30, and are shown in the far right column of Figure 3.36 from 0° up to 2° of input rotation. The calculations are continued up to the full 360° of rotation. Results are best represented graphically, as shown in Figure 3.35. Maximum and minimum values can also be easily

Inputs:

Length O_2A	6	inches
Length AB	10	inches
Offset e	1.75	inches
Input speed of θ_2	50	rpm

Slider position

Maximum x_B	15.904	inches
located at	353.7	degrees
Minimum x_B	3.597	inches
located at	154.1	degrees

Stroke   12.307 inches

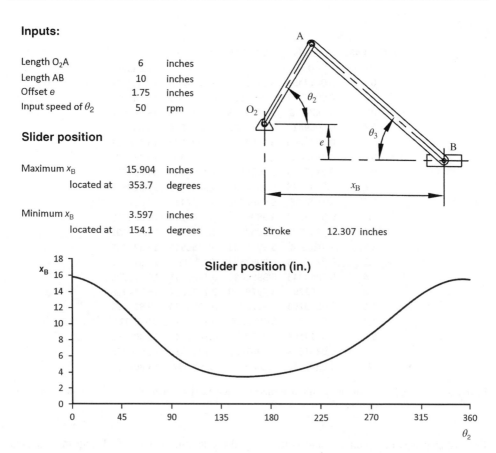

Figure 3.35   Screenshot from spreadsheet position analysis for an offset slider–crank mechanism

determined. The calculations in Figure 3.35 show that a maximum location of 15.904 inches is obtained at 353.7° rotation of the input, and a minimum location of 3.597 inches at 154.1°.

3.11.2 Distance Formula to Solve for Output Angle

Another approach to get a quick calculation of output angle utilizes the distance formula and the goal seek command in Excel. Referring to Figure 3.24, the coordinates of point A are defined using the length of the crank and the input angle.

$$a_x = (AO_2)\cos \theta_2$$
$$a_y = (AO_2)\sin \theta_2 \tag{3.59}$$

The x and y coordinates of point B can be defined using the length of the output link and the output angle.

$$b_x = (BO_4)\cos \theta_4 + O_2O_4$$
$$b_y = (BO_4)\sin \theta_4 \tag{3.60}$$

| θ_2 | | | | x_B |
degrees	radians	$O_2A \cos \theta_2$	$O_2A \sin \theta_2$	inches
0	0	6	0	15.845684
0.1	0.0017453	5.9999909	0.010472	15.843808
0.2	0.0034907	5.9999634	0.0209439	15.841902
0.3	0.005236	5.9999178	0.0314158	15.839966
0.4	0.0069813	5.9998538	0.0418876	15.838001
0.5	0.0087266	5.9997715	0.0523592	15.836006
0.6	0.010472	5.999671	0.0628307	15.833981
0.7	0.0122173	5.9995522	0.073302	15.831926
0.8	0.0139626	5.9994151	0.0837731	15.829841
0.9	0.015708	5.9992598	0.0942439	15.827727
1	0.0174533	5.9990862	0.1047144	15.825583
1.1	0.0191986	5.9988943	0.1151847	15.823409
1.2	0.020944	5.9986841	0.1256545	15.821205
1.3	0.0226893	5.9984557	0.136124	15.818972
1.4	0.0244346	5.9982089	0.1465931	15.816709
1.5	0.0261799	5.9979439	0.1570617	15.814416
1.6	0.0279253	5.9976607	0.1675298	15.812093
1.7	0.0296706	5.9973592	0.1779975	15.80974
1.8	0.0314159	5.9970394	0.1884646	15.807358
1.9	0.0331613	5.9967013	0.1989311	15.804946
2	0.0349066	5.996345	0.209397	15.802504

Figure 3.36 Sample calculations from spreadsheet position analysis for an offset slider–crank mechanism

The distance between points A and B must equal the length of link AB. Using the distance formula,

$$AB = \sqrt{(b_x - a_x)^2 + (b_x - a_x)^2} \tag{3.61}$$

We can now use Excel to quickly calculate output angle. Figure 3.37 shows the general setup of the file using input data from Example Problem 3.3. The x- and y-coordinate locations for point A are calculated using Equation 3.59 and the coordinate location for point B is calculated using Equation 3.60. Note that the location of point B is based on a trial value of the output angle (shown as 90°). The distance between points A and B is calculated using Equation 3.61. If the value of the output angle were correct, the distance (13.716) would match the input length of AB (12). They do not match indicating that our trial value is incorrect.

Trial and error could be used trying new values of the output angle until the distance d matched the length of link AB. However, to determine the correct value for the output angle, we can use the goal seek command in Excel.

3.11.3 Computer Solutions Using MATLAB® and MathCAD

Though Excel is a very useful tool for position analysis, other software packages allow for more versatility. Chapter 6 will focus on building modules in MATLAB® and MathCAD for a complete kinematic analysis of linkage mechanisms, including position analysis.

Input			Location of point A	
O_2O_4	19		A_x	5.734064
O_2A	7		A_y	4.015035
AB	12			
O_4B	7.5		**Trial value of output angle**	
			θ_4	90 degrees
θ_2	35	degrees		
			Location of point B	
			B_x	19
			B_y	7.5
			Distance between A and B	
			d	13.71605

Figure 3.37 Excel file setup for determining output angle

3.12 Review and Summary

This chapter focused on the first step to complete linkage analysis, which is position analysis. Graphical methods were presented first to develop the foundation. Graphical and analytical methods were presented. Some information, such as Freudenstein's equation, serves as essential information for linkage synthesis presented in Chapter 5.

Problems

P3.1 For the linkage mechanism shown in Figure 3.38, (a) determine the number of links, (b) label all full and half joints, and (c) determine mobility.

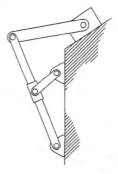

Figure 3.38

P3.2 Determine the mobility for each mechanism shown in Figure 1.38.

P3.3 Determine the mobility for each mechanism shown in Figure 1.20a.

P3.4 Determine the mobility for each mechanism shown in Figure 1.39.

P3.5 Given a four-bar linkage with the dimension below, determine if the linkage is Grashof or non-Grashof and sketch the linkage in toggle positions. Would rearranging the links change the Grashof condition? $O_2O_4 = 6$, $O_2A = 2$, $AB = 4.5$, and $BO_4 = 3.75$.

P3.6 Calculate the Grashof condition for a four-bar mechanism with link lengths of 2, 9, 6, and 4.5 inches. Will this mechanism contain change points?

P3.7 A four-bar crank–rocker mechanism has the two fixed pivot points located 19 inches apart. The coupler is 13 inches long. Determine the possible length values of the crank if the rocker is 10 inches long.

P3.8 Calculate the transmission angle and output angle for the mechanism shown in Figure 3.39.

Answer: $\mu = 142.81°$, $\theta_4 = 145.70°$

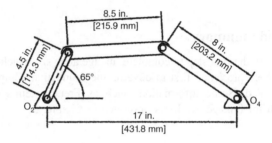

Figure 3.39

P3.9 Calculate the transmission angle and output angle for the mechanism shown in Figure 3.40.

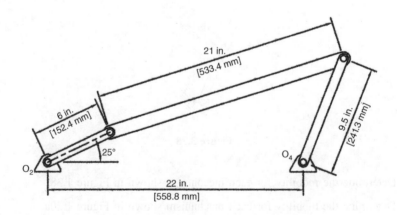

Figure 3.40

P3.10 Calculate the transmission angle and output angle for the mechanism shown in Figure 3.41.

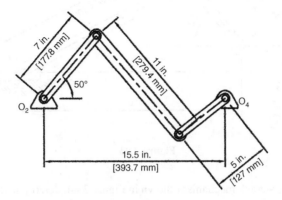

Figure 3.41

P3.11 Calculate the transmission angle and output angle for the mechanism shown in Figure 3.42.

Answer: $\mu = 83.45°$, $\theta_4 = 104.95°$

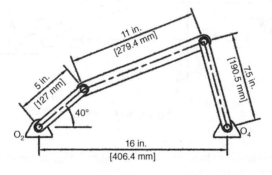

Figure 3.42

P3.12 A four-bar slider–crank mechanism has a crank that is 2.5 inches long, a coupler that is 4 inches long, and no offset. At what value(s) of input angle will the slider position (x_B) be equal to 6 inches?

P3.13 Determine the horizontal position of the slider for the position shown in Figure 3.43.

Answer: 22.07 in.(560.6 mm)

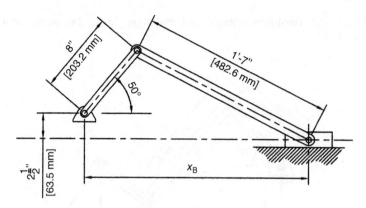

Figure 3.43

P3.14   For the slider–crank mechanism shown in Figure 3.44, determine (a) the position x_B of the slider for $\theta_2 = 30°$, (b) the total distance the slider will travel in one cycle, and (c) the time ratio of working stroke to return stroke.

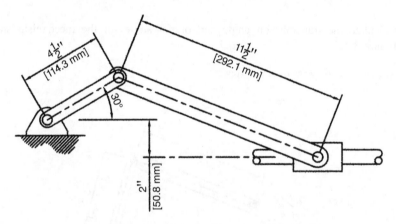

Figure 3.44

P3.15   For the slider–crank mechanism shown in Figure 3.45, determine the location x_B of the slider for the position shown.

P3.16   For the four-bar linkage mechanism shown in Figure 3.39, determine the coordinate position of points A and B.

P3.17   For the four-bar linkage mechanism shown in Figure 3.40, determine the coordinate position of points A and B.

P3.18   For the four-bar linkage mechanism shown in Figure 3.42, determine the coordinate position of points A and B.

P3.19   For a linkage mechanism with dimensions given in Figure 3.39, use Freudenstein's equation to determine output angle(s) for the input angle shown.

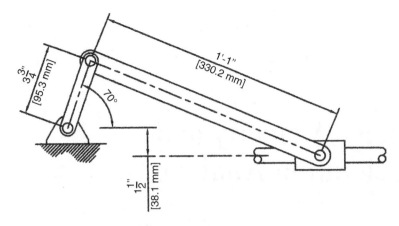

Figure 3.45

P3.20 A four-bar mechanism has the dimensions $O_2O_4 = 8$, $O_2A = 5.5$, $AB = 7$, and $O_4B = 4$ (all dimensions in inches). Determine any possible output angle for an input angle of $62°$ using Freudenstein's equation.

P3.21 For a linkage mechanism with dimensions given in Figure 3.42, use Freudenstein's equation to determine output angle(s) for the input angle shown.

P3.22 A four-bar mechanism has dimensions $O_2O_4 = 14$, $O_2A = 5$, and $BO_4 = 6$ (all dimensions in inches). The mechanism gives output angles of $122.66°$ or $-158.70°$ when the input angle is $42°$. (a) Use Freudenstein's equation to determine the length of the coupler. (b) Will values for θ_4 always be real numbers for this mechanism? Explain your answer.

P3.23 A four-bar mechanism has the dimensions $O_2O_4 = 23$, $O_2A = 9.75$, $AB = 19$, and $O_4B = 13$ (all dimensions in inches). Determine any possible output angle for an input angle of $10°$ using Freudenstein's equation.

P3.24 For a linkage mechanism with dimensions given in Figure 3.41, use Freudenstein's equation to determine output angle(s) for the input angle shown.

P3.25 A four-bar mechanism has the dimensions $O_2O_4 = 20$, $O_2A = 9$, $AB = 17$, and $O_4B = 10$ (all dimensions in mm). Determine any possible output angle for an input angle of $12°$ using Freudenstein's equation. Calculate the transmission angle using geometrical methods.

P3.26 Develop an Excel spreadsheet, similar to that illustrated in Figure 3.35, to calculate the output angles over an entire cycle of a crank–rocker mechanism. The inputs should be all link lengths. The spreadsheet should include a check for Grashof's criterion. Include calculated output angle for all input angles from $0°$ to $360°$ in $1°$ increments. Provide a plot of output angle versus input angle.

Further Reading

Paul, B. (1979) *Kinematics and Dynamics of Planar Machinery*, Prentice Hall, New Jersey.
Hrones, J.A. and Nelson, G.L. (1951) *Analysis of the Four-Bar Linkage*, MIT Technology Press, Cambridge, MA.

4

Linkage Velocity and Acceleration Analysis

4.1 Introduction

Chapter 3 discussed several methods of linkage position analysis. The next step in linkage analysis would be to calculate velocity and acceleration of critical points of interest within the linkage mechanism. Much of the material in this chapter builds on kinematic knowledge established in engineering dynamics courses and the fundamental concepts of velocity and acceleration, which were reviewed in Chapter 2. If additional information on the fundamental concepts is required, the student is encouraged to refer to textbooks on engineering dynamics.

Velocity analysis is important for finding stored kinetic energy and times required to perform certain tasks. Acceleration is critical for finding forces in the system. Velocity and acceleration analysis has become increasingly important in machine analysis due to the higher speeds associated with modern machinery. This chapter will develop several different methods for determining velocity and acceleration of linkage mechanisms. There are advantages and disadvantages to each method, so it is useful to be familiar with several methods. Graphical methods will be presented in which vectors can be drawn to scale (by hand or with CAD software) to determine magnitudes and directions. Velocity analysis with the aid of instant centers of zero velocity will also be presented.

Velocity and acceleration analysis can be very difficult and tedious even for relatively basic linkage mechanisms. It is not possible to give one specific process that will work best for solving any type of velocity and acceleration problem. The general solution process is to start with the link with the most given velocity and acceleration data. Once equations are developed for that link, you can move progressively through the mechanism. Complete velocity and acceleration analysis often requires solving simultaneous sets of equations. Generally, velocity analysis is completed prior to starting the acceleration analysis. This chapter, therefore, is separated to demonstrate methods of velocity prior to methods of acceleration.

Also, it is often desired to develop plots representing velocity and acceleration through an entire cycle of motion. Having complete velocity and acceleration information at only one position of the mechanism is of little use to an engineer doing a complete mechanism design. Doing a complete and thorough analysis is more suited for computer methods. Therefore,

Machine Analysis with Computer Applications for Mechanical Engineers, First Edition. James Doane.
© 2016 John Wiley & Sons, Ltd. Published 2016 by John Wiley & Sons, Ltd.
Companion Website: www.wiley.com/go/doane0215

concepts of utilizing computers for velocity and acceleration analysis will be beneficial. Computer methods will be presented in Chapter 6.

4.2 Finite Displacement: Approximate Velocity Analysis

We will begin our focus on a relatively quick and simple means of determining approximate velocities. In complex systems, this approach may be necessary because a true analytical solution may be very difficult (or simply too time consuming). The approximate methods can also serve as a verification of more detailed velocity analysis equations. Although not presented in this chapter, this process can also be continued to determine an approximate acceleration by doing the second derivative numerically.

The method of finite displacement is a graphical approach that requires drawing the mechanism in a few different orientations. Therefore, this method is best suited for computer drafting software. Velocity is estimated using numerical differentiation, which was discussed in Chapter 2. Central difference method is preferred due to improved accuracy. The process requires drawing the mechanism in the position where the velocity information is required, as well as a position before and after that location.

To illustrate the procedure, consider the mechanism shown in Figure 4.1. A velocity analysis is required for the mechanism in the position shown in Figure 4.1a, where the input angle is defined as θ_2. The output angle in the initial position is measured as θ_4. To use central difference method of numerical differentiation, we need the mechanism drawn in two other positions. One of the positions will be with a small counterclockwise rotation of the driver link, as shown in Figure 4.1b, causing a new input angle of $\theta_2 + \delta$. This position results in a modified output angle of $\theta_4 + \Delta\theta_4$. The other position, not shown, will be a small clockwise rotation of the driver link giving an input angle of $\theta_2 - \delta$. For all three positions, the output angle is determined graphically. Once the angular location of the output link is known for each position, the central difference method can be used to determine the output angular velocity (and acceleration if desired) at the original position.

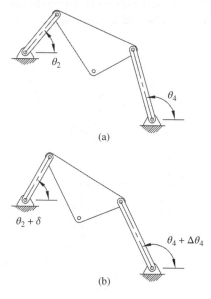

(a)

(b)

Figure 4.1 Process of finite difference for velocity analysis

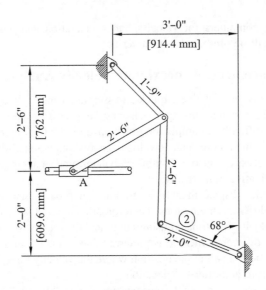

Figure 4.2

Example Problem 4.1

Link 2 of the mechanism shown in Figure 4.2 is rotating clockwise at 2.4 rad/s at the instant shown. Use finite displacement to approximate the velocity of the slider A in the position shown.

Solution: Figure 4.3 shows the mechanism in two offset positions with reference to the original position (labeled position 0).

Position 1 is the mechanism with link 2 rotated 4° counterclockwise from the original position, and position 2 is with link 2 rotated 4° clockwise from the original position (note that 4° offsets were used for clarity, but smaller offsets could also be used). The position of the slider for each of the two offset positions is labeled. It can be seen that in position 1, the slider is offset

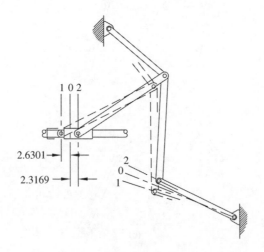

Figure 4.3

Table 4.1   Data for Example Problem 4.1

Position	θ_2 (°)	(rad)	Position of A (in.)	Time (s)	Velocity of A (in./s)
1	72	1.2566	−2.6301	0	
0	68	1.1868	0	0.029	85.29
2	64	1.1170	2.3169	0.058	

a distance of 2.6301 inches to the left of the original position. Also, in position 2 the slider is 2.3169 inches to the right of the original position.

To determine the velocity in inches/second, we need to know the time required to rotate the 4° increments:

$$\Delta t = \left(\frac{1\ s}{2.4\ \text{rad}}\right)(4°)\left(\frac{2\pi\ \text{rad}}{360°}\right) = 0.029\ s$$

Table 4.1 summarizes the data.

The velocity of the slider is determined from

$$v_A \approx \frac{\Delta s}{\Delta t} \approx \frac{2.3169 - (-2.6301)}{0.058}$$

Answer: 85.29 in./s

4.3   Instantaneous Centers of Rotation

4.3.1   Number of Instant Centers

The concepts of instant center of rotation were discussed in Chapter 2. Instant centers can be very helpful for linkage velocity analysis. Linkage mechanisms contain multiple instant centers. The number of instant centers in a linkage mechanism can be determined from the expression for the number of combinations of n things taken k at a time.

$$C_k^n = \frac{n!}{k!(n-k)!} \tag{4.1}$$

For instant centers, we are interested in the combination of n things (n representing the number of links) taken 2 at a time. Using $k = 2$, Equation 4.1 becomes

$$C_2^n = \frac{n!}{2!(n-2)!} = \frac{n(n-1)(n-2)(n-3)\cdots 1}{2(n-2)(n-3)\cdots 1} \tag{4.2}$$

Canceling terms gives the final equation to determine the number of instant centers in a mechanism with n links:

$$C = \frac{n(n-1)}{2} \tag{4.3}$$

As a general example, a four-bar mechanism will have six instant centers: $\left(C = \frac{4(4-1)}{2} = 6\right)$.

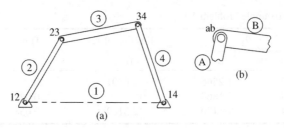

Figure 4.4 (a) Locations of primary instant centers. (b) Instant center notation

4.3.2 Primary Instant Centers

Now that the number of instant centers has been determined, we will discuss procedures for finding the location of each instant center. The primary instant centers, which are those located at the permanent connections between links (link joints), will be discussed first. The instant centers at pin connections can easily be located by inspection. Figure 4.4a shows the four primary centers for a general four-bar mechanism. Instant centers 12 and 14 are commonly called fixed centers because they remain fixed to the frame.

The numbering system for instant centers will be that the center for two links moving relative to each other is specified by the two numbers (or letters) identifying the links. An example is shown in Figure 4.4b for the joint between links A and B with the center labeled ab. It is also customary for the label to be in numerical or alphabetical order (label is ab not ba).

4.3.3 Kennedy–Aronhold Theorem

Once the primary instant centers are identified, the remaining centers can be located using the Kennedy–Aronhold theorem. To illustrate the theorem, consider three rigid bodies in plane motion (connected or not connected). The number of instant centers can be determined from Equation 4.3 to be $C = n(n-1)/2 = 3(3-1)/2 = 3$. The Kennedy–Aronhold theorem states that these three instant centers will lie on the same straight line.

4.3.4 Locating Instant Centers for Typical Four-Bar Mechanisms

We will now look at examples of the process of finding all instant centers for a four-bar mechanism. From Equation 4.3, we know that a four-bar mechanism ($n = 4$) will have six instant centers, four of which will be primary instant centers. A four-bar mechanism is shown in Figure 4.5a with the primary centers labeled.

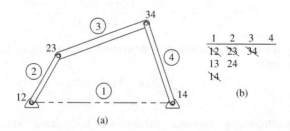

Figure 4.5 (a) Four-bar mechanism with primary instant centers labeled. (b) Table of instant centers for the mechanism

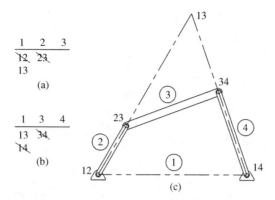

Figure 4.6 Steps for locating instant center 13. (a) Auxiliary table for combination with 2. (b) Auxiliary table for combination with 4. (c) Location of instant center 13

It is helpful to keep track of located instant centers in an organized way, especially for large linkage mechanisms. One way is with the use of a table as shown in Figure 4.5b. The top row of the table consists of link numbers in order (including the stationary ground link). The column of numbers below any link is the number at the top of the column combined with all link numbers to the right of that column. The values in the table will represent all possible instant centers. We can cross out the primary instant centers because they have been located. We can see that the remaining instant centers are 13 and 24, both of which will be located using the Kennedy–Aronhold theorem.

We will first locate instant center 13. To consider 1 and 3 in combination with 2, we make an auxiliary table of centers, as shown in Figure 4.6a. The auxiliary table has 1, 2, and 3 in the first row and the columns below are built the same as before.

Based on the Kennedy–Aronhold theorem, the table indicates that instant center 13 must be on the same line as instant centers 12 and 23. Similarly, a second auxiliary table for 1 and 3 in combination with 4 is developed, as shown in Figure 4.6b, which indicates that instant center 13 must be on the same line as centers 13 and 34. Therefore, instant center 13 will be at the intersection point of the two lines. The center is shown in Figure 4.6c.

The steps for finding the last remaining instant center are outlined in Figure 4.7. An auxiliary table for 2 and 4 in combination with 1 is shown in Figure 4.7a, and the table for 2 and 4 in

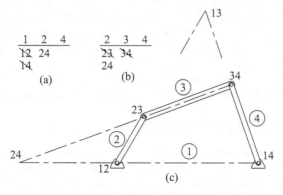

Figure 4.7 Steps for locating instant center 24. (a) Auxiliary table for combination with 1. (b) Auxiliary table for combination with 3. (c) Location of instant center 24

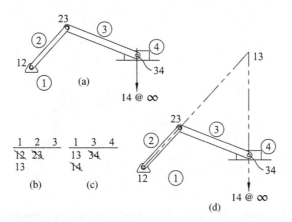

Figure 4.8 Slider–crank instant centers. (a) Primary instant centers. (b) Auxiliary table for combination with 2. (c) Auxiliary table for combination with 4. (d) Location of instant center 13

combination with 3 is shown in Figure 4.7b. The two auxiliary tables show that instant center 24 must lay on the line of centers 12 and 14 and the line of centers 23 and 34. Figure 4.7c shows instant center 24 located at the intersection of the lines.

4.3.5 Locating Instant Centers for Slider–Crank Mechanisms

A very similar process exists for finding the instant centers for a slider–crank mechanism. As shown in Figure 4.8a, three of the primary instant centers (12, 23, and 34) can be located as before. The primary center for the sliding joint (14) is not as obvious. A slider can be considered as a rocker with the length of link 4 being infinite. The consideration is possible because as the rocker length increases, the arc of motion starts to become very flat and begins to approach a straight line. Therefore, the instant center 14 will be located at infinity along a line perpendicular to the slider motion, as shown in Figure 4.8a.

Once the primary centers are located, we will again use the Kennedy–Aronhold theorem to locate the remaining instant centers. Figure 4.8b and c show the two auxiliary tables required to locate instant center 13. It can be determined that instant center 13 will be located at the intersection of the line passing through centers 12 and 23 and the line passing through centers 14 and 34. The final location of instant center 13 is shown in Figure 4.8d.

Next, we will locate instant center 24. From the auxiliary table in Figure 4.9a, we can determine that center 24 must be located on the line through centers 23 and 34, which is easily located. Then using the auxiliary table in Figure 4.9b, we determine that center 24 must be on the line through centers 12 and 14. However, instant center 14 is located at infinity. Based on the concept of parallel lines crossing at infinity, the line through 12 and 14 will pass through point 14 and be parallel to the line through 34 pointing toward instant center 14.

4.3.6 Locating Instant Centers for Other Mechanisms

The previous sections discussed the procedures for determining instant center locations for basic four-bar mechanisms. The concepts introduced can be used to find instant centers for mechanisms with more than four links. As a general example, consider the six-bar mechanism shown in Figure 4.10a, which will have a total of 15 instant centers. Seven of the instant centers are primary instant centers, as shown in Figure 4.10b, and can be located by inspection.

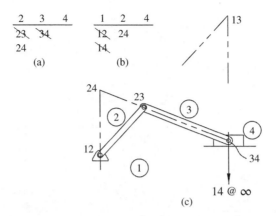

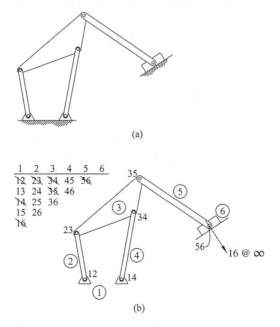

Figure 4.9   Slider–crank instant centers. (a) Auxiliary table for combination with 1. (b) Auxiliary table for combination with 3. (c) Location of instant center 24

The remaining eight instant centers can be located using the Kennedy–Aronhold theorem. Figure 4.11 shows the process for locating the remaining instant centers along with the necessary auxiliary tables. Note that instant center 36 is located off the page.

4.3.7 Velocity Analysis Using Instant Centers

Once the instant centers have been located, they can be utilized to determine velocity magnitude and direction of points within the system. As an illustrative example, consider the slider–crank mechanism shown in Figure 4.12. The location of instant center 13, which is labeled as point *I* in the figure, can be determined using methods discussed in Section 4.3.5.

Figure 4.10   (a) Six-bar mechanism. (b) Primary instant center locations

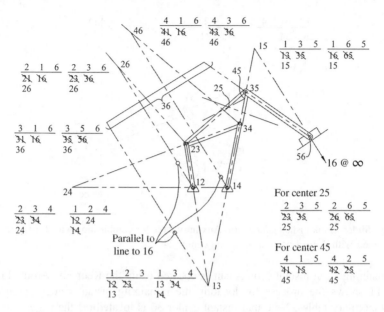

Figure 4.11   Instant centers for six-bar mechanism

The crank rotates at a constant speed ω_2. The velocity magnitude of point A is then $v_A = (O_2A)\omega_2$. The direction of the velocity vector can be determined based on the direction of ω_2 and will be perpendicular to the line from point A to the instant center.

Now we will utilize the instant center 13. At the position shown in Figure 4.12, the coupler (link 3) moves as if it were in pure rotation about instant center 13. Therefore, the rotation speed

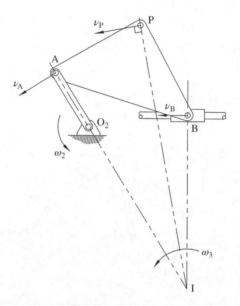

Figure 4.12   Determining velocities using instant centers

of link 3 will be

$$\omega_3 = \frac{v_A}{(IA)} \tag{4.4}$$

The quantity IA is the distance from the instant center to point A. The velocity magnitude of any point on link 3 can now be determined using the rotational speed.

$$\omega_3 = \frac{v_A}{(IA)} = \frac{v_P}{(IP)} = \frac{v_B}{(IB)} \tag{4.5}$$

The following example will better illustrate the process.

Example Problem 4.2

The 6.5 inch radius flywheel shown in Figure 4.13 is rotating at a constant speed of 140 rpm counterclockwise. What is the velocity of block B at the instant shown?

Solution: Converting the rotational speed of the flywheel gives

$$\omega = 140\,\frac{\text{rev}}{\text{min}} \left(\frac{2\pi\,\text{rad}}{\text{rev}}\right)\left(\frac{1\,\text{min}}{60\,\text{s}}\right) = 14.66\,\frac{\text{rad}}{\text{s}}$$

The flywheel is in pure rotation. The velocity magnitude of point A can be determined from

$$v_A = r\omega = 6.5\,\text{in.}\left(14.66\,\frac{\text{rad}}{\text{s}}\right) = 95.3\,\frac{\text{in.}}{\text{s}}$$

The direction of the velocity vector will be perpendicular to the radial line, as shown in Figure 4.14.

Next we must locate the instant center. The velocity of point B will be restricted to a horizontal line. The instant center, which is illustrated in Figure 4.15, can be located following the procedures discussed in Section 4.3.5. The angle between AB and the horizontal can be determined using law of sines:

$$\frac{\sin 60}{15} = \frac{\sin \alpha}{6.5} \Rightarrow \alpha = 22.04°$$

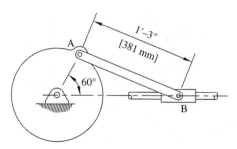

Figure 4.13

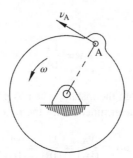

Figure 4.14

All other angles can then be determined. Law of sines can again be used to get distances to the instant center:

$$\frac{\sin 30}{15} = \frac{\sin 67.96}{r_A} = \frac{\sin 82.04}{r_B}$$

$$r_A = 27.81 \quad r_B = 29.71$$

Velocity of point B can be determined using the rotational velocity of the connecting rod:

$$\omega_3 = \frac{v_A}{r_A} = \frac{v_B}{r_B}$$

$$\frac{95.3}{27.81} = \frac{v_B}{29.71}$$

Answer: $v_B = 101.8 \dfrac{\text{in.}}{\text{s}}$

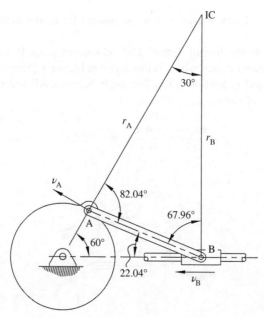

Figure 4.15

4.4   Graphical Velocity Analysis

4.4.1   Basic Concepts

Next we will move into a graphical method for velocity analysis using vector polygons. Velocity analysis can be accomplished fairly easily using vector polygons, especially with the aid of computer drafting software. One major disadvantage would be that it is a slow process and not useful for analysis of multiple positions of a mechanism. The visual nature of this method can help in understanding and can serve as a good method of checking more complex analytical solutions.

The solution steps presented assume that the lengths of all links are known, the angular positions of all links are known, and the input angular velocity is known. Consider the four-bar mechanism shown in Figure 4.16a. We need to find the angular velocities of links three and four. We are also interested in the velocity of point C on the coupler.

Begin the analysis of the link with the most known information. Typically, this will be the driving link because its angular velocity is generally given. The magnitude of the velocity of point A is determined from

$$v_A = (AO_2)\omega_2 \tag{4.6}$$

The vector direction, as shown in Figure 4.16a, will be perpendicular to the line AO_2. The direction also agrees with the counterclockwise rotational velocity direction. The vector for the velocity of point A can now be drawn to some convenient scale to begin a velocity polygon shown in Figure 4.16c.

Next we consider the velocity of point B. The direction of the angular velocity of link 4 is unknown; therefore, the direction of the velocity vector for point B cannot be directly determined. However, it is known that link 4 is in pure rotation. Therefore, the velocity vector for point B must be perpendicular to line BO_4, as shown in Figure 4.16a.

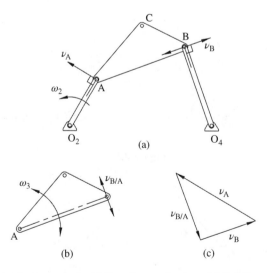

Figure 4.16   Graphical velocity analysis. (a) Four-bar mechanism. (b) Relative velocity of point B with respect to point A. (c) Velocity polygon

The last velocity term required is the relative velocity of point B with respect to point A, as illustrated in Figure 4.16b. Again, the direction of the angular velocity of link 3 is unknown. However, it is known that the relative velocity must be perpendicular to line AB because the relative velocity is based on link 3 rotating about a fixed point A.

Figure 4.16c is the vector polygon representing the velocity difference equation $\vec{v}_B = \vec{v}_A + \vec{v}_{B/A}$. Because the velocity polygon must close, the directions of the velocity vectors $\vec{v}_B$ and $\vec{v}_{B/A}$ can now be determined. With the vector polygon drawn to a convenient scale based on the known length of $\vec{v}_A$, the remaining velocity vector lengths can be determined graphically.

This process is very quick and easy in CAD software, but can also be done to some acceptable level of accuracy by hand drawings. This general process can be repeated to determine velocity magnitudes of other points such as point C on the coupler. Once the velocity of point B has been determined, the rotational velocity of link 4 is calculated by $\omega_4 = v_B/(BO_4)$. Similarly, the rotational velocity of the coupler is determined from $\omega_3 = v_{B/A}/(AB)$.

Example Problem 4.3

The crank in the slider–crank mechanism shown in Figure 4.17 rotates at a constant rate of 160 rpm clockwise. Determine the angular velocity of the connecting rod and the velocity of the slider for the position shown using the vector polygon method.

Solution: Converting the rotational speed of the crank gives

$$\omega_2 = 160\,\frac{rev}{min}\left(2\pi\,\frac{rad}{rev}\right)\left(\frac{1\ min}{60\ s}\right) = 16.75\,\frac{rad}{s}$$

The velocity of point A can be determined based on the crank's pure rotation at a constant speed. The direction will be perpendicular to the crank, as shown in Figure 4.18a.

$$v_A = (AO_2)\omega_2 = 4\ in.\left(16.75\,\frac{rad}{s}\right) = 67\,\frac{in.}{s}$$

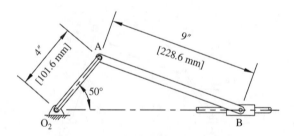

Figure 4.17

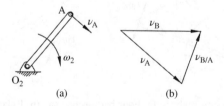

(a)   (b)

Figure 4.18

The velocity polygon, as shown in Figure 4.18b, can now be constructed to scale. The polygon is constructed based on the relative velocity equation.

$$\vec{v}_B = \vec{v}_A + \vec{v}_{B/A}$$

The velocity vector representing the velocity of point B with respect to point A will act perpendicular to the connecting rod. The magnitude of that velocity is measured from the velocity polygon.

$$v_{B/A} = 45.8 \frac{\text{in.}}{\text{s}}$$

The relative velocity can be used to determine the rotational velocity of the connecting rod:

$$v_{B/A} = (AB)\omega_3$$

$$(9 \text{ in.})\omega_3 = 45.8 \frac{\text{in.}}{\text{s}}$$

Answer: $\omega_3 = 5.1 \dfrac{\text{rad}}{\text{s}}$ (ccw)

The velocity of the slider is also measured from the velocity polygon.

Answer: $v_B = 67 \dfrac{\text{in.}}{\text{s}} \rightarrow$

Example Problem 4.4

Link O_2A in the mechanism shown in Figure 4.19 rotates counterclockwise at a constant rate of 50 rpm. Using the velocity vector polygon method, determine the rotational velocity of link O_4B.

Solution: Converting the rotation speed to radians per second gives

$$\omega_2 = 50 \frac{\text{rev}}{\text{min}} \left(2\pi \frac{\text{rad}}{\text{rev}} \right) \left(\frac{1 \text{ min}}{60 \text{ s}} \right) = 5.236 \frac{\text{rad}}{\text{s}}$$

The magnitude of the velocity at point A is

$$v_A = (AO_2)\omega_2 = 76.2 \text{ mm} \left(5.236 \frac{\text{rad}}{\text{s}} \right) = 399 \frac{\text{mm}}{\text{s}}$$

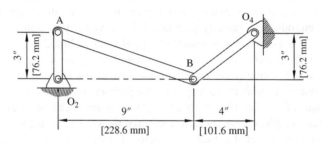

Figure 4.19

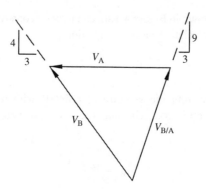

Figure 4.20

The velocity direction at A will be horizontal to the left, as shown in Figure 4.20. The velocity of point B is determined by constructing a velocity polygon for the relative velocity equation.

$$\vec{v}_B = \vec{v}_A + \vec{v}_{B/A}$$

The direction of the velocities will be known. The velocity of point B will act perpendicular to line O_4B, and the relative velocity of point B with respect to A will act perpendicular to line AB. The complete velocity polygon is shown in Figure 4.20. The length of the vector representing the velocity of B is measured to give

$$v_B = 460 \frac{mm}{s}$$

The length of link 4 will be 5 inches (127 mm). The rotational velocity of link 4 is given as

$$\omega_4 = \frac{v_B}{O_4} = \frac{460 \ mm/s}{127 \ mm}$$

Answer: $\omega_4 = 3.62 \frac{rad}{s}$ (cw)

4.4.2 Component Method

The component method is based on the concept that two points on the same rigid body will have equal components of velocity in the direction along a line joining the two points. In the basic four-bar mechanisms, such as the one shown in Figure 4.21a, the component method is a simple way to determine the velocity of point B if the velocity of point A is known. The velocity of point A is known and is directed perpendicular to link 2.

The process of the component method is illustrated in Figure 4.21b for determining the velocity of point B. The velocity vector at A needs to be broken into components: one parallel to line AB and one perpendicular to line AB. The parallel component of $\vec{v}_A$ must equal the parallel component of $\vec{v}_B$. The final velocity vector of point B can be determined using the parallel component length and the known direction of the velocity of B (perpendicular to link 4). The process can be expanded to find the velocity of other points on the coupler, which will be illustrated in an example problem.

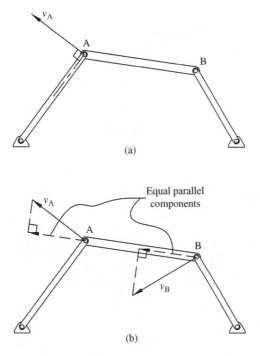

Figure 4.21   Component method for velocity analysis

Example Problem 4.5

Solve Example Problem 4.4 using the component method.

Solution: The velocity of A was determined from Example Problem 4.4 to be

$$v_A = 399 \frac{mm}{s}$$

Figure 4.22a shows the velocity vector at A along with the projection of that vector in the direction along line AB. The projected vector $v_{A_{per}}$ is then copied to point B, as shown in Figure 4.22b. That vector is then used to determine the length of the velocity vector at B, which acts perpendicular to line O_4B. The length of the vector representing the velocity of B is measured to give

$$v_B = 460 \frac{mm}{s}$$

The rotational velocity of link 4 is given as

$$\omega_4 = \frac{v_B}{BO_4} = \frac{460 \, mm/s}{127 \, mm}$$

Answer: $\omega_4 = 3.62 \frac{rad}{s}$ (cw)

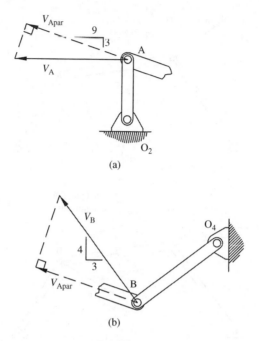

(a)

(b)

Figure 4.22

4.4.3 Parameter Studies

Consider an in-line slider–crank mechanism, as shown in Figure 4.23. The length of the crank will be defined as R and the length of the crank as C. As the crank rotates, the slider will move a distance, defined as x, from the toggle position. Using methods similar to those in Chapter 3, the distance x can be defined in terms of the input angle:

$$x = R(1 - \cos \theta_2) + C\left(1 - \sqrt{1 - \left(\frac{R}{C}\right)^2 \sin^2 \theta_2}\right) \qquad (4.7)$$

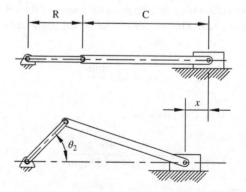

Figure 4.23 Slider position for an in-line slider-crank mechanism

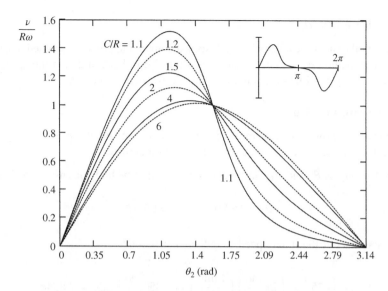

Figure 4.24 Normalized slider velocity

For a crank rotation speed of ω, the velocity of the slider can be determined by differentiating position with respect to time:

$$v = R\omega \sin \theta_2 \left(1 + \frac{R}{C} \frac{\cos \theta_2}{\sqrt{1 - (R/C)^2 \sin^2 \theta_2}} \right) \tag{4.8}$$

We can now get a dimensionless form by dividing by $R\omega$:

$$\frac{v}{R\omega} = \sin \theta_2 \left(1 + \frac{R}{C} \frac{\cos \theta_2}{\sqrt{1 - (R/C)^2 \sin^2 \theta_2}} \right) \tag{4.9}$$

Equation 4.9 can be plotted, as shown in Figure 4.24, for different ratios of coupler length to crank length.

Figure 4.24 shows normalized slider velocity for half of the crank cycle (180° of rotation). For rotation values in the second half of the cycle, the velocity values become negative as shown in the insert.

4.5 Analytical Velocity Analysis Methods

4.5.1 Introduction

This section will present three general approaches for analytical velocity analysis: vector method, loop-closure method, and differentiation of position method. Chapter 2 covered basic concepts of velocity analysis and should be referenced for additional information.

4.5.2 Vector Method

The first method of analytical velocity analysis is a common approach discussed in rigid body dynamics. The method uses relative velocities discussed in Section 2.5. Although Chapter 2 demonstrated the general process, another example is provided below.

Example Problem 4.6

The mechanism shown in Figure 4.25 has a constant input speed of $\omega_2 = 75$ rpm counterclockwise. Calculate ω_3 and ω_4 for the position shown.

Solution: The angles θ_3 and θ_4 can be determined using methods discussed in Chapter 3.

$$\theta_3 = 15.80° \qquad \theta_4 = 96.97°$$

We must first convert the input speed to radians per second.

$$\omega_2 = 75\frac{\text{rev}}{\text{min}}\left(2\pi\frac{\text{rad}}{\text{rev}}\right)\left(\frac{1\text{ min}}{60\text{ s}}\right) = 7.854\frac{\text{rad}}{\text{s}} \Rightarrow \vec{\omega}_2 = 7.854\hat{k}$$

The velocity of point A in pure rotation about point O_2 is

$$\vec{v}_A = \vec{\omega}_2 \times \vec{r}_2 = 7.854\hat{k} \times \left(6\cos(40)\hat{i} + 6\sin(40)\hat{j}\right) = -30.29\hat{i} + 36.10\hat{j}$$

The velocity of point B can be expressed using relative velocity:

$$\vec{v}_B = \vec{v}_A + \vec{v}_{B/A} = \vec{v}_A + \vec{\omega}_3 \times \vec{r}_3$$

$$\vec{v}_B = -30.29\hat{i} + 36.10\hat{j} + \omega_3\hat{k} \times \left(15\cos(15.8)\hat{i} + 15\sin(15.8)\hat{j}\right)$$

$$\vec{v}_B = -30.29\hat{i} + 36.10\hat{j} - 4.08\omega_3\hat{i} + 14.43\omega_3\hat{j}$$

$$\vec{v}_B = (-30.29 - 4.08\omega_3)\hat{i} + (36.10 + 14.43\omega_3)\hat{j} \tag{1}$$

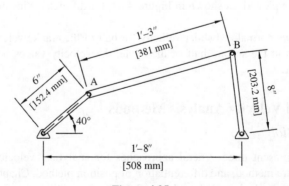

Figure 4.25

The velocity of point B can also be expressed in pure rotation about point O_4:

$$\vec{v}_B = \vec{\omega}_4 \times \vec{r}_4$$

$$\vec{v}_B = \omega_4 \hat{k} \times \left(8\cos(96.97)\hat{i} + 8\sin(96.97)\hat{j}\right)$$

$$\vec{v}_B = -7.94\omega_4 \hat{i} - 0.97\omega_4 \hat{j} \qquad (2)$$

The components of Equations 1 and 2 must be equal.

$$\begin{cases} -30.29 - 4.08\omega_3 = -7.94\omega_4 \\ 36.10 + 14.43\omega_3 = -0.97\omega_4 \end{cases}$$

Solving the set of equations gives

Answer:
$$\vec{\omega}_3 = -2.67\hat{k}$$
$$\vec{\omega}_4 = 2.44\hat{k}$$

4.5.3 Loop-Closure Method

Consider the four-bar mechanism shown in Figure 4.26 with the link nodes expressed with position vectors. The closed vector polygon yields

$$\vec{R}_1 + \vec{R}_2 + \vec{R}_3 + \vec{R}_4 = 0 \qquad (4.10)$$

The velocity equation is determined by differentiating with respect to time:

$$\dot{\vec{R}}_1 + \dot{\vec{R}}_2 + \dot{\vec{R}}_3 + \dot{\vec{R}}_4 = 0 \qquad (4.11)$$

If link 1 is the stationary ground link then $\dot{\vec{R}}_1 = 0$. If all the joints are revolute joints, as they are in the figure, the velocity equation becomes

$$\vec{\omega}_2 \times \vec{R}_2 + \vec{\omega}_3 \times \vec{R}_3 + \vec{\omega}_4 \times \vec{R}_4 = 0 \qquad (4.12)$$

For a planar mechanism, the cross products give the following equation:

$$-\left(\omega_2 R_{2_y} + \omega_3 R_{3_y} + \omega_4 R_{4_y}\right)\hat{i} + \left(\omega_2 R_{2_x} + \omega_3 R_{3_x} + \omega_4 R_{4_x}\right)\hat{j} = 0 \qquad (4.13)$$

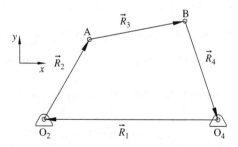

Figure 4.26 Linkage mechanism with closed position vector polygon

The vector equation can be separated into two scalar equations by noting that all $\hat{i}$ terms and all $\hat{j}$ terms must equal zero.

$$\begin{cases} \omega_3 R_{3_y} + \omega_4 R_{4_y} = -\omega_2 R_{2_y} \\ \omega_3 R_{3_x} + \omega_4 R_{4_x} = -\omega_2 R_{2_x} \end{cases} \tag{4.14}$$

The equations can be expressed in matrix form with ω_2 being the known input.

$$\begin{bmatrix} R_{3_y} & R_{4_y} \\ R_{3_x} & R_{4_x} \end{bmatrix} \begin{bmatrix} \omega_3 \\ \omega_4 \end{bmatrix} = \begin{bmatrix} -\omega_2 R_{2_y} \\ -\omega_2 R_{2_x} \end{bmatrix} \tag{4.15}$$

The unknown angular velocities can now be expressed using Cramer's rule:

$$\omega_3 = \frac{\begin{vmatrix} -\omega_2 R_{2_y} & R_{4_y} \\ -\omega_2 R_{2_x} & R_{4_x} \end{vmatrix}}{\begin{vmatrix} R_{3_y} & R_{4_y} \\ R_{3_x} & R_{4_x} \end{vmatrix}} = -\omega_2 \left(\frac{R_{2_y} R_{4_x} - R_{4_y} R_{2_x}}{R_{3_y} R_{4_x} - R_{4_y} R_{3_x}} \right) \tag{4.16}$$

$$\omega_4 = \frac{\begin{vmatrix} R_{3_y} & -\omega_2 R_{2_y} \\ R_{3_x} & -\omega_2 R_{2_x} \end{vmatrix}}{\begin{vmatrix} R_{3_y} & R_{4_y} \\ R_{3_x} & R_{4_x} \end{vmatrix}} = -\omega_2 \left(\frac{R_{3_y} R_{2_x} - R_{2_y} R_{3_x}}{R_{3_y} R_{4_x} - R_{4_y} R_{3_x}} \right) \tag{4.17}$$

Example Problem 4.7

Solve Example Problem 4.6 using the loop closure method.

Solution: First, we will calculate the angular velocity of link 3 using Equation 4.16.

$$\omega_3 = -\omega_2 \left(\frac{R_{2_y} R_{4_x} - R_{4_y} R_{2_x}}{R_{3_y} R_{4_x} - R_{4_y} R_{3_x}} \right)$$

The x and y components of each position vector need to be calculated:

$$R_{2_x} = 6 \cos 40 = 4.60 \qquad R_{2_y} = 6 \sin 40 = 3.86$$

$$R_{3_x} = 15 \cos 15.8 = 14.43 \qquad R_{3_y} = 15 \sin 15.8 = 4.08$$

$$R_{4_x} = 8 \cos 96.97 = -0.97 \qquad R_{4_y} = 8 \sin 96.97 = 7.94$$

Using the above values along with the rotational velocity of link 2 of 7.854 rad/s, the angular velocity of link 3 is

$$\omega_3 = -7.854 \left(\frac{3.86(-0.97) - 7.94(4.60)}{4.08(-0.97) - 7.94(14.43)} \right)$$

Answer: $\omega_3 = -2.67 \dfrac{\text{rad}}{\text{s}}$

Next we can calculate the rotational velocity of link 4 using Equation 4.17:

$$\omega_4 = -\omega_2 \left(\frac{R_{3_y} R_{2_x} - R_{2_y} R_{3_x}}{R_{3_y} R_{4_x} - R_{4_y} R_{3_x}} \right) = -7.854 \left(\frac{4.08(4.60) - 3.86(14.43)}{4.08(-0.97) - 7.94(14.43)} \right)$$

Answer: $\omega_4 = 2.44 \dfrac{\text{rad}}{\text{s}}$

4.5.4 Differentiation of Position Coordinate Equation

In some problems it is possible to develop a position equation, which can be differentiated to obtain a velocity equation. Typically, geometrical constraints of points on a body are used to develop position equations. Although this procedure is effective, it will generally be limited to simple cases. The process is best described with the use of an example.

Example Problem 4.8

Slider A in the mechanism shown in Figure 4.27 moves at a constant rate of 2.1 ft/s downward. Determine the velocity of slider B as a function of the angle θ.

Solution: The distance y can be determined as functions of θ.

$$y = 30 \cos \theta$$

The time derivative gives the vertical velocity of slider A.

$$\dot{y} = -2.1 \frac{\text{ft}}{\text{s}} = -30 \sin \theta \dot{\theta} \quad \Rightarrow \quad \dot{\theta} = \frac{2.1}{30 \sin \theta} = \frac{0.07}{\sin \theta}$$

The distance x can also be determined as a function of θ.

$$x = 30 \sin \theta$$

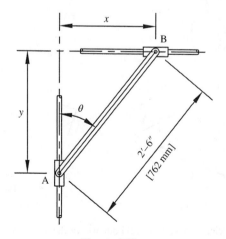

Figure 4.27

The time derivative gives

$$\dot{x} = 30 \cos \theta \dot{\theta} = 30 \cos \theta \frac{0.07}{\sin \theta}$$

Answer: $\dot{x} = 2.4 \cot \theta$

4.6 Graphical Acceleration Analysis Methods

Now that processes of velocity analysis have been explained, we can move to acceleration analysis. We will begin with graphical methods using vector polygons. Much like the graphical method of velocity analysis discussed in Section 4.4, acceleration analysis can also be performed graphically. However, graphical acceleration analysis is more complex because the directions of acceleration vectors are more difficult to visualize and often need to be assumed. Graphical analysis methods can be very useful in complex mechanisms because true analytical solutions may be very difficult. Graphical methods can also serve as a way of verifying the results of analytical methods.

Consider the linkage mechanism shown in Figure 4.28a. It is desired to determine the angular acceleration of links 3 and 4 using graphical methods. We will assume that a complete velocity analysis has been completed; therefore, the angular velocities of all links are known. The process will again typically start with link 2 because it is the link with the most known information. For a given rotational velocity and acceleration of link 2, point A will have both a normal and tangential component of acceleration as shown. Because link 2 is in pure rotation, the magnitude of the normal component of the acceleration of point A will be $(a_A)_n = (AO_2)\omega_2^2$

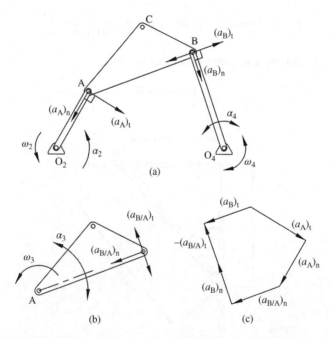

Figure 4.28 Graphical acceleration analysis. (a) Four-bar mechanism. (b) Relative acceleration of point B with respect to point A. (c) Acceleration vector polygon

and its direction will be toward the center of rotation. The tangential component will have a magnitude $(a_A)_t = (AO_2)\alpha_2$ and will be perpendicular to line AO_2 in the direction relating to the direction of α_2. The two vector components can be drawn at a convenient scale to start a vector polygon, as shown in Figure 4.28c.

Next we move to point B and consider the relative acceleration of point B with respect to point A. The relative acceleration, as shown in Figure 4.28b, is based on pure rotation about point A. Therefore, the relative acceleration will again have a normal and tangential component. The normal component $\left(\vec{a}_{B/A}\right)_n$ will have a known direction toward point A. The magnitude is also known because ω_3 was determined in a previous velocity analysis.

Link 4 is also in pure rotation, and the acceleration of point B will have both a normal and a tangential component. The direction of the normal component will be oriented toward the fixed rotation point. The magnitude of the normal component is also known due to previous velocity analysis. The tangential component will be oriented perpendicular to line BO_4, although the actual direction and magnitude cannot be determined until the direction and magnitude of α_4 are known.

We can now continue construction of the vector polygon shown in Figure 4.28c. The relative acceleration equation $\vec{a}_B = \vec{a}_A + \vec{a}_{B/A}$ will be used to determine the unknown acceleration values. Writing each acceleration vector in terms of normal and tangential components gives

$$(\vec{a}_B)_t + (\vec{a}_B)_n = (\vec{a}_A)_t + (\vec{a}_A)_n + \left(\vec{a}_{B/A}\right)_t + \left(\vec{a}_{B/A}\right)_n \qquad (4.18)$$

All information is known about the normal and tangential vectors for the acceleration of point A. Both vectors can be drawn to start the polygon shown in Figure 4.28c. Magnitude and direction of $\left(\vec{a}_{B/A}\right)_n$ are known so that vector is added to the polygon. The normal component of the acceleration of point B also has known magnitude and direction, so it can be added to the polygon. However, that vector needs to be subtracted because it appears on the left-hand side of Equation 4.18. To subtract that vector, simply draw it in the opposite direction. The two unknown tangential accelerations, $\left(\vec{a}_{B/A}\right)_t$ and $(\vec{a}_B)_t$, can be added to the polygon. The directions and magnitudes can be determined based on the fact that the vector polygon must close. Since the polygon is drawn to scale, the length of the unknown vectors will give their magnitude.

Once the polygon is complete, the angular accelerations of links 3 and 4 can be calculated. The angular acceleration of link 3 is determined using the relative acceleration of point B with respect to point A to give $\alpha_3 = \left(a_{B/A}\right)_t/(AB)$. Link 4 is in pure rotation about the fixed pivot, so the angular acceleration is determined from $\alpha_4 = (a_B)_t/(BO_4)$.

Example Problem 4.9

Figure 4.29 shows a four-bar mechanism. Link 2 is rotating at 90 rpm (ccw) and is slowing down at a rate of 12.5 rad/s². Determine the angular acceleration of link 4 for the position shown.

Solution: Converting the rotational speed of the crank gives

$$\omega_2 = 90\,\frac{\text{rev}}{\text{min}}\left(2\pi\,\frac{\text{rad}}{\text{rev}}\right)\left(\frac{1\ \text{min}}{60\ \text{s}}\right) = 9.42\,\frac{\text{rad}}{\text{s}}$$

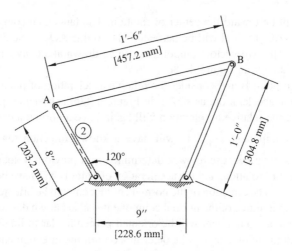

Figure 4.29

The velocity magnitude of point A is

$$v_A = r_2\omega_2 = 8 \text{ in.}\left(9.42\frac{\text{rad}}{\text{s}}\right) = 75.36\frac{\text{in.}}{\text{s}}$$

The velocity polygon, as discussed in Section 4.4.1, can now be constructed with the direction of the velocity of point A directed perpendicular to link 2. Figure 4.30a shows the polygon. The unknown velocities are measured from the polygon to give

$$v_{B/A} = 73.0\frac{\text{in.}}{\text{s}} \qquad v_B = 88.8\frac{\text{in.}}{\text{s}}$$

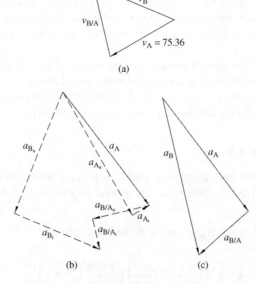

Figure 4.30

Acceleration information about link 2 can be determined.

$$a_{A_n} = r_2 \omega_2^2 = 8 \text{ in.} \left(9.42 \frac{\text{rad}}{\text{s}} \right)^2 = 709.9 \frac{\text{in.}}{\text{s}^2}$$

$$a_{A_t} = r_2 \alpha_2 = 8 \text{ in.} \left(12.5 \frac{\text{rad}}{\text{s}^2} \right) = 100.0 \frac{\text{in.}}{\text{s}^2}$$

Both accelerations can be drawn to begin the acceleration polygon, as shown in Figure 4.30b. The normal component of the acceleration of point B is also known.

$$a_{B_n} = r_4 \omega_4^2 = r_4 \left(\frac{v_B}{r_4} \right)^2 = \frac{v_B^2}{r_4}$$

$$a_{B_n} = \frac{\left(88.8 \frac{\text{in.}}{\text{s}} \right)^2}{12 \text{ in.}} = 657.1 \frac{\text{in.}}{\text{s}^2}$$

This acceleration term can now be added to the acceleration polygon in a direction along link 4. Similarly, the normal component of the acceleration of point B with respect to A is

$$a_{B/A_n} = \frac{v_{B/A}^2}{r_3} = \frac{\left(73.0 \frac{\text{in.}}{\text{s}} \right)^2}{18 \text{ in.}} = 296.1 \frac{\text{in.}}{\text{s}^2}$$

Both tangential components are unknown, but the directions are known. Laying out the tangential vectors on the polygon will give the directions, as shown in Figure 4.30b. The final acceleration polygon is shown in Figure 4.30c. The acceleration terms are measured from the polygon to give

$$a_{B/A} = 334.9 \frac{\text{in.}}{\text{s}^2} \qquad a_B = 808.6 \frac{\text{in.}}{\text{s}^2}$$

The angular acceleration of link 4 can now be calculated.

$$\alpha_4 = \frac{a_{B_t}}{r_4}$$

The tangential acceleration of point B is measured from the polygon shown in Figure 4.30b to be 471.2 in./s².

$$\alpha_4 = \frac{471.2 \frac{\text{in.}}{\text{s}^2}}{12 \text{ in.}}$$

Answer: $\alpha_4 = 39.3 \frac{\text{rad}}{\text{s}^2}$ (cw)

4.7   Analytical Acceleration Analysis Methods

The vector method for acceleration builds on the vector method for velocity developed in Section 4.5.2. Relative acceleration equations were discussed in Section 2.5 and will be used here to determine accelerations within the mechanism.

Example Problem 4.10

Using the results for the mechanism in Example Problem 4.6 calculate α_3 and α_4 for the position shown.

Solution: The acceleration of point A in pure rotation about point O_2 is

$$\vec{a}_A = \left(\vec{a}_A\right)_n + \left(\vec{a}_A\right)_t = \vec{\omega}_2 \times \vec{v}_A + \vec{\alpha}_2 \times \vec{r}_2$$

From the problem statement for Example Problem 4.6, we know that $\vec{\omega}_2 = 7.854\hat{k}$ and $\vec{\alpha}_2 = 0$. The velocity of point A was calculated to be $\vec{v}_A = -30.29\hat{i} + 36.10\hat{j}$.

$$\vec{a}_A = 7.854\hat{k} \times \left(-30.29\hat{i} + 36.10\hat{j}\right) = -283.53\hat{i} - 237.90\hat{j}$$

The acceleration of point B can be expressed using relative acceleration equations.

$$\vec{a}_B = \vec{a}_A + \left(\vec{a}_{B/A}\right)_n + \left(\vec{a}_{B/A}\right)_t = \vec{a}_A + \vec{\omega}_3 \times \left(\vec{\omega}_3 \times \vec{r}_3\right) + \vec{\alpha}_3 \times \vec{r}_3$$

From Example Problem 4.6, $\vec{\omega}_3 = -2.67\hat{k}$.

$$\vec{a}_B = -283.53\hat{i} - 237.90\hat{j} + -2.67\hat{k} \times \left(-2.67\hat{k} \times \left(15\cos(15.8)\hat{i} + 15\sin(15.8)\hat{j}\right)\right)$$
$$+ \alpha_3\hat{k} \times \left(15\cos(15.8)\hat{i} + 15\sin(15.8)\hat{j}\right)$$

$$\vec{a}_B = -283.53\hat{i} - 237.90\hat{j} + -2.67\hat{k} \times \left(10.90\hat{i} - 38.54\hat{j}\right) - 4.08\alpha_3\hat{i} + 14.43\alpha_3\hat{j}$$

$$\vec{a}_B = -283.53\hat{i} - 237.90\hat{j} + -102.89\hat{i} - 29.12\hat{j} - 4.08\alpha_3\hat{i} + 14.43\alpha_3\hat{j}$$

$$\vec{a}_B = (-386.42 - 4.08\alpha_3)\hat{i} + (-267.05 + 14.43\alpha_3)\hat{j} \tag{1}$$

The acceleration of point B can also be expressed in pure rotation about point O_4.

$$\vec{a}_B = \vec{\omega}_4 \times \left(\vec{\omega}_4 \times \vec{r}_4\right) + \vec{\alpha}_4 \times \vec{r}_4$$

From Example Problem 4.6 $\vec{\omega}_4 = 2.44\hat{k}$.

$$\vec{a}_B = 2.44\hat{k} \times \left(2.44\hat{k} \times \left(8\cos(96.97)\hat{i} + 8\sin(96.97)\hat{j}\right)\right)$$
$$+ \vec{\alpha}_4 \times \left(8\cos(96.97)\hat{i} + 8\sin(96.97)\hat{j}\right)$$

$$\vec{a}_B = 5.79\hat{i} - 47.40\hat{j} - 7.94\alpha_4\hat{i} - 0.97\alpha_4\hat{j}$$

$$\vec{a}_B = (5.79 - 7.94\alpha_4)\hat{i} + (-47.40 - 0.97\alpha_4)\hat{j} \tag{2}$$

The components of Equations 1 and 2 must be equal.

$$\begin{cases} -386.42 - 4.08\alpha_3 = 5.79 - 7.94\alpha_4 \\ -267.05 + 14.43\alpha_3 = -47.40 - 0.97\alpha_4 \end{cases}$$

Solving the set of equations gives

Answer: $\begin{aligned} \vec{\alpha}_3 &= 11.49\hat{k} \\ \vec{\alpha}_4 &= 55.26\hat{k} \end{aligned}$

4.8 Kinematic Analysis of Linkage Mechanisms with Moving Slides

4.8.1 Sliding Motion

Kinematic analysis of mechanisms containing sliding connections becomes a little bit more complex. Before exploring full linkage mechanisms, we need to develop the basic ideas of sliding motion. Figure 4.31 shows two rigid bodies. Rigid body 1 and 2 are in sliding contact. The axis along the line of contact is the tangential axis, and the normal axis is perpendicular to the line of contact. Each rigid body is moving in the directions shown, but remains in contact along the sliding surface.

A point P is shown at the intersection of the normal and tangential axes. However, there are two points actually located at the intersection. Let P_1 represent point P on rigid body 1, and let P_2 represent point P on rigid body 2. The total velocity of point P_1 will depend on the motion of rigid body 1, and the total velocity of point P_2 will depend on the total motion of rigid body 2. However, the normal component of both velocities must be equal, $\vec{v}_{1_n} = \vec{v}_{2_n}$, for the two rigid bodies to remain in contact.

Next consider the mechanism shown in Figure 4.32a. Link 3 is a slider, which is pinned to link 2 at point P. The slider moves along the long axis of link 4. The input rotary motion of link 2 is then transferred to link 4 through the slider. The typical problem would consist of finding the output rotational velocity of link 4 given the rotational velocity of link 2.

Again, we will have coincidence points at location P, now on three links P_2, P_3, P_4. The velocity of point P_2 can be determined using only link 2, as shown in Figure 4.32b. The magnitude of the velocity will be

$$v_{P_2} = r_2\omega_2 \tag{4.19}$$

and the direction will be perpendicular to the long axis of the link as shown. Similarly, the velocity of point P_4 is shown in Figure 4.32c. The magnitude is unknown, but the direction can be determined from inspection.

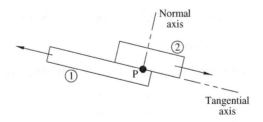

Figure 4.31 Sliding motion between two rigid bodies

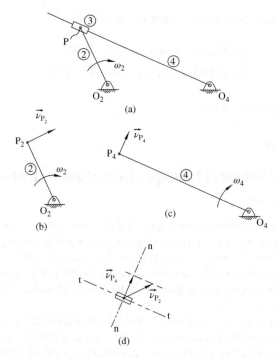

Figure 4.32 (a) Mechanism with sliding joint. (b) Velocity of P_2. (c) Velocity of P_4. (d) Sliding motion

The sliding motion of link 3 along link 4 is shown in Figure 4.32d. Line t–t is the tangential axis and line n–n is the normal axis. The known velocity vector $\vec{v}_{P_2}$ is added at point P. The normal components of $\vec{v}_{P_2}$ and $\vec{v}_{P_4}$ must be equal, which allows for a graphical determination of the velocity of point P_4. The rotational velocity of link 4 can now be determined by referring back to Figure 4.32c.

Example Problem 4.11

The linkage mechanism shown in Figure 4.33 is driven by link 2, which is rotating clockwise at a constant rate of 12 rad/s at the instant shown. Determine the rotational speed of link 4 for the position shown.

Solution: For the instant shown, we know the output angle is 30°. The angular position of link 2 as well as the distance from O_2 to the slider can be determined from geometry. The distance from O_2 to the slider is determined from the law of cosines.

$$r_2^2 = 508^2 + 355.6^2 - 2(508)(355.6)\cos 30$$

$$r_2 = 267.6\,\text{mm}$$

The angle of link 2, measured from the positive x-axis, can be determined using law of sines:

$$\frac{\sin \theta_2}{508} = \frac{\sin 30}{267.6}$$

$$\theta_2 = 108.37°$$

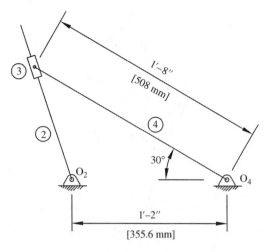

Figure 4.33

These dimensions are shown in Figure 4.34. The velocity of point P_2, shown in Figure 4.34, can be determined based on the rotational speed of link 2.

$$v_{P_2} = r_2\omega_2 = 0.2676\,\mathrm{m}\left(12\frac{\mathrm{rad}}{\mathrm{s}}\right) = 3.2\frac{\mathrm{m}}{\mathrm{s}}$$

The direction will be perpendicular to link 2 as shown. Next, we consider the slider, as shown in Figure 4.35. The angle a can be determined from

$$a = 180 - 30 - 108.37 = 41.63°$$

Based on the geometry, angle b will be equal to angle a. The magnitude of the velocity of point P_4 is

$$v_{P_4} = \frac{v_{P_2}}{\cos b} = \frac{3.2}{\cos 41.63} = 4.28\frac{\mathrm{m}}{\mathrm{s}}$$

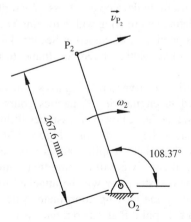

Figure 4.34

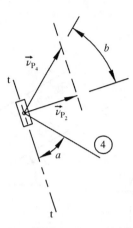

Figure 4.35

The rotation velocity of link 4 can now be determined.

$$\omega_4 = \frac{v_{P_4}}{r_4} = \frac{4.28\,\dfrac{m}{s}}{0.508\ m}$$

Answer: $\omega_4 = 8.4\dfrac{rad}{s}$ (cw)

Kinematic analysis of mechanisms with sliding motion can also be solved graphically with the use of velocity and acceleration vector polygons. The process is illustrated in Chapter 11 to determine forces. See Example Problem 11.5 for more information on the graphical method.

4.8.2 Motion Relative to Rotating Axes

The dynamics review in Chapter 2 applied kinematics using nonrotating reference frames to describe the motion. Previous sections of this chapter also focus on planar mechanism without relative sliding between links. Because sliding contact between two nongrounded links occurs frequently in machines, it is useful to develop a process to describe motion of a point using a reference frame that is translating and rotating with a rotating body. Sliding contact problems can be solved using methods previously discussed in Section 4.8.1, but the solution becomes more elegant with the use of a rotating reference frame (especially for more complex problems).

We will now consider motion relative to rotating axes. The concept of moving reference frames is generally covered in engineering dynamics courses, so this may serve as a review. The discussion presented here, however, assumes little previous knowledge of the process. First, we will consider the general case. Figure 4.36a shows a rigid body with points A and B in a fixed *X–Y* plane (capital letters representing the fixed coordinate system). For the most general case, we will assume that point A moves independent of point B. A moving coordinate system *x–y* will be attached to the rigid body at point B (lower case letters representing the moving coordinate system). The moving coordinate system moves in translation with point B and also rotates with the angular velocity ω of the rigid body.

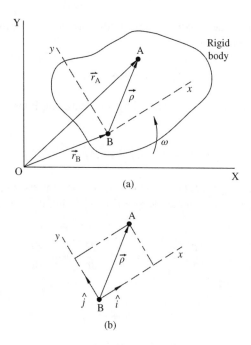

Figure 4.36 (a) Position vectors using rotating coordinate system. (b) Unit vectors to locate point A

The motion of point A can now be expressed from the moving coordinate system. The position of point A can be defined as

$$\vec{r}_A = \vec{r}_B + \vec{\rho} \tag{4.20}$$

where $\vec{\rho}$ is the vector representing the position of point B with respect to point A. If we use unit vectors $\hat{i}$ and $\hat{j}$ attached to the moving coordinate system, as shown in Figure 4.36b, Equation 4.20 becomes

$$\vec{r}_A = \vec{r}_B + \left(x\hat{i} + y\hat{j}\right) \tag{4.21}$$

The velocity of point A can now be determined by differentiation of Equation 4.21 with respect to time. Keep in mind that the unit vectors $\hat{i}$ and $\hat{j}$ rotate with the rigid body, so the time derivatives will not be zero. Using the product rule, the time derivative gives

$$\vec{v}_A = \vec{v}_B + \dot{x}\hat{i} + x\frac{d\hat{i}}{dt} + \dot{y}\hat{j} + y\frac{d\hat{j}}{dt} \tag{4.22}$$

We can simplify the expression. First, the time derivatives of the unit vectors become

$$x\frac{d\hat{i}}{dt} + y\frac{d\hat{j}}{dt} = \vec{\omega} \times x\hat{i} + \vec{\omega} \times y\hat{j} = \vec{\omega} \times \left(x\hat{i} + y\hat{j}\right) = \vec{\omega} \times \vec{\rho} \tag{4.23}$$

Next, we define $\dot{x}\hat{i} + \dot{y}\hat{j}$ as the velocity relative to the moving x–y coordinate system $\dot{\vec{\rho}}_{\text{rel}}$. The final velocity equation becomes

$$\vec{v}_A = \vec{v}_B + \vec{\omega} \times \vec{\rho} + \dot{\vec{\rho}}_{\text{rel}} \tag{4.24}$$

Acceleration of point A can be determined by differentiating Equation 4.24 with respect to time:

$$\vec{a}_A = \vec{a}_B + \dot{\vec{\omega}} \times \vec{\rho} + \vec{\omega} \times \dot{\vec{\rho}} + \frac{d\dot{\vec{\rho}}_{\text{rel}}}{dt} \tag{4.25}$$

The $\dot{\vec{\omega}}$ term is simply the angular rotation of the moving coordinate system, which is attached to the rigid body. Next, we will examine the $\vec{\omega} \times \dot{\vec{\rho}}$ term. Using the unit vectors of the moving coordinate system,

$$\dot{\vec{\rho}} = \frac{d}{dt}\left(x\hat{i} + y\hat{j}\right) = \dot{x}\hat{i} + x\frac{d\hat{i}}{dt} + \dot{y}\hat{j} + y\frac{d\hat{j}}{dt} \tag{4.26}$$

Using Equation 4.23,

$$\dot{\vec{\rho}} = \dot{x}\hat{i} + \dot{y}\hat{j} + \vec{\omega} \times \vec{\rho}$$

$$\dot{\vec{\rho}} = \dot{\vec{\rho}}_{\text{rel}} + \vec{\omega} \times \vec{\rho} \tag{4.27}$$

Therefore, $\vec{\omega} \times \dot{\vec{\rho}}$ becomes

$$\vec{\omega} \times \dot{\vec{\rho}} = \vec{\omega} \times \left(\dot{\vec{\rho}}_{\text{rel}} + \vec{\omega} \times \vec{\rho}\right) = \vec{\omega} \times (\vec{\omega} \times \vec{\rho}) + \vec{\omega} \times \dot{\vec{\rho}}_{\text{rel}} \tag{4.28}$$

Substituting Equation 4.28 into Equation 4.25 gives

$$\vec{a}_A = \vec{a}_B + \dot{\vec{\omega}} \times \vec{\rho} + \vec{\omega} \times (\vec{\omega} \times \vec{\rho}) + \vec{\omega} \times \dot{\vec{\rho}}_{\text{rel}} + \frac{d\dot{\vec{\rho}}_{\text{rel}}}{dt} \tag{4.29}$$

The last term can now be examined:

$$\frac{d\dot{\vec{\rho}}_{\text{rel}}}{dt} = \frac{d}{dt}\left(\dot{x}\hat{i} + \dot{y}\hat{j}\right) = \ddot{x}\hat{i} + \dot{x}\frac{d\hat{i}}{dt} + \ddot{y}\hat{j} + \dot{y}\frac{d\hat{j}}{dt} \tag{4.30}$$

We will define the terms $\ddot{x}\hat{i} + \ddot{y}\hat{j}$ as the relative acceleration vector $\ddot{\vec{\rho}}_{\text{rel}}$:

$$\frac{d\dot{\vec{\rho}}_{\text{rel}}}{dt} = \dot{x}\frac{d\hat{i}}{dt} + \dot{y}\frac{d\hat{j}}{dt} + \ddot{\vec{\rho}}_{\text{rel}}$$

$$\frac{d\dot{\vec{\rho}}_{\text{rel}}}{dt} = \vec{\omega} \times \left(\dot{x}\hat{i} + \dot{y}\hat{j}\right) + \ddot{\vec{\rho}}_{\text{rel}} = \vec{\omega} \times \dot{\vec{\rho}}_{\text{rel}} + \ddot{\vec{\rho}}_{\text{rel}} \tag{4.31}$$

Substituting Equation 4.31 into Equation 4.29 gives

$$\vec{a}_A = \vec{a}_B + \dot{\vec{\omega}} \times \vec{\rho} + \vec{\omega} \times (\vec{\omega} \times \vec{\rho}) + \vec{\omega} \times \dot{\vec{\rho}}_{\text{rel}} + \vec{\omega} \times \dot{\vec{\rho}}_{\text{rel}} + \ddot{\vec{\rho}}_{\text{rel}} \tag{4.32}$$

The term $\vec{\omega} \times \dot{\vec{\rho}}_{\text{rel}}$ appears twice, so they can be collected to give the final acceleration equation:

$$\vec{a}_A = \vec{a}_B + \dot{\vec{\omega}} \times \vec{\rho} + \vec{\omega} \times (\vec{\omega} \times \vec{\rho}) + 2\vec{\omega} \times \dot{\vec{\rho}}_{\text{rel}} + \ddot{\vec{\rho}}_{\text{rel}} \qquad (4.33)$$

4.8.3 Coriolis Component

The term $2\vec{\omega} \times \dot{\vec{\rho}}_{\text{rel}}$ in the final acceleration Equation 4.33 is known as the Coriolis acceleration named after the French scientist Gaspard-Gustave Coriolis (1792–1843). Machine analysis problems with moving slides are more difficult and the Coriolis component tends to be a source of confusion for students. This section will prove that the Coriolis component exists, and it will hopefully clarify the total acceleration equation. Consider the rotating link shown in Figure 4.37a. The link rotates at a constant speed ω. The slider moves along the link with a constant linear velocity v_s. Assuming the link was not rotating, after some small time Δt the slider would move a distance d_1 as shown in Figure 4.37b. Assuming the slider is fixed to the rotating link, after time Δt the slider will move along an arc distance d_2 as shown in Figure 4.37c. Logically, the total motion would be the summation of the two motions. Figure 4.37d shows that the combined effect would get the slider to point p. However, the

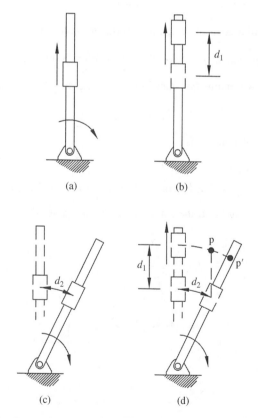

(a)   (b)

(c)   (d)

Figure 4.37 (a) Rotating link with slider. (b) Motion due to slider alone. (c) Motion due to rotating link alone. (d) Final location of the slider

actual location of the slider must be at point p′. The additional acceleration required to move the slider from point p to point p′ is the Coriolis acceleration.

Example Problem 4.12

Consider the rotating disk shown in Figure 4.38, which rotates at a constant rate of 7 rad/s (ccw). At the instant shown, the radial distance to point A is $r = 4$ in.. The moving coordinate system x–y rotates with the disk. Determine the absolute velocity and acceleration of point A for (a) $v = 0$ and (b) $v = 2.5\dfrac{\text{in.}}{\text{s}}$.

Solution: First we consider part (a). Although the motion of point A becomes pure rotation, which is easy to solve, we will use Equations 4.24 and 4.33 in order to illustrate the process and better understand the individual terms of the velocity and acceleration equations. First we will determine the absolute velocity of point A. Note that the origin of the moving coordinate system is now at point O (Figure 4.36 used point B, so the equations also used point B).

$$\vec{v}_A = \vec{v}_O + \vec{\omega} \times \vec{\rho} + \dot{\vec{\rho}}_{rel}$$

For our example, the origin of the rotating coordinate system remains at the center of the disk.

$$\vec{v}_O = 0$$

The vector locating point A in the rotating coordinate system is

$$\vec{\rho} = x\hat{i} + y\hat{j} = 4\cos 40\hat{i} + 4\sin 40\hat{j} = 3.06\hat{i} + 2.57\hat{j}$$

Since the position of A remains constant in the rotating coordinate system,

$$\dot{\vec{\rho}}_{rel} = \dot{x}\hat{i} + \dot{y}\hat{j} = 0$$

The absolute velocity of point A is

$$\vec{v}_A = 0 + \vec{\omega} \times \vec{\rho} + 0$$
$$\vec{v}_A = 7\hat{k} \times \left(3.06\hat{i} + 2.57\hat{j}\right) = -17.99\hat{i} + 21.42\hat{j}$$

The velocity vector is relative to the rotating coordinate system.

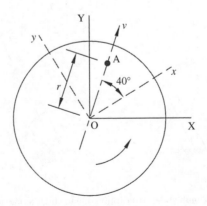

Figure 4.38

Next, we determine the absolute acceleration of point A for part (a).

$$\vec{a}_A = \vec{a}_O + \dot{\vec{\omega}} \times \vec{\rho} + \vec{\omega} \times (\vec{\omega} \times \vec{\rho}) + 2\vec{\omega} \times \dot{\vec{\rho}}_{rel} + \ddot{\vec{\rho}}_{rel}$$

The moving coordinate system origin remains fixed to the disk center and rotates at a constant speed giving

$$\vec{a}_O = 0 \qquad \dot{\vec{\omega}} = 0$$

Again, point A is stationary in the rotating coordinate system:

$$\dot{\vec{\rho}}_{rel} = \ddot{\vec{\rho}}_{rel} = 0$$

The absolute acceleration is

$$\vec{a}_A = 0 + 0 \times \vec{\rho} + \vec{\omega} \times (\vec{\omega} \times \vec{\rho}) + 2\vec{\omega} \times 0 + 0$$

$$\vec{a}_A = \vec{\omega} \times (\vec{\omega} \times \vec{\rho}) = 7\hat{k} \times (-17.99\hat{i} + 21.42\hat{j}) = -149.94\hat{i} - 125.93\hat{j}$$

The acceleration vector is relative to the rotating coordinate system. The orientation of both the velocity and acceleration vectors are shown in Figure 4.39a.

Answer for part (a): $\begin{aligned} \vec{v}_A &= -17.99\hat{i} + 21.42\hat{j} \quad (\text{in./s}) \\ \vec{a}_A &= -149.94\hat{i} - 125.93\hat{j} \quad (\text{in./s}^2) \end{aligned}$

Next, we consider part (b) when the particle is in motion relative to the rotating disk. As with the previous case, the rotating coordinate system origin remains fixed to the disk center and rotates at a constant rate.

$$\vec{v}_O = 0 \qquad \vec{a}_O = 0 \qquad \dot{\vec{\omega}} = 0$$

Point A is now moving relative to the rotating axes.

$$\dot{\vec{\rho}}_{rel} = \dot{x}\hat{i} + \dot{y}\hat{j} = 2.5 \cos 40\hat{i} + 2.5 \sin 40\hat{j} = 1.91\hat{i} + 1.61\hat{j}$$

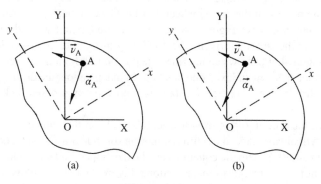

(a)   (b)

Figure 4.39

The absolute velocity of point A is

$$\vec{v}_A = 0 + \vec{\omega} \times \vec{\rho} + \dot{\vec{\rho}}_{rel}$$

$$\vec{v}_A = -17.99\hat{i} + 21.42\hat{j} + 1.91\hat{i} + 1.61\hat{j} = -16.08\hat{i} + 23.03\hat{j}$$

Point A is moving with a constant velocity in the rotating coordinate system.

$$\ddot{\vec{\rho}}_{rel} = 0$$

The absolute acceleration of point A is

$$\vec{a}_A = 0 + 0 \times \vec{\rho} + \vec{\omega} \times (\vec{\omega} \times \vec{\rho}) + 2\vec{\omega} \times \dot{\vec{\rho}}_{rel} + 0$$

$$\vec{a}_A = \vec{\omega} \times (\vec{\omega} \times \vec{\rho}) + 2\vec{\omega} \times \dot{\vec{\rho}}_{rel}$$

$$\vec{a}_A = -149.94\hat{i} - 125.93\hat{j} + 2(7\hat{k}) \times (1.91\hat{i} + 1.61\hat{j})$$

$$\vec{a}_A = -149.94\hat{i} - 125.93\hat{j} - 22.54\hat{i} + 26.74\hat{j}$$

$$\vec{a}_A = -172.48\hat{i} - 99.19\hat{j}$$

Again, the velocity and acceleration vectors are relative to the rotating coordinate system. The orientations are shown in Figure 4.39b. The acceleration does not pass through the disk center due to the Coriolis acceleration component.

Answer for part (b): $\begin{aligned} \vec{v}_A &= -16.08\hat{i} + 23.03\hat{j} \quad \text{(in./s)} \\ \vec{a}_A &= -172.48\hat{i} - 99.19\hat{j} \quad \text{(in./s}^2) \end{aligned}$

4.8.4 Geneva Mechanisms

Another common mechanism utilizing sliding joints is an indexing device known as a Geneva mechanism. The Geneva mechanism is used for intermittent circular output motion from continuous circular input motion. The motion is illustrated for a general Geneva mechanism in Figure 4.40. Figure 4.40a shows the driving wheel and the driven wheel. The driving wheel contains a pin that creates the positive drive. The driven wheel contains some number of slots. The figure shows an example with six slots but the minimum is three slots. The number of slots will determine the motion of the driven wheel. For the Geneva mechanism shown, the driving wheel rotates counterclockwise. Figure 4.40a shows the point in time when the pin enters slot 1. Once the pin enters the slot, the driven wheel will begin to rotate in a direction opposite the driving wheel. Figure 4.40b shows an intermediate position of motion. In Figure 4.40c, the pin has left slot 1, which has now indexed one position. In that position, the driving wheel is still in motion, but the driven wheel is held in a stationary position by a raised locking plate on the driving wheel.

As mentioned, the motion of the driven wheel depends on the number of slots it contains. The dwell time will always exceed the time of motion (or indexing) for a Geneva mechanism, as shown in Figure 4.40. Kinematic equations can be developed for Geneva mechanisms, but it is also convenient to use graphical representations. Figure 4.41 shows plots of Geneva wheel kinematics including driven wheel displacement, velocity, and acceleration.

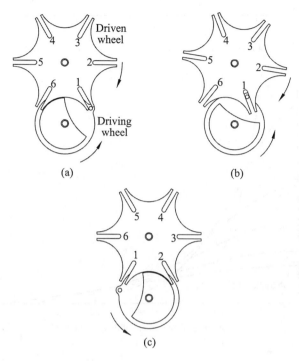

Figure 4.40 Geneva mechanism. (a) Pin entering slot 1. (b) Intermediate position. (c) Dwell period

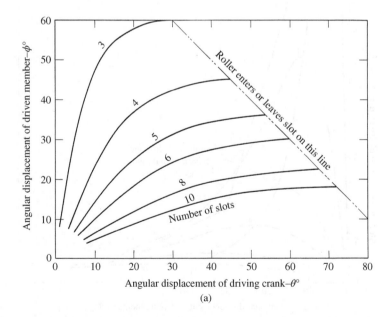

Figure 4.41 Geneva wheel kinematics. (a) Angular displacement of external Geneva. Reproduced from Tuttle, Mechanisms for Engineering Design, John Wiley & Sons, © 1967 (b) Angular velocity of external Geneva. Reproduced from Tuttle, Mechanisms for Engineering Design, John Wiley & Sons, © 1967 (c) Angular acceleration of external Geneva. Reproduced from Tuttle, Mechanisms for Engineering Design, John Wiley & Sons, © 1967

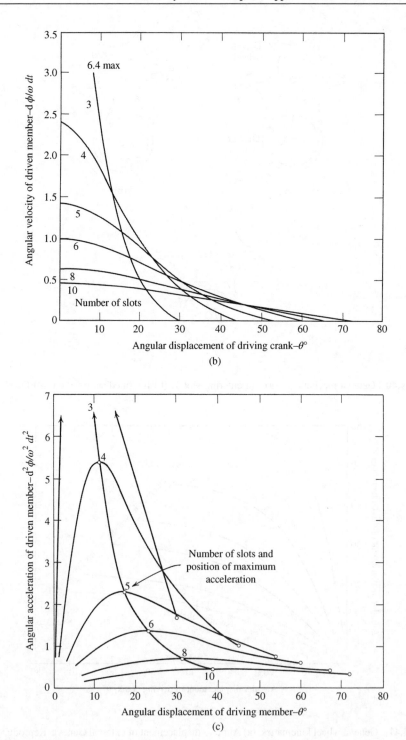

Figure 4.41   (*Continued*)

4.9   Review and Summary

Velocity and acceleration analysis is a vital, yet tedious, part of machine analysis. Many methods exist for calculating velocity and acceleration, but for complete kinematic analysis of a mechanism, computer methods (presented in Chapter 6) are often preferred. This chapter provided graphical and analytical procedures for kinematic analysis. Thorough and accurate kinematic analysis is essential to continue kinetic analysis of linkage mechanisms. Force analysis, covered in Chapter 11, will build on the kinematics developed in this chapter.

Problems

P4.1   Figure 4.42 shows two positions of a linkage mechanism. The driver (link 2) rotates clockwise at a constant rate of 65 rpm. Referring to the positions shown, for a rotation of 25° of the input, the output rotates 37.4°. Determine the approximate angular velocity of the output link?

Answer: $\omega_{\text{out}} \approx 97.4$ rpm

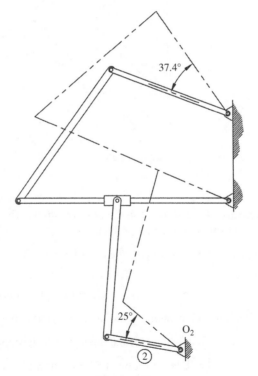

Figure 4.42

P4.2   The mechanism shown in Figure 4.43 has a 4 inch crank O_2A that is rotating at 1.8 rad/s clockwise at the instant shown. The distance from O_4 to B is 7.5 inches and points B and C lie on the same vertical line in the position shown. Use finite displacement to

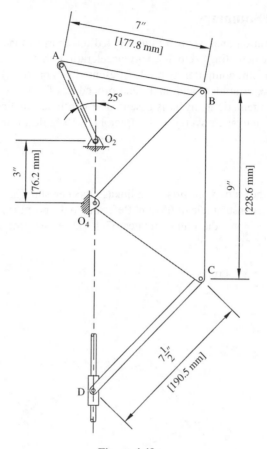

Figure 4.43

estimate the velocity of the slider D in the position shown. (Note that it will be very useful to use CAD software for this problem.)

P4.3 Use finite displacement to estimate the angular velocity of link O_4–B–C from Problem P4.2.

P4.4 Use finite displacement to estimate the acceleration of the slider from Problem P4.2.

P4.5 Locate and label all instant centers for the linkage mechanism shown in Figure 4.44.

P4.6 Locate and label all instant centers for the linkage mechanism shown in Figure 4.45.

P4.7 At the instant shown, the slider in Problem P4.6 is known to be moving to the left at 3 inches/s. Use instant center 13 to determine the velocity magnitude and the direction of the midpoint of the coupler.

Answer: $v = 70\,\dfrac{\text{in.}}{\text{s}}\,\angle 21.07°$

P4.8 Locate and label all instant centers for the linkage mechanism shown in Figure 4.46.

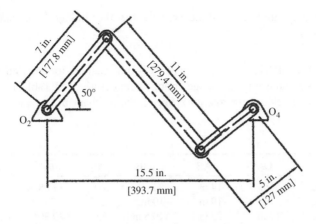

Figure 4.44

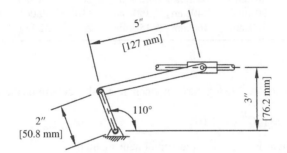

Figure 4.45

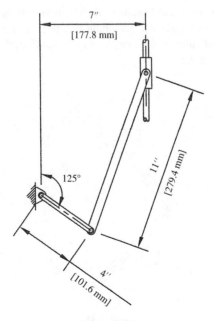

Figure 4.46

P4.9 Locate and label all instant centers for the linkage mechanism shown in Figure 4.43.

For Problems P4.10–P4.16 use the vector polygon method to determine the rotational velocities of links 3 and 4 for the linkage arrangement described in Table 4.2 (positive values of ω_2 indicate counterclockwise rotation). Sketch the mechanism to scale.

Table 4.2

Problem	O_2O_4	O_2A	AB	O_4B	θ_2	θ_3	θ_4	ω_2
P4.10	14 in.	5 in.	14 in.	8 in.	40°	17.08°	66.32°	80 rpm
P4.11	20 in.	7 in.	19 in.	10 in.	110°	8.33°	111.07°	−100 rpm
P4.12	25 in.	11 in.	22 in.	10.5 in.	−35°	49.30°	99.01°	200 rpm
P4.13	8 in.	10 in.	17 in.	8.5 in.	130°	1.49°	72.43°	40 rpm
P4.14	100 mm	40 mm	65 mm	70 mm	60°	28.92°	109.27°	−75 rpm
P4.15	240 mm	90 mm	140 mm	190 mm	110°	19.00°	136.76°	−150 rpm
P4.16	300 mm	130 mm	160 mm	150 mm	26°	32.18°	108.56°	90 rpm

P4.17 Solve Problem P4.11 using the component method to determine the velocity of point B.

Answer: $v_B = 140 \dfrac{\text{in.}}{\text{s}} \angle 21.07°$

P4.18 Solve Problem P4.12 using the component method.

P4.19 Solve Problem P4.16 using the component method.

P4.20 For the mechanism shown in Figure 4.47, the output link (link 4) is rotating at 85 rpm clockwise in the position shown. The input link (link 2) is 50% longer than the output link and is perpendicular to the coupler in the position shown. What is the rotation speed of the input link?

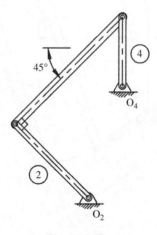

Figure 4.47

P4.21 Solve Problem P4.10 using the vector method.

Answer:
$$\omega_3 = -1.75 \frac{\text{rad}}{\text{s}}$$
$$\omega_4 = 2.69 \frac{\text{rad}}{\text{s}}$$

P4.22 The four-bar mechanism shown in Figure 4.48 has $\theta_3 = 12.283°$ and $\theta_4 = 100.492°$ in the position shown. Link 2 is rotating clockwise with a constant speed of 80 rpm. Use the vector method to determine the rotational velocity of links 3 and 4.

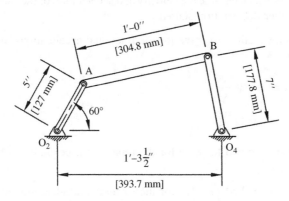

Figure 4.48

P4.23 Solve Problem P4.14 using the vector method.

Answer:
$$\omega_3 = 3.72 \frac{\text{rad}}{\text{s}}$$
$$\omega_4 = -2.35 \frac{\text{rad}}{\text{s}}$$

P4.24 The crank in the mechanism shown in Figure 4.46 is rotating at a constant rate of 120 rpm clockwise. Analytically determine the velocity of the slider.

P4.25 The crank in the mechanism shown in Figure 4.44 is rotating at a constant rate of 90 rpm counterclockwise. Use the loop-closure method to determine the angular velocities of links 3 and 4.

P4.26 Solve Problem P4.15 using the loop-closure method.

P4.27 Graphically determine the rotational acceleration of link 4 for Problem P4.21.

P4.28 The crank for Problem P4.11 is accelerating at a rate of 14 rad/s². Graphically determine the rotational acceleration of link 3.

P4.29 Use the vector polygon approach to determine the rotational acceleration of links 3 and 4 for Problem P4.16.

P4.30 The crank in Figure 4.45 is rotating at a constant rate of 90 rpm clockwise. Determine the acceleration of the slider for the position shown.

Answer: $\vec{a}_B = 13.3\hat{i}\ \dfrac{\text{in.}}{\text{s}}$

P4.31 Solve Problem P4.28 using the vector method.

P4.32 Solve Problem P4.29 using the vector method.

P4.33 An in-line slider–crank mechanism has a crank length of 3 inches and a connecting rod length of 11 inches. The crank rotates at a rate of 95 rpm clockwise. At a point when the crank is at 90°, it begins to decelerate at a constant rate and comes to a complete stop after 12 rotations. Determine the velocity and acceleration of the slider at a position one revolution after deceleration started.

P4.34 Use the vector method to solve for the rotational accelerations of links 3 and 4 for Problem P4.13.

Answer:
$$\alpha_3 = -1.28\,\dfrac{\text{rad}}{\text{s}^2}$$
$$\alpha_4 = -9.11\,\dfrac{\text{rad}}{\text{s}^2}$$

P4.35 Use the vector method to solve for the rotational accelerations of links 3 and 4 for Problem P4.14.

P4.36 Use the vector method to solve for the rotational accelerations of links 3 and 4 for Problem P4.15.

P4.37 Link O_2A if the mechanism shown in Figure 4.49 is rotating at a rate of 5.2 rad/s clockwise and accelerating at a rate of 9 rad/s². Determine the angular velocity and acceleration of the right-angled output link.

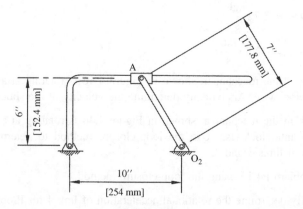

Figure 4.49

P4.38 Link O_2A shown in Figure 4.50 is 5 inches (127 mm) and rotates at a constant rate of 18 rad/s counterclockwise. Determine the angular velocity and acceleration of link O_3B for the instant shown.

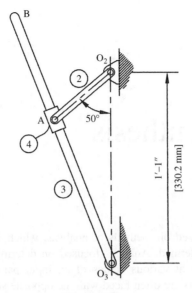

Figure 4.50

Further Reading

Greenwood, D.T. (1988) *Principles of Dynamics*, Prentice-Hall, New Jersey.
Tuttle, S.B. (1967) *Mechanisms for Engineering Design*, John Wiley & Sons, Inc., New York.

5

Linkage Synthesis

5.1 Introduction

The previous chapters focused on methods of analysis, which involve scenarios when the linkage geometry is fully defined. Analysis focused on determining kinematics of a fully defined linkage mechanism at various values of an input parameter. Though analysis is critically important, engineers are often faced with the opposite scenario. Engineers typically have the challenge of specifying linkage geometry to generate a desired motion. The ultimate task is generally to combine analysis with synthesis to develop the final linkage arrangement.

This chapter mainly focuses on methods of linkage synthesis (design). There are many methods for linkage synthesis, although the two main categories are graphical synthesis and analytical synthesis. Even though analytical synthesis is more accurate and allows for such things as optimization, graphical methods will be discussed first. Graphical methods are more visual, which allow for easier understanding of the general synthesis process. The graphical methods presented in this chapter will use basic hand drafting methods, although computer-aided design (CAD) software could be utilized for greater accuracy and speed.

5.1.1 Synthesis Classifications

The three main categories in kinematic synthesis are path generation, function generation, and motion generation. In path generation, the goal is for a specified point on the mechanism to trace a specific curve or path. For example, a point on the coupler may need to develop a specific coupler curve (as discussed in Section 3.5). In addition, it is often critical to have the point move at a specific rate, meaning that the velocity and acceleration along the path are often of importance. The Hrones and Nelson atlas (see Section 3.5.3) is a great tool to aid in path generation.

Function generation deals with a direct correlation or relationship, such as a mathematical relationship, between input and output motion (commonly input angle versus output angle). The input would move to locate x (over some defined range) and the output would move based on a function $y = f(x)$. Function generators are essentially rudimentary analog computing devices and were used in several military applications before electronic computers. Cam mechanisms, which are discussed in Chapter 9, are great function generators. Four-bar mechanisms can also be designed to serve as function generators and are generally less

Machine Analysis with Computer Applications for Mechanical Engineers, First Edition. James Doane.
© 2016 John Wiley & Sons, Ltd. Published 2016 by John Wiley & Sons, Ltd.
Companion Website: www.wiley.com/go/doane0215

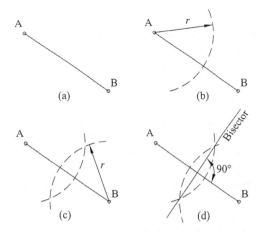

Figure 5.1   Steps for drawing a perpendicular bisector

expensive than cam mechanisms. Freudenstein's equation is a very useful tool for function generator synthesis, and the procedures are discussed in detail in Section 5.7.

Motion generation is the last category, and it will be a major focus of this chapter. Motion generation involves the planar motion of a line such that it moves through a set of successive locations. Unlike path generation, the positions of the line as well as the angular orientations are important. Linkage mechanisms are typically used for motion generation. The required method of motion generation synthesis depends on the number of desired locations for the line.

5.1.2   *Essential Engineering Geometry and Drafting*

Performing graphical linkage synthesis by hand requires certain fundamental drafting skills. A critical skill would be the ability to construct a perpendicular bisector. Bisecting a line will divide the line into two equal parts. A perpendicular bisector draws a perpendicular line through the midpoint of the original line. For hand drafting techniques, all that is required for drawing a perpendicular bisector is a compass and a straightedge. The steps required are illustrated in Figure 5.1. The line to be bisected is shown in Figure 5.1a as line AB. Using a compass, draw an arc of radius *r* centered on point A, as shown in Figure 5.1b. The radius value *r* is arbitrary but must be greater than half the distance AB. Draw an arc of the same radius *r* now centered on point B, as shown in Figure 5.1c. Draw a line that passes through the intersection points of the two arcs, as shown in Figure 5.1d. The line drawn through the intersections will be the perpendicular bisector.

5.2   Synthesis

5.2.1   *Type Synthesis*

Type synthesis deals with the process of determining the kinds of components that should be used in a mechanism for a particular motion. The components could be various types of links, cams, gears, and so on. Some of the considerations for determining types of components may be cost, manufacturability, ease of maintenance, space constraints, or material availability. Although these considerations are very important, they are outside the scope of this text. The other considerations deal with kinematics, mainly what are possible configurations of components that give the desired motion characteristics. Though type synthesis may sound like a very simple task, it can be very difficult. The process can be difficult for students, but it will generally improve with experience.

5.2.2 Number Synthesis

It is often desired to have a linkage mechanism with a specific mobility (degrees of freedom). In number synthesis, the goal is to determine possible combinations of the number and type of links (binary, ternary, quaternary, etc.) and joints required to give that desired mobility. The equations for mobility (e.g., Kutzbach's equation developed in Section 3.2) are very useful for number synthesis. Kutzbach's equation is provided below with a modification of using j to represent joint number:

$$M = 3(N - 1) - 2j \tag{5.1}$$

For a mobility of 1,

$$\begin{aligned} 1 &= 3(N - 1) - 2j \\ 2j &- 3N + 4 = 0 \end{aligned} \tag{5.2}$$

Equation 5.2 is known as the Grübler criterion. As with Kutzbach's equation, the Grübler criterion has no information about the arrangement of links. Any combination of link and joint numbers that satisfies Equation 5.2 will maintain a single-degree-of-freedom requirement.

5.2.3 Dimensional Synthesis

The last task in the design process is dimensional synthesis, which is the process of determining the part dimensions required for a desired motion or task. Compared to type synthesis, dimensional synthesis relies less on ingenuity or personal experience. Methods for dimensional synthesis are more systematic and scientific. Determining the profile of a cam to generate a specific motion is an example of dimensional synthesis (see Chapter 9). This chapter will focus on methods of linkage dimensional synthesis, which is when a specific motion is desired and the goal is to determine the proper dimensional proportions of the links to provide that motion. With linkage mechanisms, it is often difficult to match the desired motion exactly, therefore approximations are common. As a basic example, many linkage mechanisms exist to generate an approximate straight line, but a true straight line is more difficult to generate.

It is common to need to move a link through a certain number of sequential locations. For this chapter, dimensional synthesis will be limited to two- and three-position dimensional synthesis. Dimensional synthesis for more than three desired positions is possible, although it is not well suited for graphical methods. Hartenberg and Denavit provide information about dimensional synthesis for more than three positions in their book *Kinematic Synthesis of Linkages*.

5.3 Two-Position Graphical Dimensional Synthesis

5.3.1 Introduction

We will first develop methods to move a link (or line segment) between two described locations. This section discusses two methods of two-position synthesis. The first method utilizes the rocker to move between the two positions in pure rotation. The resulting mechanism will be a Grashof crank–rocker mechanism. The second method moves the line with the coupler. Utilizing the coupler will often require the addition of two supplementary links to serve as the driver to move the mechanism between the two desired positions. Therefore, the final result is a six-bar mechanism.

5.3.2 Using Rocker Motion from a Double Rocker Mechanism

Consider a case for designing a mechanism that will convert a rocking oscillatory input to a rocking oscillatory output, as shown in Figure 5.2a. Both fixed pivot point locations are known (therefore, the length of link 1 is known) and the length of link 4 is also known. The two required positions of the output link 4 are known and shown in Figure 5.2a. Link 4 will move point B to point B' by moving through an angle θ_4. The other known information is the angle positions of the input link, which result in an input angle of θ_2 to give the required output angle θ_4. The length of links 2 and 3 must be determined.

To solve for the required length of link 2, we must use an inversion with link 2 as the stationary link in position 1. Use a construction line from O_2 to B', as shown in Figure 5.2b, and rotate that line about O_2 through the angle θ_2 so that B' moves to point c. Once point

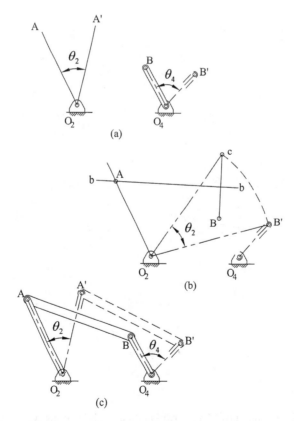

Figure 5.2 Two-position synthesis for a double rocker mechanism

c is located, draw another construction line connecting point c and point B. Draw the perpendicular bisector, labeled as line bb in the figure, of that construction line. The intersection point of line bb and the original position of link 2 will be point A and will now define the length of link 2. Figure 5.2c shows the completed linkage mechanism in both required positions.

5.3.3 Using Rocker Motion from a Crank–Rocker Mechanism

Two-position synthesis using the rocker will give a crank–rocker mechanism, but there will not be only one unique solution. The rocker pivot is uniquely defined, but the remaining links can have many different layouts. Therefore, the process often requires several iterations to develop an acceptable mechanism. The steps are rather simple and straightforward; however, the first result is typically not the best result. As one example, the first result may not satisfy Grashof's criterion, which is required for the mechanism to be a crank–rocker.

The general design procedure is best explained with an illustrative example. As shown in Figure 5.3a, start with line AB that needs to move to the location A′B′. This will be done using the rocker (link 4). Draw a construction line from A to A′ and find the perpendicular bisector of that line. The bisector is labeled as line aa in Figure 5.3b. Next, draw a construction line from B

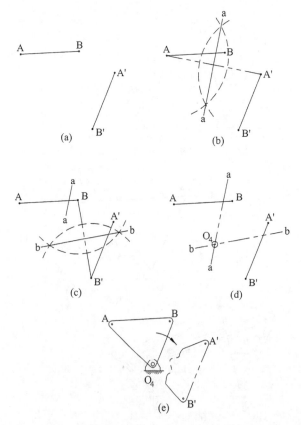

Figure 5.3 Steps for locating pivot O_4 for two-position synthesis using rocker output. (a) Required positions. (b) Bisector for AA′. (c) Bisector for BB′. (d) Location of O_4. (e) Final rocker

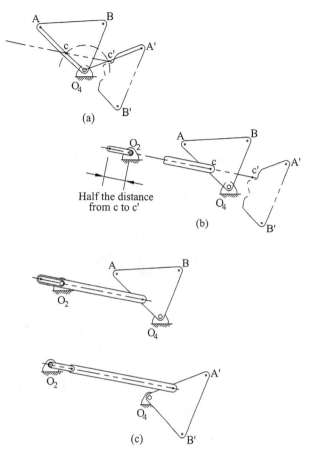

Figure 5.4 (a) Connection point for coupler. (b) One possible location of O_2. (c) Final mechanism in toggle positions

to B'. Find the perpendicular bisector, which is labeled as line bb in Figure 5.3c. As shown in Figure 5.3d, the intersection of lines aa and bb will be the location of the fixed pivot O_4. Link 4 can then be drawn. Figure 5.3e shows link 4 in the two desired positions.

Now that link 4 is fully defined, the remaining links can now be located. Pick a point on the link in both of the desired positions that will serve as the connection point between links 3 and 4. The connection point is somewhat arbitrary, but one possible position is shown as points c and c' in Figure 5.4a. For convenience, point c was located on line AO_4. An arc centered at O_4 can be drawn through point c, and point c' will be located where the arc crosses $A'O_4$. Draw a line that passes through both points c and c'. Arbitrarily, locate pivot O_2 on that line to complete link 1 (ground link). Referring to Figure 5.4b, link 2 can be drawn and its length will be half the distance between c and c'. Link 3 (coupler) can be drawn connecting links 2 and 4. The mechanism is shown in Figure 5.4c in both toggle positions. The Grashof condition must be checked to ensure the crank can complete a full revolution. If the Grashof condition is not satisfied, pivot O_2 will need to be repositioned along the line through points c and c'.

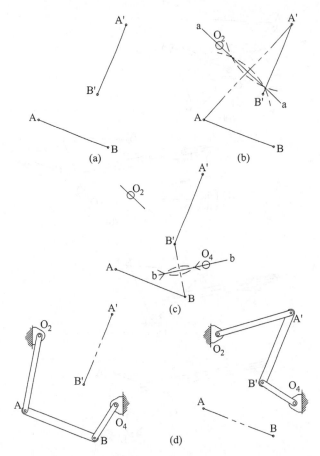

Figure 5.5 Steps for two-position synthesis using coupler motion. (a) Required positions. (b) Bisector for AA′ and possible location of O_2. (c) Bisector for BB′ and possible location of O_4. (d) Final mechanism in both positions

5.3.4 Using Coupler Motion

Two-position dimensional synthesis can also be accomplished using the coupler (link 3). The steps are illustrated in Figure 5.5. The two desired positions are shown in Figure 5.5a as line AB and line A′B′. The first step is to draw a construction line from A to A′ and find the perpendicular bisector, as shown in Figure 5.5b. Fixed pivot O_2 can be located at any convenient location along the bisector line aa. Next, create a construction line from B to B′. Figure 5.5c shows the perpendicular bisector as line bb, and the fixed pivot O_4 can be located at any convenient location along the bisector. Figure 5.5d shows the completed mechanism in the two desired positions.

As with two-position synthesis utilizing the rocker, use of the coupler for two-position synthesis will not result in only one solution. The engineer will have some flexibility to locate the fixed pivot points in convenient locations, but the locations are constrained to the two specific perpendicular bisector lines. Modifications may need to be made to the initial design to optimize the performance of the mechanism.

5.3.5 Adding a Driver Dyad

Figure 5.5 showed steps for developing a four-bar mechanism to move the coupler between two desired locations. However, the current mechanism is not constrained to only move between the two locations. It is common, therefore, to add two additional links to serve as a driving mechanism. The addition of the two links will result in a six-bar mechanism. The procedure for adding the two driving links, or driver dyad, is very similar to the procedure described in Figure 5.4 to complete the two-position synthesis using the rocker.

To illustrate the addition of a driver dyad, we will continue the example illustrated in Figure 5.5. One of the links in pure rotation (link 2 or link 4) will serve as the connection point for the driver dyad. Figure 5.6a shows the process of designing the driver dyad to connect to link 2 and move it between the two desired positions. A point along link 2 is arbitrarily picked as the connection point and the driver dyad is located using the same procedures illustrated in Figure 5.4. The complete six-bar mechanism is shown in Figure 5.6b in both toggle positions.

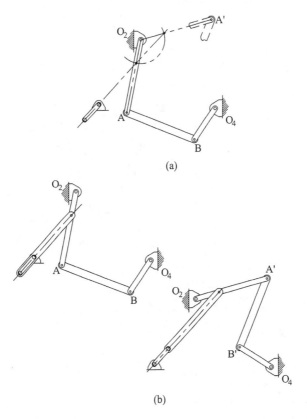

(a)

(b)

Figure 5.6 (a) Process for adding a driver dyad. (b) Complete six-bar mechanism in both toggle positions

5.4 Three-Position Graphical Dimensional Synthesis

5.4.1 Using Coupler Motion

In general, three-position synthesis cannot be obtained with the use of the rocker. In order to use rocker output, the three positions would need to be on a common arc so that all positions could be reached with pure rotation from a single pivot point. Because that is not typically the case, three-position synthesis will almost always utilize the coupler. We will begin with a very similar approach to that used for two-position synthesis with the coupler. The major difference is that for three-position synthesis, the pivot points are uniquely defined.

An illustrative example will again be used. As shown in Figure 5.7a, start with line AB that you need to move to the two locations A′B′ and A″B″. This motion will be done using the coupler (link 3). Draw a construction line from A to A′ as illustrated in Figure 5.7b. Find the perpendicular bisector of that line. Repeat the process for a construction line from A′ to A″. The fixed pivot O_2 will be located at the intersection of those two perpendicular bisectors as shown. Referring to Figure 5.7c, the next step is to draw a construction line from B to B′. Find the

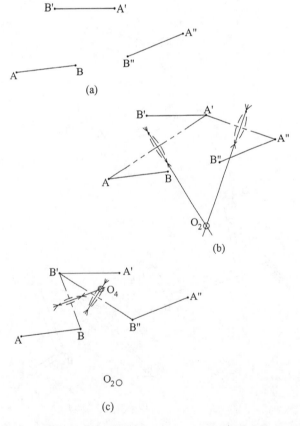

Figure 5.7 Steps for three-position synthesis using coupler motion. (a) Required positions. (b) Bisectors and location of O_2. (c) Bisectors and location of O_4

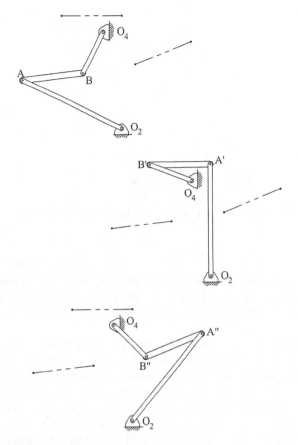

Figure 5.8 Completed three-position synthesis with mechanism shown in all required positions

perpendicular bisector of the construction line. Repeat the process for the construction line from B' to B". The fixed pivot O_4 will be located at the intersection of the two perpendicular bisectors.

The completed linkage mechanism is shown in Figure 5.8 in the three required positions. As with two-position synthesis, it may be required to add a driver dyad to move the mechanism through the desired positions.

5.4.2 Fixed Pivot Point Locations Defined

The method described in Section 5.4.1 can often result in fixed pivot point locations that are not acceptable or practical. Often acceptable locations of fixed pivots are limited due to the physical geometry of the machine. Therefore, this section will discuss an inversion method to achieve three-position synthesis with predefined locations of the fixed pivots.

As an example, consider the material handling application shown in Figure 5.9. The fabricated part must be gripped at points A and B, lifted off the machine frame, and placed on the conveyor rack. Therefore, line AB will need to move through A'B' to a final location

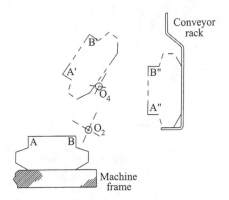

Figure 5.9 Three desired positions with inconvenient fixed pivot locations

A″B″. Following the procedure described in Section 5.4.1 will result in fixed pivot points O_2 and O_4 in the locations shown in Figure 5.9, which are not very ideal. It would be more convenient to have the pivot locations somewhere on the machine frame, as shown in Figure 5.10.

Once the ideal location of the fixed pivots has been determined, we begin the process of designing the mechanism. Defining the coupler by points A and B resulted in unacceptable fixed pivot locations, so we must move the coupler point such that the resulting fixed pivots match those shown in Figure 5.10. The first step of this process is shown in Figure 5.11a. This step will result in two intermediate points labeled O_2' and O_4'. The two points will be determined using the location of line A′B′ relative to location AB. Start by drawing an arc centered at A′ that passes through O_4. The arc is shown having radius r_A'. Draw the same arc now centered at point A. Next, draw an arc centered at B′ that passes through O_4, which is shown having radius

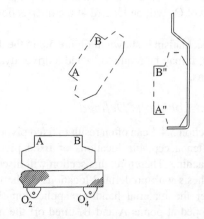

Figure 5.10 Preferred fixed pivot point locations

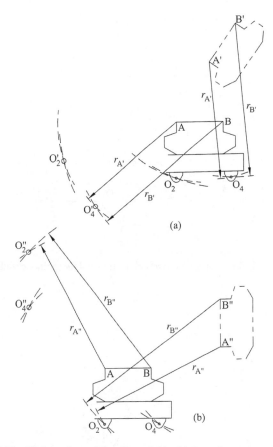

Figure 5.11 (a) Locating points O_2' and O_4'. (b) Locating points O_2'' and O_4''

r_B'. Draw that same arc now centered at B. The two arcs will intersect at the location that is defined as O_4'.

The process is now used to locate point O_2'. Arcs centered at A' and B' are now drawn at radii values to pass through O_2 and transferred to points A and B. The intersection will locate point O_2'. Figure 5.11b shows the steps for locating points O_2'' and O_4''. The steps are the same as those described for O_2' and O_4', but position A"B" is now used as the reference.

The next step is to locate the two connection points on the coupler. First, draw a construction line from O_2 to O_2' and from O_2' to O_2''. Draw perpendicular bisectors for both construction lines and locate the intersection point, which is labeled as point c in Figure 5.12. Point c will serve as a connection point and a link will connect between O_2 (because the O_2 points were used to locate point c) and c. Similarly, draw construction lines using the O_4 points and the perpendicular bisectors will intersect at the point labeled d. Point d will be the other coupler point, and it will connect to O_4.

Once the two connection points have been located, the final mechanism can be drawn. The final mechanism is shown in Figure 5.13 in all three required positions.

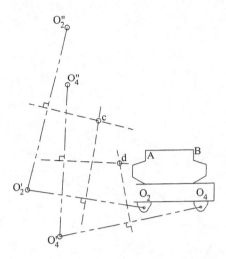

Figure 5.12   Procedure for locating the two connection points (points c and d) on the coupler

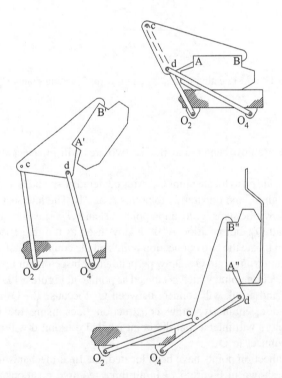

Figure 5.13   Final mechanism in all three required positions

5.5   Approximate Dwell Linkage Mechanisms

A dwell mechanism is one in which a continuous input motion is converted to an intermittent output motion. Dwell mechanisms are fairly common and have numerous applications. Industrial applications include loading or unloading parts, or transporting a part to a machine and holding it in place for a manufacturing process. Because of simplicity of design, cams are the typical choice for dwell mechanisms (see Chapter 9). However, linkage mechanisms can alternatively be used to provide approximate periods of dwell. The use of linkage mechanisms for dwell has advantages of being less expensive to manufacture and maintain. However, the design process for dwell linkage mechanisms is not as straightforward as cam design.

The general procedure for obtaining approximate dwell in a linkage mechanism is to add an output dyad to a four-bar mechanism, utilizing a portion of its coupler curve with an approximate circular arc, as shown in Figure 5.14. The final result will be a six-bar mechanism (six links is the minimum required for a linkage dwell mechanism). The basic premise for the dwell mechanism is that a node located at the center of the circular arc segment will remain relatively stationary, while point P moves along the approximate arc. Therefore, actual dwell time will depend on the length of the approximate circular arc. Initial designs may need optimization to improve the dwell characteristics.

The Hrones and Nelson atlas can be an invaluable tool for finding an appropriate four-bar mechanism with an approximate circular arc. Other useful coupler curves are those that are symmetrical. A symmetrical coupler curve will exist when lengths AB, BP, and BO_4 shown in Figure 5.14 are all equal.

The first step in designing a linkage mechanism with single dwell is to find a four-bar mechanism that generates a suitable coupler curve, as illustrated in Figure 5.14. We will begin with a procedure that requires picking a coupler curve that has an approximate circular arc of an appropriate size to give the required dwell duration. As an illustrative example, say we need a dwell that occurs as the crank rotates through 80°. The mechanism shown in Figure 5.15 has an approximate circular arc from P to P′ while the crank rotates through 80°. The linkage dimensions are defined in the figures, but the proportions shown can be scaled up or down without affecting the shape of the coupler curve.

The coupler curve from P to P′ generates an approximate arc, and the center of that arc must now be located. Draw two chords on the circular arc, such as those shown in Figure 5.16 as lines aP and bP′. Draw the perpendicular bisectors for both chords. The intersection, labeled as point D, will be the approximate center location for the approximate circular arc segment PP′.

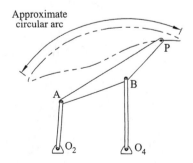

Figure 5.14 Coupler curve with approximate circular arc

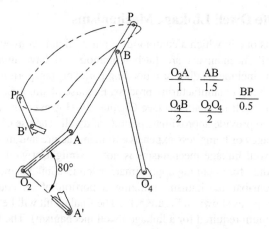

$$\frac{O_2A}{1} \quad \frac{AB}{1.5}$$

$$\frac{O_4B}{2} \quad \frac{O_2O_4}{2} \quad \frac{BP}{0.5}$$

Figure 5.15 Linkage mechanism for a dwell during 80° of crank rotation

The dwell will be achieved based on the fact that point D will remain relatively stationary, while point P moves to P′ along the approximate circular arc. Link 5 will then connect from point P to point D, as shown in Figure 5.17.

The motion of slider D is shown in Figure 5.18 in a plot versus crank rotation starting at the beginning of the dwell period. It can be seen that the slider is in an approximate dwell (approximately stationary) during the first 80° on input motion. It is often necessary to try several possible coupler curves to find a final design that is optimal.

Double dwell can also be approximated with linkage mechanisms, although the process will not be illustrated here. Using an approach similar to that described for single dwell, the four-bar mechanism would now need two circular arcs with different centers and similar direction of concavity. Double dwell can also be developed using a four-bar mechanism with two approximate straight line segments. Though double dwell can be approximated with a linkage mechanism, cams are often better suited for such an application.

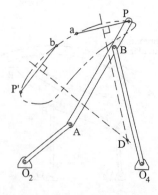

Figure 5.16 Method for locating approximate center of arc PP′

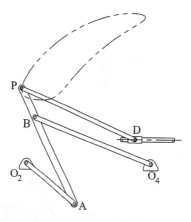

Figure 5.17 Completed single dwell linkage mechanism

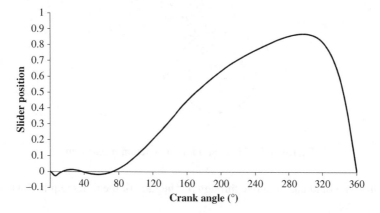

Figure 5.18 Motion of slider D with respect to crank rotation

5.6 Quick Return Mechanisms

5.6.1 Time Ratio

A quick return mechanism, as introduced in Chapter 1, is one in which the output link moves more quickly on the return stroke compared with the work stroke. There are many common applications of quick return mechanisms, but in general they are used when a high-load work stroke and a low-load return stroke are desired. The ratio of the work stroke to the return stroke is known as the time ratio (TR). To illustrate the concept of time ratio, let us examine the procedure to determine the time ratio for a defined quick return mechanism. Consider the crank–rocker mechanism shown in Figure 5.19a. As the crank rotates, the mechanism reach toggle position 1, which is shown in Figure 5.19b. From the first toggle position, the crank will rotate an angle α to reach the second toggle position. The crank will then rotate an angle β to return to the first toggle position.

For the linkage arrangement shown, the angle α is less than the angle β. Also, note that $\alpha + \beta = 360°$. Assuming the crank is rotating at a constant rate, it will take less time to move

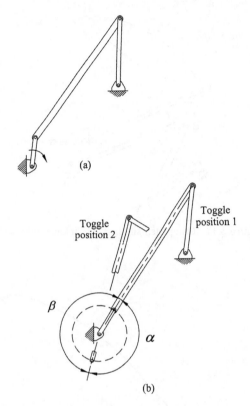

Figure 5.19 Time ratio for quick return mechanism

through α than it would take to move through β. The ratio of the times required will define the linkage time ratio:

$$\text{TR} = \frac{\alpha}{\beta} \tag{5.3}$$

Another angle will be needed for construction, which is defined by

$$\delta = |180 - \alpha| \tag{5.4}$$

The movement through α would typically represent the return stroke (quick return), and movement through β would represent the work stroke. There are multiple linkage configurations that could be used to obtain a particular time ratio. Also, it is often desirable to have relatively constant velocity during the work stroke. Therefore, it is often necessary to develop multiple possible configurations to optimize performance of a quick return mechanism.

5.6.2 Quick Return Crank–Rocker Mechanism

The layout of a crank–rocker mechanism discussed in Section 5.3.3 can be adjusted to give a specified quick return. As the output link oscillates between the two toggle positions, the crank will rotate at a constant speed. However, the crank fixed pivot point location is modified such that the crank will move through different arc lengths for both the working stroke and return stroke.

The mechanism can again be designed graphically using hand drafting techniques, although the required tools must be expanded to include a protractor to measure angular values. CAD software can also be used to simplify the process and improve accuracy of the design. The best way to illustrate the design process is with an example.

Example Problem 5.1

Design a quick return mechanism with a time ratio of working stroke to return stroke of approximately 1.2:1 with 50° of rocker output motion.

Solution: The first step is to draw the rocker in the two toggle positions at the required 50° of rotation, as shown in Figure 5.20. Fixed pivot O_4 is now located, but it is necessary to locate fixed pivot O_2. Line aa is drawn (at any convenient angle) through the end node of the rocker in one toggle position as shown.

The geometry can now be determined based on the required time ratio of 1.2:1. Using Equation 5.3, some trial and error results in the angles

$$\alpha = 163 \qquad \beta = 197$$

giving a time ratio of 1.208. Using Equation 5.4, the construction angle is

$$\delta = |180 - 163| = 17$$

A second line bb can now be drawn through the end node in the other toggle position. Line bb will be drawn at an angle δ to line aa. Fixed pivot O_2 will be located at the intersection of lines aa and bb, as shown in Figure 5.21.

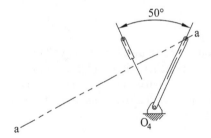

Figure 5.20

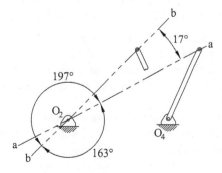

Figure 5.21

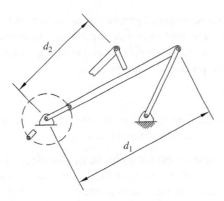

Figure 5.22

The final step is to determine the length of the crank. Two measurements, d_1 and d_2, are required. It can be seen in Figure 5.22 that distance d_1 is the sum of the crank and coupler lengths and distance d_2 is the coupler length minus the crank length.

Solving the two equations simultaneously will yield the crank and coupler lengths required:

$$L_{coupler} + L_{crank} = d_1$$
$$L_{coupler} - L_{crank} = d_2$$

5.6.3 Whitworth Mechanism

The next configuration of quick return mechanism we will examine is known as a Whitworth mechanism. The Whitworth mechanism is developed from an inversion of a slider–crank mechanism where links 2 and 4 make complete revolutions. Whitworth mechanisms are relatively easy to construct and have the advantage of higher time ratios. The construction is illustrated in Example Problem 5.2.

Example Problem 5.2

Design a Whitworth quick return mechanism with a total stroke of 4 inches and a time ratio of working stroke to return stroke of 2.6:1. Use a drive crank that is 1.5 inches long.

Solution: Locate point O_2 and draw a circle with a radius of 1.5 (crank length) about the fixed pivot O_2, as shown in Figure 5.23. Draw the crank in the two extreme positions for the required time ratio. For the case of 2.6:1, the work stroke will be through 260° and the return stroke will be through 100°. Locate pivot O_4 at the midpoint of the line connecting the ends of the two crank positions as shown.

The remaining links can now be drawn, as shown in Figure 5.24. The length BO_4 will be half the required stroke, which for this example will be 2 inches. Link BP can be any convenient length longer than the length BO_4.

The slider motion is shown in Figure 5.25 versus crank angle. The extreme positions of the slider occur at toggle positions when AB and BP are collinear.

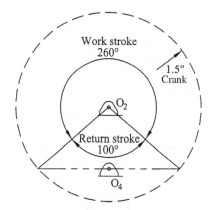

Figure 5.23

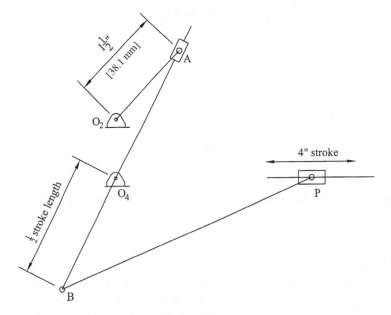

Figure 5.24

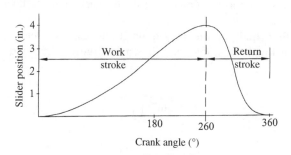

Figure 5.25

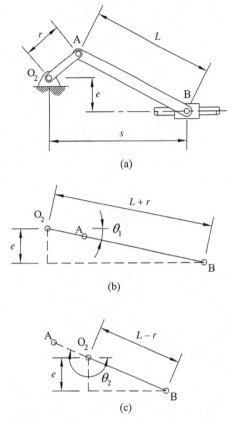

Figure 5.26 (a) Offset slider–crank mechanism. (b) Toggle position showing θ_1. (c) Toggle position showing θ_2

5.6.4 Offset Slider–Crank Mechanism

For a slider–crank mechanism with zero offset, the work stroke equals the return stroke. An offset slider–crank mechanism, however, is a quick return mechanism. Methods for determining the position of a slider crank have already been discussed in detail in Chapter 3. Determining the time ratio for an offset slider–crank mechanism simply requires the position of the mechanism in both toggle positions. Consider the general offset slider–crank mechanism shown in Figure 5.26a.

The angle of the crank in the toggle positions can be determined from simple trigonometry. Both toggle positions are shown in Figure 5.26b and c. The values are calculated using

$$\theta_1 = \sin^{-1}\left(\frac{e}{L+r}\right)$$

$$\theta_2 = \pi + \sin^{-1}\left(\frac{e}{L-r}\right)$$

(5.5)

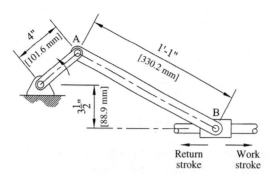

Figure 5.27

The piston will be moving forward when $\theta_1 < \theta < \theta_2$ and moving backward when $\theta_2 < \theta < 2\pi + \theta_1$. The time ratio will then be

$$TR = \frac{\theta_2 - \theta_1}{2\pi - (\theta_2 - \theta_1)} \tag{5.6}$$

Example Problem 5.3

For the offset slider–crank mechanism shown in Figure 5.27, determine the time ratio of working stroke to return stroke.

Solution: Figure 5.28 shows the mechanism in both toggle positions. The crank angle for each toggle position is calculated using Equation 5.5.

$$\theta_1 = \sin^{-1}\left(\frac{e}{L+r}\right) = \sin^{-1}\left(\frac{3.5}{13+4}\right) = 11.88°$$

$$\theta_2 = \pi + \sin^{-1}\left(\frac{e}{L-r}\right) = \pi + \sin^{-1}\left(\frac{3.5}{13-4}\right) = 202.89°$$

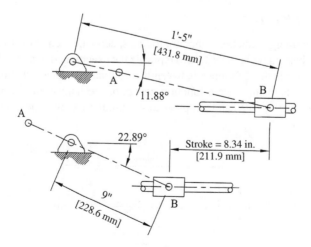

Figure 5.28

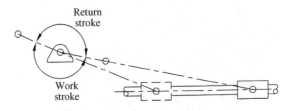

Figure 5.29

Figure 5.29 shows the overlay of both toggle positions to illustrate the work stroke and return stroke. The time ratio can be determined using Equation 5.6:

$$TR = \frac{\theta_2 - \theta_1}{2\pi - (\theta_2 - \theta_1)} = \frac{3.54 - 0.21}{2\pi - (3.54 - 0.21)} = 1.13$$

The time ratio can be confirmed graphically. The return stroke angle can be determined to be 169° and the work stroke angle is 191°. The time ratio is then calculated as

$$TR = \frac{191}{169} = 1.13$$

Answer: TR = 1.13

5.7 Function Generation

5.7.1 Introduction

Another classification of linkage synthesis is function generation, which involves a direct mathematical relationship between the input and the output. Typically a functional relationship, such as $y = \sin x$, is desired over some set range of input values. Function generators can also serve as mechanical converters. For example, function generators could be developed that will convert temperature from Celsius to Fahrenheit. Different methods exist for developing function-generating mechanisms. This section will focus on a method utilizing Freudenstein's equation, which was developed in Chapter 3.

5.7.2 Accuracy Points

If we plan on developing a function generator to fit a specific mathematical function over some set range, the mechanism will not be able to replicate the exact function and will have structural error. Structural error is the difference between the true mathematical function and the values obtained by the mechanism. Once a function generator mechanism has been developed that approximates the function, we can then plot the structural error over the interval. Figure 5.30

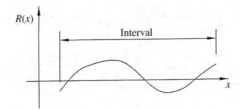

Figure 5.30 Structural error

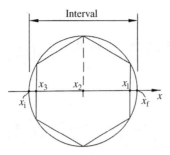

Figure 5.31 Graphical location of three accuracy points (x_1, x_2, x_3) using Chebyshev spacing

shows a typical example of a plot of structural error $R(x)$. The locations of zero error, where the plot crosses the horizontal axis, are known as precision points or accuracy points.

5.7.3 Chebyshev Spacing

The location of the accuracy points can be optimized such that the structural error magnitude is minimized over the entire interval. It will be discovered later that when using Freudenstein's equation to develop the function generator, we will be limited to three accuracy point locations. One method for optimizing the location of those accuracy points is known as Chebyshev spacing. The spacing can be visualized by drawing a circle centered on the interval with a radius equal to one-half of the interval, as shown in Figure 5.31. The two locations where the circle crosses the x-axis will define the initial point of the interval x_i and the final point of the interval x_f. Inscribe a regular polygon of $2n$ sides (where n is the number of required accuracy points) so that its sides are perpendicular to the x-axis. For the case of three accuracy points, we need a six-sided polygon, as shown in Figure 5.31. The last step is to project the vertices to the x-axis to determine accuracy points, which are labeled as x_1 through x_3 in the figure.

The location of the accuracy points can also be determined mathematically. The general equation for n accuracy points is given by

$$x_j = \frac{x_f + x_i}{2} + \frac{x_f - x_i}{2} \cos\left(\frac{\pi(2j-1)}{2n}\right) \qquad j = 1, 2, 3 \tag{5.7}$$

Example Problem 5.4

Find the three Chebyshev precision points over a range $0.25 \leq x \leq 1.5$.

Solution: For our example, $x_i = 0.25$ and $x_f = 1.5$. Using Equation 5.7 with $n = 3$,

$$x_j = \frac{x_f + x_i}{2} + \frac{x_f - x_i}{2} \cos\left(\frac{\pi(2j-1)}{2(3)}\right) \qquad j = 1, 2, 3$$

For $j = 1$,

$$x_1 = \frac{1.5 + 0.25}{2} + \frac{1.5 - 0.25}{2} \cos\left(\frac{\pi}{6}\right) = 1.4163$$

For $j = 2$

$$x_2 = \frac{1.5 + 0.25}{2} + \frac{1.5 - 0.25}{2} \cos\left(\frac{3\pi}{6}\right) = 0.875$$

For $j = 3$

$$x_3 = \frac{1.5 + 0.25}{2} + \frac{1.5 - 0.25}{2} \cos\left(\frac{5\pi}{6}\right) = 0.334$$

$$x_1 = 1.4163$$

Answer: $x_2 = 0.875$

$$x_3 = 0.334$$

5.7.4 Freudenstein's Equation: Four-Bar Linkage

Freudenstein's equation was developed in Chapter 3 and used as an analysis tool to determine output angles for a given input angle. Freudenstein's equation can be used as a powerful design tool for function generation. The typical circumstance would be that a four-bar linkage mechanism is required such that three given angular positions of the crank will result in the three mathematically prescribed locations of the follower. The task is to find the proper link lengths to provide the desired motion. The process will not provide the situation of only one possible solution. Some trial and error is typically required to develop an acceptable solution.

To illustrate the process, let us develop a function generator to fit a specific example function. For our example, it is desired to have a function generator to match the function

$$y = \frac{1}{x^{5/3}} \tag{5.8}$$

over the range $1 \leq x \leq 3$. First, we need to know the location of our accuracy points. Our desired interval is $1 \leq x \leq 3$, and that gives our initial and final values of x to be $x_i = 1$ and $x_f = 3$, respectively. Using Equation 5.7 with $n = 3$, the accuracy points will be at the following locations:

$$x_1 = \frac{3+1}{2} + \frac{3-1}{2} \cos\left(\frac{(2-1)\pi}{6}\right) = 2.866$$

$$x_2 = \frac{3+1}{2} + \frac{3-1}{2} \cos\left(\frac{(4-1)\pi}{6}\right) = 2.000 \tag{5.9}$$

$$x_3 = \frac{3+1}{2} + \frac{3-1}{2} \cos\left(\frac{(6-1)\pi}{6}\right) = 1.134$$

We can then write Freudenstein's equation for each precision point. Writing Freudenstein's equation (developed in Section 3.9) for a general precision point i,

$$K_1 \cos\theta_{4i} - K_2 \cos\theta_{2i} + K_{3i} = \cos(\theta_2 - \theta_4)_i \tag{5.10}$$

The equation can be written for all three precision points, which can be expressed in matrix form:

$$
\begin{bmatrix}
\cos\theta_{41} & -\cos\theta_{21} & 1 \\
\cos\theta_{42} & -\cos\theta_{22} & 1 \\
\cos\theta_{43} & -\cos\theta_{23} & 1
\end{bmatrix}
\begin{bmatrix}
K_1 \\
K_2 \\
K_3
\end{bmatrix}
=
\begin{bmatrix}
\cos(\theta_2 - \theta_4)_1 \\
\cos(\theta_2 - \theta_4)_2 \\
\cos(\theta_2 - \theta_4)_3
\end{bmatrix}
\tag{5.11}
$$

Recall from Chapter 3 that the K values in Freudenstein's equation contain all link lengths. Therefore, if Equation 5.11 can be solved to give all three K values, we will know the link lengths required.

You may notice that we have values of x for precision point locations and values of θ are required for Equation 5.11. To develop the relationship between function values x and y and joint angles θ_2 and θ_4, we must determine the initial and final values for θ_2 and θ_4. These initial and final angle values will determine the range of motion of the input and output links. The initial and final angle values are picked arbitrarily. It is best to sketch out the desired layout, and pick the initial and final angle values based on the desired layout. The initial and final angle values can be changed later to optimize performance. After some optimization, it was decided to range θ_2 from 10° to 80° and θ_4 from 90° to 160° for our particular example. Angles of links 2 and 4 also need to be computed for each precision point using linear interpolation. For our first precision point located at $x = 1.1340$,

$$
\frac{(\theta_2)_i - (\theta_2)_f}{x_i - x_f} = \frac{(\theta_2)_i - \theta_2}{x_i - x}
\tag{5.12}
$$

$$
\frac{10 - 80}{1 - 3} = \frac{10 - \theta_2}{1 - 1.1340} \Rightarrow \theta_2 = 14.69°
$$

The value of y at that precision point is

$$
y = \frac{1}{x^{5/3}} = \frac{1}{1.1340^{5/3}} = 0.811
\tag{5.13}
$$

Linear interpolation is used again to find the angle of link 4 at that precision point:

$$
\frac{(\theta_4)_i - (\theta_4)_f}{y_i - y_f} = \frac{(\theta_4)_i - \theta_4}{y_i - y}
\tag{5.14}
$$

$$
\frac{90 - 160}{1 - 0.1602} = \frac{90 - \theta_4}{1 - 0.8110} \Rightarrow \theta_4 = 105.75
$$

This process is repeated for the other accuracy point locations. Table 5.1 summarizes the results for the initial and final values as well as each precision point location.

Table 5.1   Values at precision points

x	θ_2	y	θ_4
1	10	1	90
1.1340	14.6891	0.8110	105.7588
2	45	0.3150	147.1020
2.8660	75.3109	0.1729	158.9431
3	80	0.1602	160

The values of θ_2 and θ_4 for each precision point can now be entered into Equation 5.11.

$$\begin{bmatrix} \cos 105.7588 & -\cos 14.6891 & 1 \\ \cos 147.1020 & -\cos 45 & 1 \\ \cos 158.9431 & -\cos 75.3109 & 1 \end{bmatrix} \begin{bmatrix} K_1 \\ K_2 \\ K_3 \end{bmatrix} = \begin{bmatrix} \cos(14.6891 - 105.7588) \\ \cos(45 - 147.1020) \\ \cos(75.3109 - 158.9431) \end{bmatrix}$$

$$\begin{bmatrix} -0.2716 & -0.9673 & 1 \\ -0.8396 & -0.7071 & 1 \\ -0.9332 & -0.2536 & 1 \end{bmatrix} \begin{bmatrix} K_1 \\ K_2 \\ K_3 \end{bmatrix} = \begin{bmatrix} -0.0187 \\ -0.2097 \\ 0.1109 \end{bmatrix}$$

(5.15)

Solving Equation 5.15 for the K values gives

$$K_1 = 0.7289$$
$$K_2 = 0.8572 \qquad\qquad (5.16)$$
$$K_3 = 1.0085$$

From Chapter 3, the following equations define the K values:

$$K_1 = \frac{O_2O_4}{O_2A}$$
$$K_2 = \frac{O_2O_4}{O_4B} \qquad\qquad (5.17)$$
$$K_3 = \frac{(O_2A)^2 - (AB)^2 + (O_4B)^2 + (O_2O_4)^2}{2(O_2A)(O_4B)}$$

There are four unknown link lengths and only three equations. We cannot get all four lengths, so we must arbitrarily set one value (e.g., set $O_2O_4 = 1$) and then solve for the other lengths. Once designed, the mechanism can then be scaled up or down to any new length of the ground link. Letting $O_2O_4 = 1$ gives

$$O_2A = \frac{O_2O_4}{K_1}$$

(5.18)

$$O_2A = \frac{1}{0.7289} = 1.3719$$

$$O_4B = \frac{O_2O_4}{K_2}$$

$$O_4B = \frac{1}{0.8572} = 1.1666 \tag{5.19}$$

$$AB = \sqrt{(O_2A)^2 + (O_4B)^2 + (O_2O_4)^2 - 2K_3(O_2A)(O_4B)} \tag{5.20}$$

$$AB = \sqrt{1.3719^2 + 1.1666^2 + 1^2 - 2(1.0085)(1.3719)(1.1666)} = 1.0074$$

The linkage mechanism can now be drawn using the calculated link lengths. Figure 5.32 shows the linkage drawn along with the axis labels for x and y. To illustrate the accuracy, it is shown in four different locations. The x pointer can be set to any value in the proposed range $1 \leq x \leq 3$ and the y pointer will point to the value determined from the equation $y = 1/x^{5/3}$.

The first figure, for example, has the x pointer located on 1.2 as an input value. The output should be $y = 1/1.2^{5/3} = 0.738$, which is the approximate location of the output pointer.

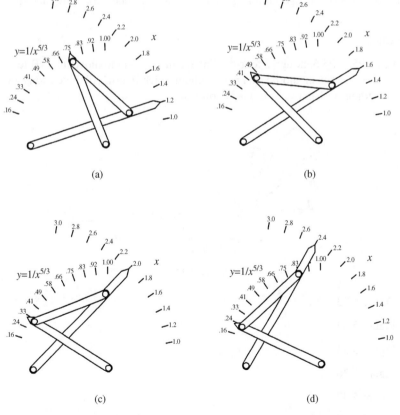

Figure 5.32 Final function generator layout for the function $y = 1/x^{5/3}$

5.7.5 Error Analysis for Function Generation

The function generators discussed in this section will provide an output that approximately matches the desired mathematical function. One source of error will simply be graphical error, which results in inaccuracies in drawing the mechanism. Graphical error can be reduced with improved drafting skill and implementation of CAD software. Mechanical error is another source of error. Mechanical error develops due to imperfections in manufacturing or excessive tolerance at connection points.

5.8 Review and Summary

This chapter discussed many common methods to graphically design linkage mechanisms to perform specific tasks. Methods presented focused on manual drafting techniques using simple equipment, but they can easily be performed with the aid of computer 2D drafting software such as AutoCAD or 3D solid modeling software. Though linkage mechanisms are not the simplest choice for dwell mechanisms, the procedures for approximate dwell linkage mechanisms were presented. It will be seen in future chapters that cams provide a more precise dwell mechanism, although they will also have disadvantages. This chapter also provided the process of designing linkage mechanisms for the function of developing a mathematical relationship between input motion and output motion. Freudenstein's equation was used to develop the required linkage geometry for four-bar linkage mechanisms as well as for slider–crank mechanisms.

Problems

For Problems P5.1–P5.6, design a four-bar linkage mechanism to move line AB to A′B′ using the rocker. The system should be designed without quick return. All dimensions are given to allow for solution by hand or with CAD software.

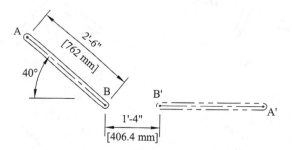

Figure 5.33

P5.1   Figure 5.33

P5.2   Figure 5.34

P5.3   Figure 5.35

P5.4   Figure 5.36

P5.5   Figure 5.37

P5.6   Figure 5.38

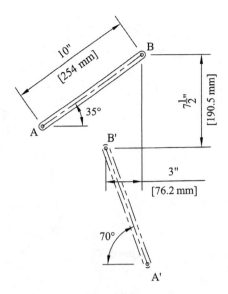

Figure 5.34

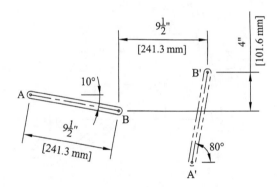

Figure 5.35

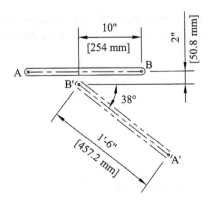

Figure 5.36

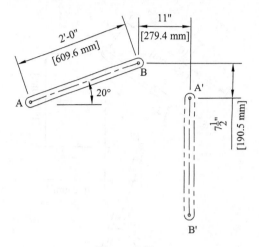

Figure 5.37

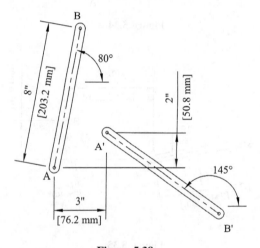

Figure 5.38

For Problems P5.7–P5.11, design a four-bar linkage mechanism to move the coupler from AB to A'B'. All dimensions are given to allow for solution by hand or with CAD software.

P5.7 Figure 5.33

P5.8 Figure 5.35

P5.9 Figure 5.36

P5.10 Figure 5.37

P5.11 Figure 5.38

P5.12 Add a driver dyad to Problem P5.7.

P5.13 Add a driver dyad to Problem P5.8.

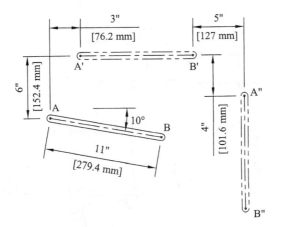

Figure 5.39

P5.14 Add a driver dyad to Problem P5.9.

P5.15 Add a driver dyad to Problem P5.10.

P5.16 Add a driver dyad to Problem P5.11.

For Problems P5.17–P5.22, design a four-bar linkage mechanism to move the coupler through the positions AB, A'B', and A"B". All dimensions are given to allow for solution by hand or with CAD software.

P5.17 Figure 5.39

P5.18 Figure 5.40

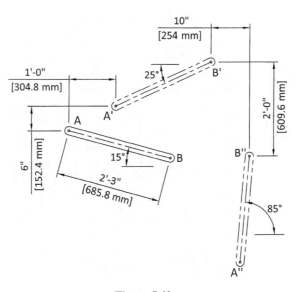

Figure 5.40

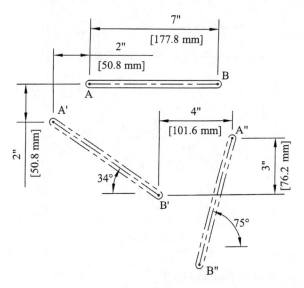

Figure 5.41

P5.19 Figure 5.41

P5.20 Figure 5.42 (ignore the shaded area)

P5.21 Figure 5.44 (ignore the fixed pivot locations shown)

P5.22 Figure 5.43

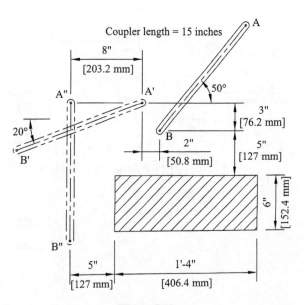

Figure 5.42

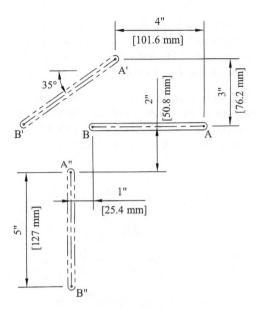

Figure 5.43

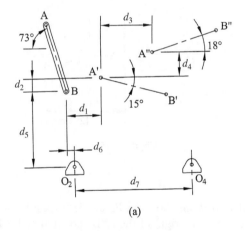

(a)

	inch	mm
d_1	9.75	247.7
d_2	3.5	88.9
d_3	14.5	368.3
d_4	6.75	171.5
d_5	20.75	527.1
d_6	2	50.8
d_7	33	838.2

(b)

Figure 5.44

P5.23 The 19-inch (482.6 mm) long coupler AB needs to move through the three positions shown in Figure 5.44a with the fixed pivot points located as shown. See Figure 5.44b for all dimensions. Use CAD software to draw the desired coupler locations. Synthesize the mechanism and draw it in the three desired locations.

P5.24 A linkage mechanism is required to allow the table surface of a drafting table to be adjustable to the approximate three positions shown in Figure 5.45. The fixed pivot points must be located within the 10-inch wide support leg. Synthesize an appropriate mechanism and designate the fixed pivot locations.

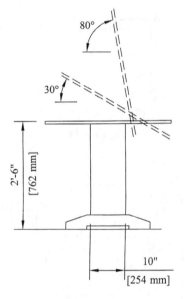

Figure 5.45

P5.25 The three coupler positions shown in Figure 5.39 must be obtained with both fixed pivot points located to the right of the $A''B''$ position and below the $A'B'$ position. Pick locations of the fixed pivot points to satisfy the conditions and design the mechanism.

P5.26 Solve Problem P5.20 with the fixed pivot points located within the shaded area.

P5.27 Solve Problem P5.23 with $d_5 = 19$ inches, $d_6 = 0$, and $d_7 = 20$ inches.

P5.28 Design a quick return crank–rocker mechanism with the output link oscillating through 70° for continuous rotation of the crank. The ratio of work to return stroke shall be 3 to 2.

P5.29 Design a quick return crank–rocker mechanism with the output link oscillating through 65° for continuous rotation of the crank. The ratio of work to return stroke shall be approximately 1.5 to 1.

For Problems P5.30–P5.35, design a Whitworth mechanism to satisfy the given conditions:

	Stroke	Time ratio	Crank length
P5.30	3.5 in.	3:1	1.5 in.
P5.31	120 mm	2.6:1	40 mm
P5.32	110 mm	3.5:1	32 mm
P5.33	4.5 in.	3:1	2 in.
P5.34	6 in.	2:1	2.5 in.
P5.35	85 mm	3.8:1	28 mm

For Problems P5.36–P5.40, determine the time ratio of working stroke to return stroke for the offset slider–crank mechanism defined. Draw a scaled model of the mechanism in both toggle positions.

	Crank length	Coupler length	Offset
P5.36	2.5 in.	9.5 in.	0.75 in.
P5.37	75 mm	240 mm	25 mm
P5.38	50 mm	130 mm	18 mm
P5.39	4 in.	11 in.	1 in.
P5.40	120 mm	350 mm	38 mm

P5.41 Five Chebyshev precision points are required over a range $1 \leq x \leq 4$. Determine the location of the precision points mathematically and represent them graphically.

P5.42 Three Chebyshev precision points are required over a range $3 \leq x \leq 7$. Determine the location of the precision points mathematically and represent them graphically.

P5.43 Synthesize a four-bar linkage mechanism to model the expression $y = x^{1.3}$ over a range $3 \leq x \leq 7$.

P5.44 Design a function generator to convert temperatures between Fahrenheit and Celsius. Define appropriate ranges for temperatures for conversion. This problem is a more open-ended design project that requires assumptions and optimizations from the student.

P5.45 Design a function generator using an input of circle diameter that gives outputs of circumference and area of the circle.

Further Reading

Hartenberg, R.S. and Denavit, J. (1964) *Kinematic Synthesis of Linkages*, McGraw-Hill, New York.
Hrones, J.A. and Nelson, G.L. (1951) *Analysis of the Four-Bar Linkage*, MIT Technology Press, Cambridge MA.
Sandor, G.N. and Erdman, A.G. (1984) *Advanced Mechanism Design: Analysis and Synthesis*, vol. 2, Prentice Hall, Upper Saddle River, NJ.

6

Computational Methods for Linkage Mechanism Kinematics

6.1 Introduction

The previous chapters have developed topics essential for performing a kinematic analysis of linkage mechanisms. Procedures for position analysis were covered in Chapter 3 and procedures for velocity and acceleration analysis were covered in Chapter 4. Position analysis methods were broken into geometrical methods (Section 3.8) and analytical methods (Section 3.9). Graphical methods are simple to visualize, but not well suited to computer programming. Some of the analytical methods presented would be useful for computer programs, but more versatile methods do exist.

Techniques presented in this chapter allow for computational methods of linkage mechanism kinematics, which are less visual but more suitable to computer applications. The intent of this chapter is to build computational methods, which allow for a faster and more thorough analysis of linkage mechanisms. Once a complete program has been created to calculate position, velocity, and acceleration, it can be expanded to do a complete kinetic analysis using material covered in Chapter 11.

Although it has been shown in previous chapters that spreadsheets are useful tools for machine analysis, most of this chapter will focus on using MATLAB® and MathCAD. It is assumed that the reader has a fundamental understanding of each program, although some basic software concepts will be introduced throughout the chapter as and when required. Other software packages could also be used to solve the problems developed in this chapter.

6.2 Matrix Review

6.2.1 Introduction

Although many engineering students will already be familiar with matrix operations, it is never a bad idea to do a little review. This section will not provide a complete coverage of matrices, but will focus on the essential content required for linkage analysis methods presented in this chapter. This section will also provide concepts of matrix operations in MATLAB® and MathCAD.

Machine Analysis with Computer Applications for Mechanical Engineers, First Edition. James Doane.
© 2016 John Wiley & Sons, Ltd. Published 2016 by John Wiley & Sons, Ltd.
Companion Website: www.wiley.com/go/doane0215

This chapter will progressively build computer code to perform kinematic analysis of linkage mechanisms. Most of the work presented in this chapter will utilize matrices, which are easily solved in programs like MATLAB® and MathCAD (Excel can also perform matrix operations). Some computational methods, such as linkage force analysis discussed in Chapter 11, will require the solution of a set of simultaneous equations. Throughout the discussions, these simultaneous equations will be represented in matrix form. This section will provide a brief summary of required topics of matrix algebra. It will later be seen that force analysis will utilize matrix methods for solving nine simultaneous equations for a basic four-bar mechanism.

6.2.2 Matrix Notation

A matrix is an array of elements. The horizontal sets of elements are called rows and the vertical sets of elements are called columns. The matrix $[A]$ shown in Equation 6.1 is an $m \times n$ matrix, which indicates it has m rows and n columns. A square matrix is one in which $m = n$. Square matrices will be used frequently in this chapter because they are important for solution of simultaneous equations. The number of equations (number of rows) will be the same as the number of unknowns (number of columns). This will be seen more in Section 6.2.4.

$$[A]_{m \times n} = \begin{bmatrix} a_{11} & a_{12} & \cdots & a_{1n} \\ a_{21} & a_{22} & \cdots & a_{2n} \\ \vdots & \vdots & \ddots & \vdots \\ a_{m1} & a_{m2} & \cdots & a_{mn} \end{bmatrix} \tag{6.1}$$

Each element in a matrix is denoted by a_{ij}, which indicates the element is in row i and column j. The diagonal (or principal diagonal) consists of the terms where $i = j$ $(a_{11}, a_{22}, a_{33}, \ldots)$. A symmetric matrix is a square matrix in which $a_{ij} = a_{ji}$ for all values of i and j. An example of a symmetric matrix would be

$$[A] = \begin{bmatrix} 1 & 2 & -4 \\ 2 & 7 & 9 \\ -4 & 9 & 3 \end{bmatrix}$$

The identity matrix is a square matrix where the diagonal terms are all equal to 1 and all other terms are 0.

$$[I] = \begin{bmatrix} 1 & 0 & 0 \\ 0 & 1 & 0 \\ 0 & 0 & 1 \end{bmatrix}$$

6.2.3 Matrix Operations

Adding or subtracting two matrices can only exist if the two matrices have the same dimensions. In other words, both matrices must have the same number of rows and the same number of columns. If matrix $[A]$ and matrix $[B]$ are to be added to create matrix $[C]$, the elements of matrix $[C]$ are determined from Equation 6.2:

$$[C] = [A] + [B]$$
$$c_{ij} = a_{ij} + b_{ij}$$

(6.2)

Subtraction is the same process, only the elements are subtracted instead of added. Addition and subtraction are both commutative, meaning that $[A] \pm [B] = [B] \pm [A]$.

In applications such as solving systems of equations, there is a very useful operation known as the determinate of a matrix. The determinate of a matrix $[A]$ is typically represented with the symbol $|A|$. The determinate of a 2×2 matrix is calculated from

$$\begin{vmatrix} a & b \\ c & d \end{vmatrix} = a \cdot d - b \cdot c$$

(6.3)

Multiplication of two matrices is a more complex process. If matrix $[C]$ is the product of matrix $[A]$ and matrix $[B]$, the elements of matrix $[C]$ are determined from Equation 6.4.

$$[C] = [A][B]$$
$$c_{ij} = \sum_{k=1}^{n} a_{ik} b_{kj}$$

(6.4)

The term n shown in the summation is the number of columns in matrix $[A]$ and the number of rows in matrix $[B]$, which must be equal. Multiplication of matrices is typically not commutative, meaning that in general $[A][B] \neq [B][A]$. Matrix multiplication is rather tedious by hand for large matrices, but is very simply accomplished in programs like MATLAB®. It should be noted that matrix division is not a defined operation. However, a matrix inverse (described next) can be used to perform a process equivalent to division.

Next we will define the matrix inverse. The inverse of a matrix is the same concept of the reciprocal of a number. For example, the reciprocal of 5 is 1/5 (or 5^{-1}). If you multiply the original number by the reciprocal, you get a value of 1, that is, $5 \times 1/5 = 1$. Consider matrix $[A]$ that is square and nonsingular (a nonsingular matrix is one in which the determinate is not equal to zero). The matrix $[A]^{-1}$ is the inverse and Equation 6.5 shows that multiplying the original matrix by the inverse will give an identity matrix:

$$[A][A]^{-1} = [I]$$

(6.5)

The main use of a matrix inverse is to have a method that represents matrix division (because dividing matrices does not exist). The procedure for this will be illustrated in Section 6.2.4. Matrix inversion can be done by hand calculations using methods such as LU decomposition, although it is tedious for large matrices. Programs such as MATLAB® and MathCAD have built-in functions to determine a matrix inverse. Spreadsheets can calculate matrix inverse, and many calculators can also perform matrix operations including matrix inverse.

Several operations have been defined. Although these operations can all be performed using hand calculations, this chapter will focus fairly heavily on MATLAB® applications. Example Problem 6.1 will illustrate how to use MATLAB® for the different operations discussed.

Example Problem 6.1

Consider the matrices $[A]$ and $[B]$ as defined below. Use MATLAB® to determine (a) $[A] + [B]$, (b) $[A][B]$, (c) $|B|$, and (d) $[A]^{-1}$.

$$[A] = \begin{bmatrix} 7 & -5 & 1 \\ 0.5 & 16 & -4 \\ 2 & -1 & 9 \end{bmatrix} \qquad [B] = \begin{bmatrix} 4 & -4 & 0 \\ 2 & 7 & 3.5 \\ -1 & 0 & 10 \end{bmatrix}$$

Solution: Before we can perform the operations, we need to define the two matrices in MATLAB®. The coefficient matrices are entered in MATLAB® as

$$A = \begin{bmatrix} 7 & -5 & 1; & 0.5 & 16 & -4; & 2 & -1 & 9 \end{bmatrix}$$
$$B = \begin{bmatrix} 4 & -4 & 0; & 2 & 7 & 3.5; & -1 & 0 & 10 \end{bmatrix}$$

Figure 6.1 shows the solution process for each question.

6.2.4 Representing Simultaneous Equations in Matrix Form

Programs such as MATLAB® and MathCAD are very useful for solving sets of simultaneous linear equations. Before discussing the actual solution process, it is important to understand how to represent a set of equations in matrix form. Consider the set of n equations shown in Equation 6.6.

$$\begin{cases} a_{11}x_1 + a_{12}x_2 + \cdots + a_{1n}x_n = b_1 \\ a_{21}x_1 + a_{22}x_2 + \cdots + a_{2n}x_n = b_2 \\ \qquad\qquad \vdots \\ a_{n1}x_1 + a_{n2}x_2 + \cdots + a_{nn}x_n = b_n \end{cases} \qquad (6.6)$$

The n unknowns are $x_1, x_2, \cdots x_n$ and can be written as a matrix $[x]$. The known coefficients a_{ij} and b_i can be expressed as matrices $[A]$ and $[B]$.

$$[A] = \begin{bmatrix} a_{11} & a_{12} & \cdots & a_{1n} \\ a_{21} & a_{22} & \cdots & a_{2n} \\ \vdots & \vdots & \ddots & \vdots \\ a_{n1} & a_{n2} & \cdots & a_{nn} \end{bmatrix}$$

$$[B] = \begin{bmatrix} b_1 \\ b_2 \\ \vdots \\ b_n \end{bmatrix} \qquad (6.7)$$

```
>> A = [7 -5 1 ; 0.5 16 -4 ; 2 -1 9];
>> B = [4 -4 0 ; 2 7 3.5 ; -1 0 10];
>> A+B

ans =

    11.0000    -9.0000     1.0000
     2.5000    23.0000    -0.5000
     1.0000    -1.0000    19.0000

>> A*B

ans =

    17.0000   -63.0000    -7.5000
    38.0000   110.0000    16.0000
    -3.0000   -15.0000    86.5000

>> det(B)

ans =

    374

>> inv(A)

ans =

     0.1386     0.0436     0.0040
    -0.0124     0.0604     0.0282
    -0.0322    -0.0030     0.1134

>>
```

Figure 6.1

The system of equations shown in Equation 6.6 can be expressed in matrix form as

$$
\begin{bmatrix}
a_{11} & a_{12} & \cdots & a_{1n} \\
a_{21} & a_{22} & \cdots & a_{2n} \\
\vdots & \vdots & \ddots & \vdots \\
a_{n1} & a_{n2} & \cdots & a_{nn}
\end{bmatrix}
\begin{bmatrix}
x_1 \\
x_2 \\
\vdots \\
x_n
\end{bmatrix}
=
\begin{bmatrix}
b_1 \\
b_2 \\
\vdots \\
b_n
\end{bmatrix}
\tag{6.8}
$$

$$[A][x] = [B]$$

Once the equations have been represented in matrix form, there are several possible solution methods for determining the unknowns. One common approach that may be familiar is known as naïve Gauss elimination, which can be improved by use of pivoting. Other common methods are LU decomposition, which comes in different forms, and an iterative method known as Gauss–Seidel. Although these methods are powerful, they tend to be tedious and time consuming to complete by hand. Development or review of these methods is beyond the scope of this text.

This chapter will focus on utilizing computer software to solve the matrix equations. Programs such as MathCAD and MATLAB® have built-in functions that perform matrix operations and allow for easy solution to simultaneous equations. Spreadsheets, such as Excel, also have the ability to perform matrix operations, although the solution process tends to be more complex.

Solution to Equation 6.8 in MATLAB® can be performed in two different methods. The first method utilizes matrix inversion and the second uses the left division operator. Though both methods work, the left division operator (\) is more efficient. The following example illustrates both procedures.

Example Problem 6.2

Solve the set of equations $\begin{cases} x_1 - x_2 + x_3 & = 2 \\ 2x_1 + 3x_2 - 2x_3 = -4 \\ 6x_1 + 2x_2 + 6x_3 = 10 \end{cases}$ using MATLAB®.

Solution: The set of equations can be expressed in matrix form.

$$\begin{bmatrix} 1 & -1 & 1 \\ 2 & 3 & -2 \\ 6 & 2 & 6 \end{bmatrix} \begin{bmatrix} x_1 \\ x_2 \\ x_3 \end{bmatrix} = \begin{bmatrix} 2 \\ -4 \\ 10 \end{bmatrix}$$

The coefficient matrices are

$$[A] = \begin{bmatrix} 1 & -1 & 1 \\ 2 & 3 & -2 \\ 6 & 2 & 6 \end{bmatrix} \qquad [B] = \begin{bmatrix} 2 \\ -4 \\ 10 \end{bmatrix}$$

The coefficient matrices are entered in MATLAB® as

$$A = \begin{bmatrix} 1 & -1 & 1; & 2 & 3 & -2; & 6 & 2 & 6 \end{bmatrix}$$
$$B = \begin{bmatrix} 2; & -4; & 10 \end{bmatrix}$$

The first solution method utilized a matrix inverse. The MATLAB® command is entered as

$$x = inv(A)^*B$$

The second method is a left division operator, which is entered as

$$x = A \backslash B$$

```
>> A = [1 -1 1 ; 2 3 -2 ; 6 2 6]        >> A = [1 -1 1 ; 2 3 -2 ; 6 2 6]

A =                                      A =
        1    -1     1                            1     -1     1
        2     3    -2                            2      3    -2
        6     2     6                            6      2     6
>> B = [2 ; -4 ; 10]                     >> B = [2 ; -4 ; 10]

B =                                      B =
        2                                       2
       -4                                      -4
       10                                      10
>> x = inv(A)*B                          >> x = A\B

x =                                      x =
        0.0625                                  0.0625
       -0.2500                                 -0.2500
        1.6875                                  1.6875

>>                                       >>
```

(a)   (b)

Figure 6.2

Both methods were done in MATLAB® and are shown in Figure 6.2, and each will generate the same result for the unknown values.

$$Answer: x = \begin{bmatrix} 0.0625 \\ -0.2500 \\ 1.6875 \end{bmatrix}$$

6.3 Position Equations

6.3.1 Introduction

The first step of linkage analysis is position analysis. Several methods for position analysis were discussed in Chapter 3. Those methods focused on both graphical and analytical methods. Although spreadsheets are a very useful tool for position analysis, it is difficult to make a spreadsheet that allows for a large variety of mechanisms. The computer example in Section 3.11.1 only allows for position analysis of a basic four-bar slider–crank mechanism. More commonly, to allow for wider versatility, computer solutions are performed using iterative methods.

6.3.2 Iterative Solution Method

Recall from the discussion on the loop closure equation (see Section 3.9.1) that we developed equations using Cartesian coordinates. From that analysis we developed Equation 3.34 for x-components and Equation 3.35 for y-components. Each is represented below, modified as residual functions of the unknown values of θ_3 and θ_4:

$$f_1(\theta_3, \theta_4) = -(O_2A)\cos\theta_2 - (AB)\cos\theta_3 + (O_4B)\cos\theta_4 + (O_2O_4)$$
$$f_2(\theta_3, \theta_4) = -(O_2A)\sin\theta_2 - (AB)\sin\theta_3 + (O_4B)\sin\theta_4$$

(6.9)

The Newton–Raphson method for multidimensional root finding will be used for finding values of θ_3 and θ_4 that result in zero residual for Equation 6.9.

The actual values of θ_3 and θ_4 are not known. The first step is to make an initial guess of $\theta_3^{(1)}$ and $\theta_4^{(1)}$ (the superscript represents iteration counter). There are values $\Delta\theta_3$ and $\Delta\theta_4$ that will cause the residual functions to be zero. Using the initial guess values and the $\Delta\theta$ terms in the residual functions gives $f_i\left(\theta_3^{(1)} + \Delta\theta_3, \theta_4^{(1)} + \Delta\theta_4\right)$, where $i = 1, 2$.

$$f_1\left(\theta_3^{(1)} + \Delta\theta_3, \theta_4^{(1)} + \Delta\theta_4\right) = -(O_2A)\cos\theta_2 - (AB)\cos\left(\theta_3^{(1)} + \Delta\theta_3\right) + (O_4B)\cos\left(\theta_4^{(1)} + \Delta\theta_4\right) + (O_2O_4)$$
$$f_2\left(\theta_3^{(1)} + \Delta\theta_3, \theta_4^{(1)} + \Delta\theta_4\right) = -(O_2A)\sin\theta_2 - (AB)\sin\left(\theta_3^{(1)} + \Delta\theta_3\right) + (O_4B)\sin\left(\theta_4^{(1)} + \Delta\theta_4\right)$$

(6.10)

Equation 6.10 gives two nonlinear equations for the unknowns $\Delta\theta_3$ and $\Delta\theta_4$. We can now use Taylor series for a function of two variables to get two linear equations.

$$f(x, y) \approx f(a, b) + (x - a)f_x(a, b) + (y - b)f_y(a, b) + \cdots$$

(6.11)

First we will approximate the first function in Equation 6.10 about the point $\left(\theta_3^{(1)}, \theta_4^{(1)}\right)$.

$$f_1\left(\theta_3^{(1)} + \Delta\theta_3, \theta_4^{(1)} + \Delta\theta_4\right) \approx f_1\left(\theta_3^{(1)}, \theta_4^{(1)}\right) + \left[\left(\theta_3^{(1)} + \Delta\theta_3\right) - \theta_3^{(1)}\right]\frac{\partial f_1}{\partial(\theta_3^{(1)} + \Delta\theta_3)}$$
$$+ \left[\left(\theta_4^{(1)} + \Delta\theta_4\right) - \theta_4^{(1)}\right]\frac{\partial f_1}{\partial(\theta_4^{(1)} + \Delta\theta_4)} = 0$$

(6.12)

Substituting Equation 6.9 gives

$$-(O_2A)\cos\theta_2 - (AB)\cos\theta_3^{(1)} + (O_4B)\cos\theta_4^{(1)} + (O_2O_4) + \Delta\theta_3\left[(AB)\sin\theta_3^{(1)}\right]$$
$$+ \Delta\theta_4\left[-(O_4B)\sin\theta_4^{(1)}\right] = 0$$

(6.13)

Rearranging Equation 6.13 gives the following linear equation with unknowns $\Delta\theta_3$ and $\Delta\theta_4$:

$$\left[(AB)\sin\theta_3^{(1)}\right]\Delta\theta_3 + \left[-(O_4B)\sin\theta_4^{(1)}\right]\Delta\theta_4$$
$$= (O_2A)\cos\theta_2 + (AB)\cos\theta_3^{(1)} - (O_4B)\cos\theta_4^{(1)} - (O_2O_4)$$

(6.14)

The process is repeated for the second function in Equation 6.10. The Taylor series gives the following:

$$f_2\left(\theta_3^{(1)} + \Delta\theta_3, \theta_4^{(1)} + \Delta\theta_4\right) \approx f_2\left(\theta_3^{(1)}, \theta_4^{(1)}\right) + \left[\left(\theta_3^{(1)} + \Delta\theta_3\right) - \theta_3^{(1)}\right]\frac{\partial f_2}{\partial(\theta_3^{(1)} + \Delta\theta_3)}$$
$$+ \left[\left(\theta_4^{(1)} + \Delta\theta_4\right) - \theta_4^{(1)}\right]\frac{\partial f_2}{\partial(\theta_4^{(1)} + \Delta\theta_4)} = 0$$

(6.15)

Substituting Equation 6.9 gives

$$-(O_2A) \sin \theta_2 - (AB) \sin \theta_3^{(1)} + (O_4B) \sin \theta_4^{(1)} + \Delta\theta_3 \left[-(AB) \cos \theta_3^{(1)} \right]$$
$$+\Delta\theta_4 \left[(O_4B) \cos \theta_4^{(1)} \right] = 0 \tag{6.16}$$

Rearranging Equation 6.16 gives the following linear equation:

$$\left[-(AB) \cos \theta_3^{(1)} \right] \Delta\theta_3 + \left[(O_4B) \cos \theta_4^{(1)} \right] \Delta\theta_4 = (O_2A) \sin \theta_2 + (AB) \sin \theta_3^{(1)} - (O_4B) \sin \theta_4^{(1)}$$
$$\tag{6.17}$$

Now Equations 6.14 and 6.17 can be solved simultaneously for the two unknowns $\Delta\theta_3$ and $\Delta\theta_4$. Arranging the equations in matrix form gives

$$
\begin{bmatrix}
(AB) \sin \theta_3^{(1)} & -(O_4B) \sin \theta_4^{(1)} \\
-(AB) \cos \theta_3^{(1)} & (O_4B) \cos \theta_4^{(1)}
\end{bmatrix}
\begin{bmatrix}
\Delta\theta_3 \\
\Delta\theta_4
\end{bmatrix}
$$
$$
=
\begin{bmatrix}
(O_2A) \cos \theta_2 + (AB) \cos \theta_3^{(1)} - (O_4B) \cos \theta_4^{(1)} - (O_2O_4) \\
(O_2A) \sin \theta_2 + (AB) \sin \theta_3^{(1)} - (O_4B) \sin \theta_4^{(1)}
\end{bmatrix}
\tag{6.18}
$$

Equation 6.18 can now be solved for $\Delta\theta_3$ and $\Delta\theta_4$, which will be in radians. An improved estimate of the unknowns can now be determined as

$$\theta_3^{(2)} = \theta_3^{(1)} + \Delta\theta_3$$
$$\theta_4^{(2)} = \theta_4^{(1)} + \Delta\theta_4 \tag{6.19}$$

The iterations continue until the estimates satisfy the residual functions within a specified tolerance.

The Newton–Raphson method may not converge if the initial guess values are too far away from the actual solution. Therefore, to get best results, the initial guess values should be estimated by graphical layout of the mechanism. The complete process is described in the next couple of examples.

Example Problem 6.3

A four-bar mechanism has the dimensions $O_2O_4 = 10$, $O_2A = 4.25$, $AB = 7.75$, and $O_4B = 5.5$ (all dimensions in inches). Determine θ_3 and θ_4 for an input angle of $52°$.

Solution: We must first make initial guesses of the two angles:

$$\theta_3^{(1)} = 40 \qquad \theta_4^{(1)} = 70$$

Next we determine the values of the residual functions using the initial guess values. Using the above values in Equation 6.9 gives

$$f_1(\theta_3, \theta_4) \quad = -(O_2A)\cos\theta_2 - (AB)\cos\theta_3 + (O_4B)\cos\theta_4 + (O_2O_4)$$

$$f_1\left(\theta_3^{(1)}, \theta_4^{(1)}\right) = -(4.25)\cos 52 - (7.75)\cos 40 + (5.5)\cos 70 + 10 = 3.3277$$

$$f_2(\theta_3, \theta_4) \quad = -(O_2A)\sin\theta_2 - (AB)\sin\theta_3 + (O_4B)\sin\theta_4$$

$$f_2\left(\theta_3^{(1)}, \theta_4^{(1)}\right) = -(4.25)\sin 52 - (7.75)\sin 40 + (5.5)\sin 70 = -3.1623$$

Equation 6.18 will be used to determine $\Delta\theta_3$ and $\Delta\theta_4$:

$$\begin{bmatrix} (AB)\sin\theta_3^{(1)} & -(O_4B)\sin\theta_4^{(1)} \\ -(AB)\cos\theta_3^{(1)} & (O_4B)\cos\theta_4^{(1)} \end{bmatrix} \begin{bmatrix} \Delta\theta_3 \\ \Delta\theta_4 \end{bmatrix} = \begin{bmatrix} -f_1\left(\theta_3^{(1)}, \theta_4^{(1)}\right) \\ -f_2\left(\theta_3^{(1)}, \theta_4^{(1)}\right) \end{bmatrix}$$

$$\begin{bmatrix} (7.75)\sin 40 & -(5.5)\sin 70 \\ -(7.75)\cos 40 & (5.5)\cos 70 \end{bmatrix} \begin{bmatrix} \Delta\theta_3 \\ \Delta\theta_4 \end{bmatrix} = \begin{bmatrix} -3.3277 \\ 3.1623 \end{bmatrix}$$

The equations can be solved by hand or by using MATLAB® as discussed in Section 6.2.4. Solving the simultaneous equations gives

$$\Delta\theta_3 = -0.4732 \text{ rad} \qquad \Delta\theta_4 = 0.1878 \text{ rad}$$

Converting to degrees gives

$$\Delta\theta_3 = -27.1099° \qquad \Delta\theta_4 = 10.7603°$$

The next set of values for the angular positions will be

$$\theta_3^{(2)} = \theta_3^{(1)} + \Delta\theta_3 = 40 - 27.1099 = 12.8901°$$

$$\theta_4^{(2)} = \theta_4^{(1)} + \Delta\theta_4 = 70 + 10.7603 = 80.7603°$$

The residual functions for the second iteration will be

$$f_1\left(\theta_3^{(1)}, \theta_4^{(1)}\right) = -(4.25)\cos 52 - (7.75)\cos 12.8901 + (5.5)\cos 80.7603 + 10 = -0.7119$$

$$f_2\left(\theta_3^{(1)}, \theta_4^{(1)}\right) = -(4.25)\sin 52 - (7.75)\sin 12.8901 + (5.5)\sin 80.7603 = -0.3507$$

giving the following set of equations:

$$\begin{bmatrix} (7.75)\sin 12.8901 & -(5.5)\sin 80.7603 \\ -(7.75)\cos 12.8901 & (5.5)\cos 80.7603 \end{bmatrix} \begin{bmatrix} \Delta\theta_3 \\ \Delta\theta_4 \end{bmatrix} = \begin{bmatrix} -0.7119 \\ -0.3507 \end{bmatrix}$$

Solving and converting to degrees gives

$$\Delta\theta_3 = 3.6749° \qquad \Delta\theta_4 = 8.6835°$$

Table 6.1

Iteration	θ_3	θ_4	$\Delta\theta_3$	$\Delta\theta_4$
1	40	70	−27.1099	10.7603
2	12.8901	80.7603	3.6749	8.6835
3	16.565	89.4438	−0.4546	−0.0943
4	16.1104	89.3495	0.0005	0.0026

The process is then repeated until convergence. Table 6.1 gives values for five iterations. More iterations could be completed to improve accuracy.

Answer: $\begin{aligned} \theta_3 &= 16.11° \\ \theta_4 &= 89.35° \end{aligned}$

Example Problem 6.4

A four-bar mechanism has the dimensions $O_2O_4 = 6$, $O_2A = 2.25$, $AB = 5$, and $O_4B = 3$ (all dimensions in inches). Determine θ_3 and θ_4 for an input angle of 275°.

Solution: We must first make initial guesses of the two angles:

$$\theta_3^{(1)} = 40 \qquad \theta_4^{(1)} = 70$$

First, we determine the values of the residual functions using the initial guess values. Using the above values in Equation 6.9 gives

$$\begin{aligned} f_1(\theta_3, \theta_4) &= -(O_2A)\cos\theta_2 - (AB)\cos\theta_3 + (O_4B)\cos\theta_4 + (O_2O_4) \\ f_1\left(\theta_3^{(1)}, \theta_4^{(1)}\right) &= -(2.25)\cos 275 - (5)\cos 40 + (3)\cos 70 + 6 = 2.9997 \\ f_2(\theta_3, \theta_4) &= -(O_2A)\sin\theta_2 - (AB)\sin\theta_3 + (O_4B)\sin\theta_4 \\ f_2\left(\theta_3^{(1)}, \theta_4^{(1)}\right) &= -(2.25)\sin 275 - (5)\sin 40 + (3)\sin 70 = 1.8466 \end{aligned}$$

Equation 6.18 will be used to determine $\Delta\theta_3$ and $\Delta\theta_4$.

$$\begin{bmatrix} (AB)\sin\theta_3^{(1)} & -(O_4B)\sin\theta_4^{(1)} \\ -(AB)\cos\theta_3^{(1)} & (O_4B)\cos\theta_4^{(1)} \end{bmatrix} \begin{bmatrix} \Delta\theta_3 \\ \Delta\theta_4 \end{bmatrix} = \begin{bmatrix} -f_1\left(\theta_3^{(1)}, \theta_4^{(1)}\right) \\ -f_2\left(\theta_3^{(1)}, \theta_4^{(1)}\right) \end{bmatrix}$$

$$\begin{bmatrix} (5)\sin 40 & -(3)\sin 70 \\ -(5)\cos 40 & (3)\cos 70 \end{bmatrix} \begin{bmatrix} \Delta\theta_3 \\ \Delta\theta_4 \end{bmatrix} = \begin{bmatrix} -2.9997 \\ -1.8466 \end{bmatrix} \Rightarrow \begin{array}{l} \Delta\theta_3 = 63.2817 \\ \Delta\theta_4 = 133.1130 \end{array}$$

The process is then repeated until convergence. Table 6.2 gives values for five iterations. More iterations could be completed to improve accuracy.

Answer: $\begin{aligned} \theta_3 &= 49.55° \\ \theta_4 &= 148.59° \end{aligned}$

Table 6.2

Iteration	θ_3	θ_4	$\Delta\theta_3$	$\Delta\theta_4$
1	40	70	63.2817	133.1130
2	103.2817	203.113	−27.4951	−90.4062
3	75.7866	112.7068	−22.3202	31.6760
4	53.4664	144.3828	−3.8145	3.9741
5	49.6519	148.3569	−0.0973	0.2286

6.3.3 MATLAB® Program Module for Calculating θ_3 and θ_4

Now that the general process of the iterative solution method has been discussed, we can focus on a computer program to perform the process for the complete cycle of a mechanism. Guess values of θ_3 and θ_4 must be defined for the first position along with all the required linkage mechanism geometry. The iterative process will be implemented to determine the actual angular positions of links 3 and 4. The crank is then moved to a different position and the process is repeated. However, the previous values of θ_3 and θ_4 can now be used as the initial guess values.

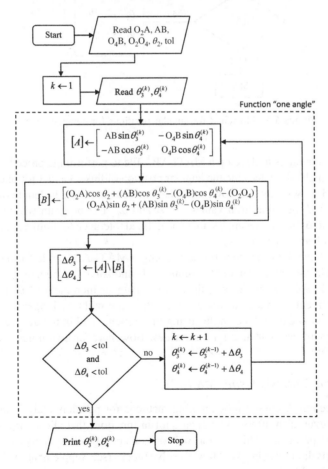

Figure 6.3 Flowchart for angular positions for single position

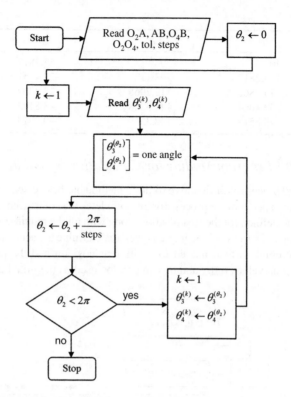

Figure 6.4 Flowchart for angular positions for complete cycle

The first requirement is to develop a MATLAB® file to perform the procedure described in Section 6.3.2, which will determine the angular position of links 3 and 4 for one defined input angle. Figure 6.3 shows a general flowchart. The required inputs are the length of each link, the input angle θ_2, and the initial trial values for θ_3 and θ_4. The program will then go through iterations until the $\Delta\theta$ terms are smaller than the defined tolerance (or until a maximum number of iterations has been reached).

Now that we have the procedure for developing a MATLAB® code for the solution at a single input angle, we can increment the input angle and solve for the new angular positions of links 3 and 4. This can be accomplished for whatever increment of θ_2 is desired. The smaller the increment, the more accurate the results, but it will take longer to solve. Define a MATLAB® function called "one angle" using the contents within the dashed box shown in Figure 6.3. We will use that function in our new flowchart for a complete cycle, which is shown in Figure 6.4.

6.3.4 Position Analysis Using MathCAD

The previous section discussed programming methods for position analysis in MATLAB®. MathCAD, however, also offers some great built-in functions that allow for fast solutions to position analysis problems. This section will demonstrate position analysis by solving the residual equations directly using MathCAD's solve blocks (see Supplementary Concept 6.1 for discussion of solve blocks).

Supplementary Concept 6.1

Solving Simultaneous Equation with MathCAD

One of the simplest ways to solve more complex simultaneous equation in MathCAD is utilizing solve blocks. The general process will be discussed here. Consider the following set of simultaneous equations:

$$4x - y^2 = -3$$
$$2x^2 = 1 + y$$

We will now solve the set of equations using MathCAD solve blocks. Before solving you must enter your guess values for each variable. You start the solve block by entering the command "given." Below that you will set the conditions, which in this case will be the above set of equations. You can enter the equations in their original form. The only thing to notice is that the equal sign is a special type (a bold equal sign) and is entered by hitting the control button and the equal sign. Once all equations have been entered, you use the Find command to solve for the variables and end the solve block. The entire process is shown in Figure 6.5. It is also important to note that solve blocks can be used for more than solving simultaneous equations. One example would be optimization, such as finding local minima or maxima of a function.

Guess values:

$$x := 0 \qquad y := 0$$

Given

$$4{\cdot}x - y^2 = -3$$

$$2{\cdot}x^2 = 1 + y$$

$$\text{Find } (x,y) = \begin{pmatrix} -0.747 \\ 0.115 \end{pmatrix}$$

Figure 6.5

To explain the process of solving linkage position with MathCAD, we will explore an illustrative example. Figure 6.6a shows the linkage mechanism to be analyzed and Figure 6.6b shows the calculations for the example problem. The link dimensions are entered along with the initial guess values for the unknown angular dimensions θ_3 and θ_4. The input speed, although not required for positional analysis, is also entered to allow for further kinematic calculations to be discussed later in this chapter. The solve block is then started by typing the word "given" and the residual equations from Equation 6.9 are entered as the constraints. The solve block is then

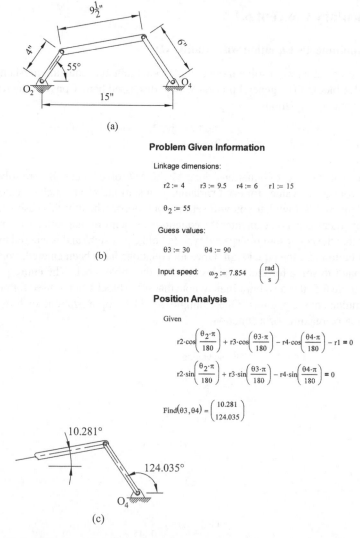

(a)

Problem Given Information

Linkage dimensions:

$$r2 := 4 \qquad r3 := 9.5 \qquad r4 := 6 \qquad r1 := 15$$

$$\theta_2 := 55$$

Guess values:

$$\theta3 := 30 \qquad \theta4 := 90$$

Input speed: $\quad \omega_2 := 7.854 \qquad \left(\dfrac{rad}{s}\right)$

Position Analysis

Given

$$r2\cdot\cos\left(\frac{\theta_2\cdot\pi}{180}\right) + r3\cdot\cos\left(\frac{\theta3\cdot\pi}{180}\right) - r4\cdot\cos\left(\frac{\theta4\cdot\pi}{180}\right) - r1 = 0$$

$$r2\cdot\sin\left(\frac{\theta_2\cdot\pi}{180}\right) + r3\cdot\sin\left(\frac{\theta3\cdot\pi}{180}\right) - r4\cdot\sin\left(\frac{\theta4\cdot\pi}{180}\right) = 0$$

$$\text{Find}(\theta3,\theta4) = \begin{pmatrix} 10.281 \\ 124.035 \end{pmatrix}$$

(b)

(c)

Figure 6.6 MathCAD solve block for position analysis. (a) Example linkage mechanism. (b) MathCAD solve block. (c) Angles of links 3 and 4

completed using the find command shown to give the actual unknown angular values. Figure 6.6c shows the two calculated angular values.

It is often desired to track coordinate data for a particular point on the coupler, such as coupler endpoints. The coordinates of points A and B represent the connection nodes for the coupler. Once the angular positions of each link are known (using the procedure from the previous sections), determining the coordinates of these two points on the coupler is a fairly simple process.

Figure 6.7   Center of mass locations for individual links

6.3.5 Linkage Center of Mass Locations

In kinetic analysis, it is often necessary to know the locations of the individual linkage center of masses. In force analysis, Newton's second law will use the acceleration values of the center of masses as well as the rotational acceleration terms. Locating the center of mass coordinates of each individual link for an arbitrary linkage position can be determined relatively easily. The process can then be repeated for several positions of the mechanism to develop coordinate locations of center of mass for a complete cycle of the mechanism.

Some vectors need to be defined before we begin our process of developing the program. Consider the mechanism shown in Figure 6.7. The center of mass of each link is labeled and located by a position vector.

The position vector locating the center of mass of link 2 is relatively easy to define. The center of mass of link 3 is given by

$$\vec{r}_{c_3} = \vec{r}_A + \vec{r}_{c_3/A} \tag{6.20}$$

The center of mass of link 4 is given by

$$\vec{r}_{c_4} = \vec{r}_{O_4} + \vec{r}_{c_4/O_4} \tag{6.21}$$

Once we understand the process for a single value of θ_2, we can develop a program to calculate the individual center of mass coordinates for a complete cycle. A MATLAB® code can be developed to automate the process. Defining the center of mass coordinates for a complete cycle requires incrementing the value of the input angle.

Individual linkage center of mass locations can then be used to find the mechanism's global center of mass location, which changes as the mechanism moves. Tracing the location of the global center can aid in balancing the mechanism. Figure 6.8 shows a plot of the x and y coordinates of the global center of mass for a crank–rocker mechanism with an 8-inch ground link, 3-inch crank, 9-inch coupler, and 6-inch rocker. The crank has a unit mass located at the midpoint. The coupler has a mass of 2.3 located 3 inches from point A. The rocker has a mass of 1.4 located 2 inches from point O_4. The individual center of mass locations were calculated at multiple points along the cycle in a spreadsheet. The total global center was then calculated and plotted.

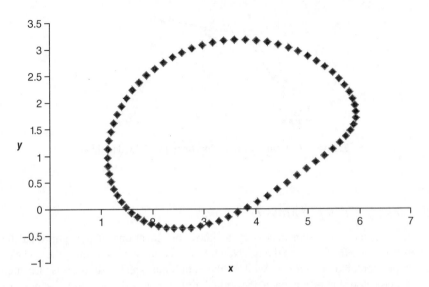

Figure 6.8   Example of the global center of mass location for a complete cycle

6.4   Velocity Analysis

6.4.1 Numerical Differentiation

Once position analysis is complete, we can focus on a complete velocity analysis. The most direct method for determining velocity is to numerically differentiate the position data. Numerical differentiation, which was discussed in Section 2.6, of position data will result in the velocity values. Numerical differentiation can be used to determine linear and angular velocities as well as accelerations. The numerical differentiation equations developed in Chapter 2 used Taylor series expansions to derive a finite difference approximation (central difference). Higher accuracy can be obtained, especially if utilizing computer software, in two different ways. The first way would be to decrease the step size, although that may be limited by the actual data. The second way to improve accuracy is by retaining higher order terms of the Taylor series.

6.4.2 Derivatives of Data Containing Errors or Noise

One issue with numerical differentiation of position data is that the data will usually contain measurement errors or noise. Differentiating data containing such errors will magnify the error. Figure 6.9 illustrates the effects of noise in the data. The solid line in Figure 6.9a is a smooth polynomial function, and its derivative (found using central difference) is shown as the solid line in Figure 6.9b. The dashed line in Figure 6.9a is the same polynomial function with some higher frequency content representing noise. Notice that the noise only slightly distorts the original line. However, the error is magnified through numerical differentiation shown as the dashed line in Figure 6.9b.

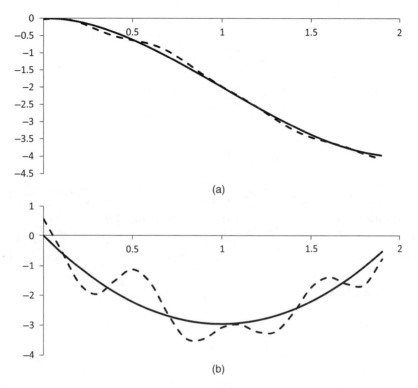

Figure 6.9 Error magnification from differentiation. (a) Function data. (b) First derivative using central difference method

If the data contain noise, the differentiation can often be improved by first filtering the position data. Curve fitting methods, such as least-squares regression, can also be used to smooth data prior to differentiation. The noise content is often a higher frequency than the desired content. Therefore, low-pass digital filters such as a Butterworth filter can be used to remove most of the noise content prior to differentiation. A cutoff frequency must be defined to represent the threshold of the filter. Frequency content higher than the cutoff frequency will be removed, although there will be distortion at frequencies near the cutoff frequency.

6.4.3 Velocity Analysis in MathCAD

We can continue to solve block process discussed in Section 6.3.3 for position analysis to solve for rotational velocities. To illustrate the process, we will again use the mechanism shown in Figure 6.6a, building from the MathCAD program shown in Figure 6.6b. The angular dimensions θ_3 and θ_4 calculated from the previous solve block will now be considered known values.

The constraint equations for the velocity solve block will be developed from velocity equations developed in Section 4.5. The velocity of point A, the joint between the crank and the coupler, can be easily determined based on the crank's pure rotational motion. Figure 6.10a shows the process for calculating the velocity of point A. Both the rotational speed of the crank

Velocity Analysis

$$\omega_{2v} := \begin{pmatrix} 0 \\ 0 \\ \omega_2 \end{pmatrix} \qquad r_2 := \begin{pmatrix} r2\cdot\cos\left(\dfrac{\theta_2\cdot\pi}{180}\right) \\ r2\cdot\sin\left(\dfrac{\theta_2\cdot\pi}{180}\right) \\ 0 \end{pmatrix} \qquad v_A := \omega_{2v} \times r_2 \qquad v_A = \begin{pmatrix} -25.734 \\ 18.019 \\ 0 \end{pmatrix}$$

(a)

$$vB1(\omega3) := v_A + \begin{pmatrix} 0 \\ 0 \\ \omega3 \end{pmatrix} \times \begin{pmatrix} r3\cdot\cos\left(\dfrac{\theta_3\cdot\pi}{180}\right) \\ r3\cdot\sin\left(\dfrac{\theta_3\cdot\pi}{180}\right) \\ 0 \end{pmatrix} \qquad vB2(\omega4) := \begin{pmatrix} 0 \\ 0 \\ \omega4 \end{pmatrix} \times \begin{pmatrix} r4\cdot\cos\left(\dfrac{\theta_4\cdot\pi}{180}\right) \\ r4\cdot\sin\left(\dfrac{\theta_4\cdot\pi}{180}\right) \\ 0 \end{pmatrix}$$

Guess values: $\omega3 := 0$ $\omega4 := 0$

Given

$$vB1(\omega3)_0 = vB2(\omega4)_0$$

$$vB1(\omega3)_1 = vB2(\omega4)_1$$

$$\text{angvel} := \text{Find}(\omega3,\omega4)$$

$$\omega_3 := \text{angvel}_0 \qquad \omega_3 = -3.374$$

$$\omega_4 := \text{angvel}_1 \qquad \omega_4 = 4.025$$

(b)

Figure 6.10   (a) Matrix calculation for $\vec{v}_A$ (b) Solve block to calculate angular velocities of the coupler and output link

and the position vector for point A with respect to O_2 are entered as vectors. The velocity of point A can then be calculated using the cross product (which is a built in function) as shown.

The remainder of the process is shown in Figure 6.10b. The constraint equations for the solve block will come from the following equations discussed in Chapter 4:

$$\vec{v}_B = \vec{v}_A + \vec{v}_{B/A} = \vec{v}_A + \vec{\omega}_3 \times \vec{r}_3 \tag{6.22}$$

$$\vec{v}_B = \omega_4 \hat{k} \times \vec{r}_4 \tag{6.23}$$

Equation 6.24 is shown in Figure 6.10b as the function vB1(ω3) and Equation 6.25 is shown as the function vB2(ω4). Guess values are then entered for both unknown rotational velocities. In this illustration both guess values are entered as zero, although other values could be entered.

The solve block is then started by typing the command "given." The two constraints are entered as scalar equations. The first constraint states that the x-components of vB1 and vB2 must be equal, and the second constraint states that the y-components must be equal. The final values for the rotational speeds can then be determined by ending the solve block with the "find" command as shown. The calculated rotational velocities will have the units of radians per second.

6.5 Acceleration Equations

6.5.1 Numerical Differentiation

The last process in our kinematic analysis focuses on acceleration. Again, the most direct method for determining the acceleration values will be numerical differentiation of the position data developed in Section 6.3. As with velocity, it is very important that the position data be smooth prior to differentiation to get acceleration. Figure 6.9b showed how small amounts of high-frequency noise in the original signal will result in large errors in the first derivative term. Those errors will again get magnified in higher derivatives. Figure 6.11 is the second derivative of the function given in Figure 6.9a. The solid line is the second derivative of the smooth polynomial function, and the dashed line is the derivative of the signal containing noise.

6.5.2 Acceleration Analysis in MathCAD

The final portion of the MathCAD program developed in Sections 6.3.3 and 6.4.3 is to determine rotational accelerations of the links. Rotational accelerations will be determined in a process very similar to that used for the rotational velocities. The problem described in Figure 6.6 will be used to illustrate the acceleration calculations. The first step calculating the acceleration of point A is shown in Figure 6.12a. The vectors for the rotational velocity of link 2 and the velocity of point A were developed in previous steps. Because link 2 rotates at a constant rate, the acceleration of point A is the cross product of $\vec{\omega}_2$ and $\vec{v}_A$.

The acceleration of point B is then developed using both link 3 (relative acceleration) and link 4 in pure rotation. Similar to the process for velocity, guess values for the rotational

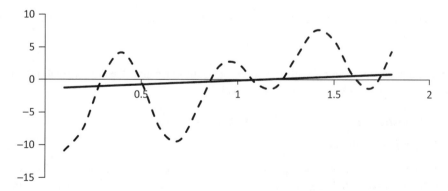

Figure 6.11 Error magnification for second derivative

Acceleration Analysis

$$a_A := \omega_{2v} \times v_A \qquad a_A = \begin{pmatrix} -141.525 \\ -202.119 \\ 0 \end{pmatrix}$$

(a)

$$aB1(\alpha 3) := a_A + \begin{pmatrix} 0 \\ 0 \\ \omega_3 \end{pmatrix} \times \left[\begin{pmatrix} 0 \\ 0 \\ \omega_3 \end{pmatrix} \times \begin{pmatrix} r3 \cdot \cos\left(\dfrac{\theta_3 \cdot \pi}{180}\right) \\ r3 \cdot \sin\left(\dfrac{\theta_3 \cdot \pi}{180}\right) \\ 0 \end{pmatrix} \right] + \begin{pmatrix} 0 \\ 0 \\ \alpha3 \end{pmatrix} \times \begin{pmatrix} r3 \cdot \cos\left(\dfrac{\theta_3 \cdot \pi}{180}\right) \\ r3 \cdot \sin\left(\dfrac{\theta_3 \cdot \pi}{180}\right) \\ 0 \end{pmatrix}$$

$$aB2(\alpha 4) := \begin{pmatrix} 0 \\ 0 \\ \omega_4 \end{pmatrix} \times \left[\begin{pmatrix} 0 \\ 0 \\ \omega_4 \end{pmatrix} \times \begin{pmatrix} r4 \cdot \cos\left(\dfrac{\theta_4 \cdot \pi}{180}\right) \\ r4 \cdot \sin\left(\dfrac{\theta_4 \cdot \pi}{180}\right) \\ 0 \end{pmatrix} \right] + \begin{pmatrix} 0 \\ 0 \\ \alpha4 \end{pmatrix} \times \begin{pmatrix} r4 \cdot \cos\left(\dfrac{\theta_4 \cdot \pi}{180}\right) \\ r4 \cdot \sin\left(\dfrac{\theta_4 \cdot \pi}{180}\right) \\ 0 \end{pmatrix}$$

Guess values:

$\alpha3 := 0 \qquad \alpha4 := 0$

Given

$$aB1(\alpha3)_0 = aB2(\alpha4)_0$$

$$aB1(\alpha3)_1 = aB2(\alpha4)_1$$

$$angacc := Find(\alpha3, \alpha4)$$

$\alpha_3 := angacc_0 \qquad \alpha_3 = -6.037$

$\alpha_4 := angacc_1 \qquad \alpha_4 = 58.748$

(b)

Figure 6.12 (a) Matrix calculation for $\vec{a}_A$. (b) Solve block to calculate angular accelerations of the coupler and output link

accelerations are used along with the solve block to determine the actual rotational accelerations. The process is shown in Figure 6.12b.

6.6 Dynamic Simulation Using Autodesk Inventor

6.6.1 Basic Concepts

The previous sections of this chapter focused on developing software code to perform kinematic analysis of linkage mechanisms. This section will discuss the basic principles of using solid modeling software to perform dynamic simulation. Focus will be on using the dynamic simulation environment in Autodesk Inventor, although other software packages offer similar analysis capabilities. It will be assumed that the reader has a basic working knowledge of solid modeling software. A vast amount of tutorial information is available online if required.

The information obtained from the dynamic simulation environment in Autodesk Inventor is based on multibody dynamics theory utilizing rigid-body components connected by joints. Joint constraints are set in the software to develop constraint equations. These joint constraints operate just as discussed in Chapter 3. If two rigid links are modeled in the software unconstrained, there would be a total of 12 degrees of freedom (3 rotational and 3 translational for each rigid body). Joint constraints are added to remove degrees of freedom. Once the system has all joint constraints applied, Lagrange's equations of motion are then used to develop a system of equations. The solution will then provide position, velocity, acceleration, and joint reaction forces.

6.6.2 Kinematic Constraints

The first step is to make an assembly drawing in Autodesk Inventor of your mechanism. To illustrate the process, we will consider a basic four-bar crank–rocker mechanism. Each of the three moving links will need to be created as parts in the software. Material properties can be included, but that is not essential for pure kinematic analysis. Though this mechanism is very basic, keep in mind that the mechanisms can become very complex. Dynamic simulation can include gears, springs, cams, and other elements and provide a complete kinematic and kinetic analysis.

After placing the three moving links into an assembly drawing, each link will have six degrees of freedom, for a total mobility of 18 (note that you can bring a fourth link and ground it so that it becomes the stationary link 1). Kinematic constraints must now be added. Each moving link can first be constrained to move in a two-dimensional plane, which reduced the degree of freedom of each link to 3 for a total mobility of 9. Joint constraints, as discussed in Chapter 3, are then added for each revolute joint.

6.6.3 Kinematic Analysis Example

The example presented in this section will follow a tutorial provided by Autodesk at the Autodesk Simulation Workshop webpage located at engineeringexploration.autodesk.com. Several tutorial sections are available that provide information on finite element analysis using Autodesk Simulation Multiphysics software. In this section we will deal with Dynamic Simulation section. Within that section, there are several modules containing videos and PowerPoint slides. For a full understanding of using Autodesk Inventor for kinematic analysis, the reader is highly encouraged to watch the videos in all modules. The tutorials focus on the general piston assembly shown in Figure 6.13. Some modules provide the essential theory behind the dynamic simulation process.

6.7 Review and Summary

This chapter expands on the concepts developed in Chapters 3 and 4 to allow for computer solutions for linkage mechanism kinematics. Though this chapter is not essential for understanding the basic concepts of linkage mechanism kinematics, it does allow for the development of computer code to aid in linkage analysis and design. The computer codes developed in

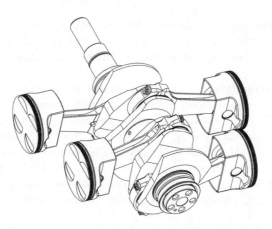

Figure 6.13 Piston assembly for dynamic simulation

this chapter are required for the development of future code based on the content presented in later chapters (e.g., code for force analysis). The implementation of the modules described in this chapter will provide the foundation of a complete kinematic analysis software package.

Problems

P6.1 Given the matrices below, determine the product $[A][B]$ using Equation 6.4. Confirm your answer by using MATLAB®. Is the product $[B][A]$ possible?

$$[A] = \begin{bmatrix} 3 & 2 & 6 \\ 2 & 1 & 7 \end{bmatrix} \qquad [B] = \begin{bmatrix} 5 & 2 & 1 & 0 & 2 \\ 3 & 9 & 6 & 1 & 7 \\ 1 & 0 & 3 & 2 & 8 \end{bmatrix}$$

P6.2 Use MATLAB® to find the inverse of the matrix $\begin{bmatrix} 3 & 1 & -2 \\ 6 & 9 & 1 \\ 1 & 4 & 12 \end{bmatrix}$.

Answer: $\begin{bmatrix} 0.493 & -0.095 & 0.09 \\ -0.336 & 0.18 & -0.071 \\ 0.071 & -0.052 & 0.1 \end{bmatrix}$

P6.3 Put the given set of equations in matrix form. Use MATLAB® to solve the set of equations.

$$\begin{cases} 2x_1 + x_2 - 2x_3 + 4x_4 & = & 25 \\ x_1 + 6x_3 + 8x_4 & = & 67.5 \\ 4x_1 + x_2 - x_3 - 3x_4 & = & -11 \\ 6x_1 + 3x_2 - 5x_3 + 9x_4 & = & 60 \end{cases}$$

P6.4 A four-bar mechanism has the dimensions of $O_2O_4 = 15$, $O_2A = 7$, $AB = 6.5$, and $O_4B = 5$. Use iterative solution method to determine θ_3 and θ_4 for an input angle of $25.5°$. Use initial guess values of $5°$ and $80°$ for θ_3 and θ_4, respectively.

P6.5   A four-bar mechanism has dimensions $O_2O_4 = 12$, $O_2A = 2.5$, $AB = 6.75$, and $BO_4 = 5$ (all dimensions in inches). Using the iterative solution method, determine $\theta_3^{(2)}$ and $\theta_4^{(2)}$ for an input angle $\theta_2 = 35$. Use $\theta_3^{(1)} = 10$ and $\theta_4^{(1)} = 100$ as the initial trial values.

P6.6   A four-bar mechanism has the dimensions $O_2O_4 = 25$, $O_2A = 12$, $AB = 12$, and $O_4B = 20$ (all dimensions in centimeters). Determine θ_3 and θ_4 for an input angle of $76°$. Use the initial trial values of $\theta_3^{(1)} = 10$ and $\theta_4^{(1)} = 150$.

$$Answer: \quad \begin{aligned} \theta_3 &= 24.36 \\ \theta_4 &= 123.94 \end{aligned}$$

P6.7   A four-bar mechanism has the dimensions $O_2O_4 = 30$, $O_2A = 9.5$, $AB = 20$, and $O_4B = 15$ (all dimensions in inches). Determine θ_3 and θ_4 for an input angle of $-40°$. The mechanism is in a crossed position. Sketch the mechanism to pick values of $\theta_3^{(1)}$ and $\theta_4^{(1)}$.

Problems P6.8 and P6.9 require the development of a computer program for iterative position analysis to determine angular positions of links 3 and 4.

P6.8   A four-bar mechanism has the dimensions given in Problem P6.6. Determine the angular positions of links 3 and 4 for input angles from 60 to 80 degrees. Plot the change in each angle versus the input angle.

P6.9   A four-bar mechanism has the dimensions given in Problem P6.4. Determine the angular positions of links 3 and 4 for input angles from 0 to 35 degrees. Plot the change in each angle versus the input angle.

P6.10   A four-bar mechanism has the dimensions $O_2O_4 = 13$, $O_2A = 10$, $AB = 8.5$, and $O_4B = 9.5$ (all dimensions in inches). Determine θ_3 and θ_4 for input angles of $50°$, $52°$, and $54°$. Sketch the mechanism to determine the initial values for iteration. Use the results to determine the angular velocity of link 4 at $52°$ if link O_2A rotates at a constant rate of 10 rad/s clockwise.

P6.11   Repeat Problem P6.10 using the input angles of $80°$, $82°$, and $84°$ to determine the angular velocities of links 3 and 4 at $82°$. Link O_2A rotates at a constant rate of 8 rad/s clockwise.

P6.12   Use the solve block function in MathCAD to solve for the angular positions and velocities of links 3 and 4 for Problem P6.5. Link 2 rotates at a constant rate of 160 rpm counterclockwise.

P6.13   Use the solve block function in MathCAD to solve for the angular positions and velocities of links 3 and 4 for Problem P6.7. Link 2 rotates at a constant rate of 300 rpm clockwise.

P6.14   Use the solve block function in MathCAD to solve for the angular positions, velocities, and accelerations of links 3 and 4 for Problem P6.6. Link 2 rotates at a constant rate of 90 rpm clockwise.

P6.15   Solve Problem P6.14 if link 2 is also accelerating at a rate of 25 rad/s^2.

Further Reading

Autodesk Simulation Workshop (2014) engineeringexploration.autodesk.com/ (accessed October 27, 2014).

Chapra, S.C. (2012) *Applied Numerical Methods with MATLAB® for Engineers and Scientists*, McGraw-Hill, New York.

Greenwood, D.T. (1988) *Principles of Dynamics*, Prentice Hall, Upper Saddle River, NJ.

7

Gear Analysis

7.1 Introduction

Many common applications in machine analysis deal with transmission of rotary motion between rotating shafts. Gears are the most common machine element used to transmit rotary motion, and they perform the task with very uniform motion. Applications range from large power gears, which are rated in thousands of horsepower, to ultraprecision gears used for high-quality instrumentation with very small power requirements.

Gears have been used throughout history. As discussed in Chapter 1, early gears were made of wood with cylindrical cogs. Gears can be traced back to Aristotle in approximately 330 BC, although other writings suggest the use of gears by the Babylonians and Egyptians as early as 1000 BC. Gears were used by the Romans and Greeks for astronomical devices and for measuring distance. Leonardo da Vinci (1452–1519) utilized gear drives in many design ideas. Water wheels became a typical industrial power source, although the rotation speed of the water wheel was very slow. Gears were used to increase rotation speed to the level needed for uses such as textile machinery. Figure 7.1 shows a brief historical timeline of the development of modern gearing. French gear theorist Philip de la Hire recommended the involute curve for gearing and studied epicycloids. Charles Camus expanded la Hire's work, and Leonard Euler developed rules of conjugate action. Methods for computing gear tooth shapes were developed by Kaestner. Other pioneers include Robert Willis and Edward Sang, who provided much work in gear teeth theory.

As discussed in Section 1.6, there are many different types of gears. The relationship of the axes of rotation is a primary factor for determining which gear type to use. Some gear types, such as spur gears and herringbone gears, are only suited for situations where the axes of rotation are parallel. It is also very common to have shafts that intersect at a 90° angle. Old applications, such as that shown in Figure 7.2, utilized cage gears. Current situations requiring intersecting axes of rotation involve a different type of gear, such as bevel gears or worm gears. Helical gears are very versatile and can operate on parallel axes (parallel helical gears) or intersecting axes (crossed helical gears).

Gears are highly standardized for benefits of interchangeability. National associations, such as the American Gear Manufacturers Association (AGMA), have standardized the fundamental gear parameters. Because of required use of set standards, the study of gears requires understanding a lot of specialized terminology. Therefore, initial sections of this chapter

Machine Analysis with Computer Applications for Mechanical Engineers, First Edition. James Doane.
© 2016 John Wiley & Sons, Ltd. Published 2016 by John Wiley & Sons, Ltd.
Companion Website: www.wiley.com/go/doane0215

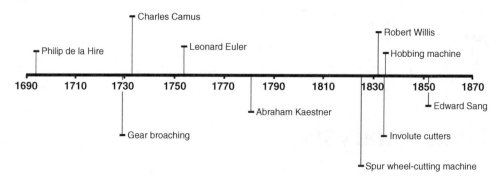

Figure 7.1   Brief historical timeline of modern gearing

Figure 7.2   Cage gear from Pantigo Windmill, Long Island, New York

focus heavily on terminology. Important requirements for tooth profiles will also be discussed with a primary focus on involute curves.

This chapter will provide a more detailed analysis of spur gears, helical gears, bevel gears, and worm gears. Analysis will cover transfer of motion from one gear to another. Though many applications require several gears in a train, it is essential to understand kinematics of single gear pairs. Analysis can then be expanded to gear trains, which will be covered in Chapter 8.

7.2   Involute Curves

7.2.1   Conjugate Profiles

The number of forms that can be used for a gear tooth profile is almost infinite. The involute form is the most common; however, before beginning to study involute curves, it is important to understand the basic idea of conjugate gear tooth action. When two rotating bodies in contact, such as the case shown in Figure 7.3, must have a constant speed ratio, the contact surfaces must be conjugate to one another.

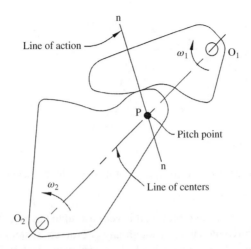

Figure 7.3 Pitch point location for conjugate surfaces

More specifically, the normals to the profiles must pass through a fixed point, known as a pitch point, on the common centerline of the two rotating bodies. It will be shown in later sections of this chapter that the speed ratio is determined by the ratio of distances from pivot points to the pitch point (ratio of O_1P to O_2P). Therefore, the pitch point must remain in a fixed position to ensure constant speed ratio.

7.2.2 Properties of Involute Curves

By far, the most common conjugate profile used to develop gear teeth is the involute of a circle. An involute curve is best visualized by tracing the path to the end of a line that is unwound from the circumference of a circle known as a base circle. Figure 7.4 shows an involute curve developed from a base circle with a radius of 1.

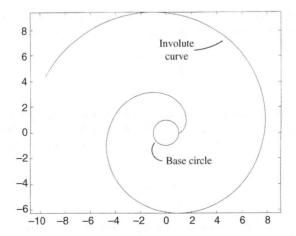

Figure 7.4 Involute curve for base circle with unit radius

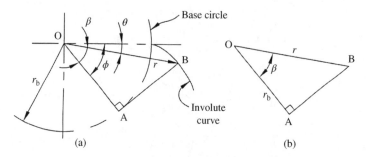

Figure 7.5   Development of involute curve equation

When trying to draw a gear tooth profile, it is very useful to have a mathematical expression to define points on an involute curve. Though any coordinate system could be utilized, it is often easiest to initially define points using polar coordinates. Referring to Figure 7.5, polar coordinates for point B on the involute are defined by radial distance r and polar angle θ. Developing a relationship between r and θ is not direct, and derivation of the polar coordinates will utilize the concept of the involute being developed from unwinding a string from the base circle.

Referring to Figure 7.5a, the length of the tangent line AB is simply the length of the string that has been unwound from the start of the involute up to point B. Therefore, the length of the tangent line that generates the involute is also equal to the length of the arc subtended by the roll angle (ϕ), giving a tangent line length of

$$AB = r_b\phi \tag{7.1}$$

Referring to Figure 7.5b, the length of the tangent line AB can also be determined from Pythagorean's theorem to be

$$AB = \sqrt{r^2 - r_b^2} \tag{7.2}$$

The right-hand side of Equations 7.1 and 7.2 must be equal, and combination of both equations gives a value of the roll angle ϕ in terms of position r and the radius of the base circle (r_b).

$$\phi = \frac{\sqrt{r^2 - r_b^2}}{r_b} \tag{7.3}$$

Also from Figure 7.5b, the angle β can be expressed in terms of the opposite and adjacent sides of the right triangle. Using Equation 7.2 to define the length AB, the tangent of the angle β is given below. The result is exactly equal to the roll angle in Equation 7.3.

$$\tan \beta = \frac{\sqrt{r^2 - r_b^2}}{r_b} = \phi \tag{7.4}$$

Ultimately, the angle of interest for defining the position of point B in polar coordinates is the polar angle (θ). Referring to Figure 7.5a, this angle can now be determined as the difference between ϕ and β. Using the result of Equation 7.4 to define the roll angle, the polar angle can then be expressed:

$$\theta = \phi - \beta = \tan \beta - \beta \tag{7.5}$$

For involute curves, the polar angle is also commonly known as the involute function of β, and is represented as follows:

$$\text{inv } \beta = \tan \beta - \beta \tag{7.6}$$

With the polar angle defined, the second term required to give polar coordinates of the point on the involute is the radial distance. The radial distance can be expressed in terms of the known base circle radius and previously determined angle β by referring to Figure 7.5b. The cosine of angle β is the ratio of base circle radius to radial position.

$$r = \frac{r_{\text{b}}}{\cos \beta} \tag{7.7}$$

Values of the radial distance (r) and the polar angle (θ) give polar coordinates of points on the involute curve, which can also be converted to Cartesian coordinate values for convenience of plotting in CAD software.

7.3 Terminology

7.3.1 Pitch Circle and Pressure Angle

The pitch circles are imaginary circles drawn through the pitch point, which was previously defined in Section 7.2.1 as part of the discussion of conjugate profiles. Because of the nature of involute curves, the pitch point will lie on the common tangent line to both base circles (line aa in Figure 7.6). The pitch circles are then drawn concentrically to the base circles passing through the pitch point. It is important to note that the pitch circle is not a physical part of the gear. Pitch circles only represent circles through the pitch point, and therefore only exist once two gears are meshed together.

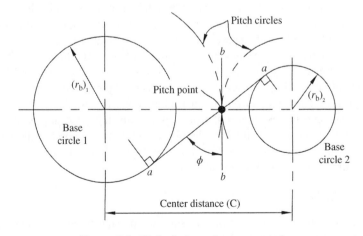

Figure 7.6 Pitch circles and pressure angle

Supplementary Concept 7.1

The Many Uses of Pitch

It is important to note that the term "pitch" can have several meanings in the discussion of gears. Some pitches are dimensions, while others are ratios. Some common examples include pitch point, pitch circle, axial pitch, circular pitch, and diametral pitch. A major reason for so many pitches is that pitches that are useful for specification of one kind of gear may not apply to other kinds of gears. Axial pitch, for example, is a very important dimension in worm gearing and has no real meaning for spur gears.

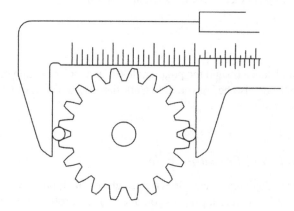

Figure 7.7   Measurement over pins

Pressure angle (ϕ), as shown in Figure 7.6, is the angle measured between the common tangent to pitch circles (line bb) and the common tangent to base circles (line aa). Pressure angles are standardized, and common nominal pressure angles are 14.5°, 20°, and 25°. Changes in center distance will affect the pressure angle due to the fact that the pitch point moves as center distance changes. Therefore, the common nominal pressure angles listed above exist only at a specific center distance. Because the involute teeth are conjugate curves, changes in center distance will not affect the speed ratio between the mating gears.

As stated, the pitch circle is an imaginary reference circle. Therefore, the diameter of the pitch circle cannot be directly measured on an existing gear. An indirect method of measuring the pitch circle of an existing gear uses measurements over pins, as illustrated in Figure 7.7. Cylindrical pins of known diameter are placed in opposing tooth spaces as shown. A measurement can then be taken over the pins. A slightly modified approach is necessary for gears with odd number of teeth where the pins are placed as opposed as possible. Tables, such as those found in *Machinery's Handbook*, can be referred to get gear dimensional information. Similar methods exist for internal gears.

7.3.2 Base Circle

Figure 7.8a shows a pitch circle with the pressure line passing through the pitch point at an angle equal to the pressure angle ϕ. The base circle is drawn so that it is tangent to the pressure

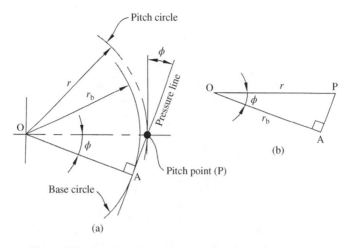

Figure 7.8 Relationship between base circle and pitch circle

line. Therefore, the base circle radius can be determined for a known pitch circle and pressure angle. It can be seen in Figure 7.8b that the ratio of base circle radius to pitch circle radius is determined by cosine of the pressure angle.

$$\frac{r_b}{r} = \cos \phi \tag{7.8}$$

In Equation 7.8, r_b is the radius of the base circle, r is the radius of the pitch circle, and ϕ is the pressure angle.

7.3.3 General Gear Tooth Terminology

All properties for involute gear teeth can be determined based on only three parameters: the number of teeth, the pressure angle, and the diametral pitch. Diametral pitch (P) is the number of teeth per inch of pitch diameter. Therefore, a large diametral pitch relates to a small tooth. Fine pitch gears are gears with a diametral pitch of 24 or greater, and course pitch gears are gears with a diametral pitch below 24. Figure 7.9 shows three gears with different diametral pitch. Each gear has a pressure angle of 20° and 12 teeth; therefore, the only difference is the number of teeth per inch of pitch diameter. Also, it is very important to note that mating gears must have the same diametral pitch.

Figure 7.10 shows a more complete comparison of gear teeth sizes for different diametral pitch values.

The diameter of the pitch circle can be determined for a given diametral pitch and number of teeth. Because diametral pitch is the number of teeth per inch along the pitch circle, dividing the number of teeth by the diametral pitch yields the diameter of the pitch circle in inches.

$$d = \frac{N}{P} \tag{7.9}$$

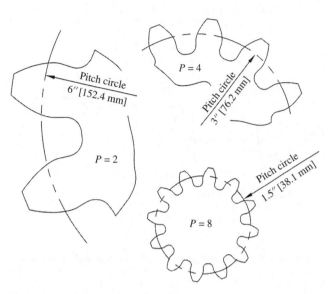

Figure 7.9 Comparison of relative size of gears with different diametral pitch (all gears shown have 12 teeth and a pressure angle $\phi = 20°$)

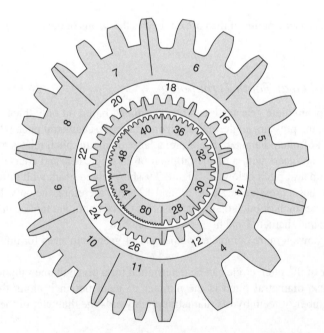

Figure 7.10 Actual sizes of gear teeth of various diametral pitches. Reproduced from Juvinall and Marshek, Machine Component Design, 5th edition, John Wiley & Sons, © 2011

Table 7.1 Gear terms specific to tooth size

Term	Description	Figures
Addendum	Radial distance representing height of gear tooth above pitch circle	Figure 7.11
Dedendum	Radial distance representing depth of gear tooth below pitch circle	
Face width	Thickness of gear measured in the direction parallel to the axis of rotation	Figure 7.12
Tooth thickness	Arc length representing thickness of one tooth along pitch circle	Figure 7.13

Note: The dedendum distance is purposely set larger than the addendum distance to create clearance, which is discussed in Section 8.3.4.

In Equation 7.9, d is the pitch diameter, N is the number of teeth, and P is the diametral pitch. Rearranging Equation 7.8 and using the calculated pitch diameter gives the radius of the base circle:

$$r_b = \frac{d}{2}\cos\phi \tag{7.10}$$

where d is the pitch diameter and ϕ is the pressure angle.

Many terms exist to describe gear teeth. Table 7.1 defines terms specific to tooth size. The total depth of a gear tooth is defined relative to the pitch circle. The addendum and dedendum represent the amount added to the pitch circle and deducted from the pitch circle, respectively.

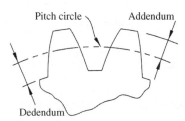

Figure 7.11

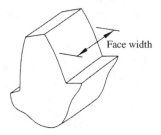

Figure 7.12

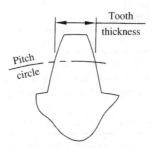

Figure 7.13

Table 7.2   Gear terms specific to tooth surfaces

Term	Description	Figures
Top land	Top surface of gear tooth	Figure 7.14
Bottom land	Bottom surface between gear teeth	
Face	Surface on side of gear tooth (contacting surface) above the pitch circle	Figure 7.15
Flank	Surface on side of gear tooth (contacting surface) below pitch circle	

Standard values for addendum and dedendum are as follows:

$$\text{addendum} = \frac{1}{P} \tag{7.11}$$

$$\text{dedendum} = \frac{1.25}{P} \quad \text{for } P < 20$$

$$\text{dedendum} = \frac{1.20}{P} + 0.002 \quad \text{for } P \geq 20 \tag{7.12}$$

In Equations 7.11 and 7.12, P is the diametral pitch. Table 7.1 also shows terminology for face width and tooth thickness. Face width is the depth of the gear in the direction along the axis. Tooth thickness is specific to the location at the pitch circle. Also, note that the tooth thickness is measured along an arc.

Table 7.2 gives terminology relating to different tooth surfaces. Top land and bottom land refer to tooth surface at the top of the tooth (at addendum circle) and space between teeth (at

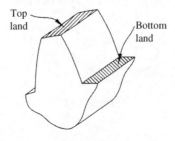

Figure 7.14

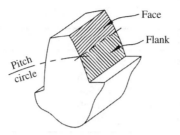

Figure 7.15

dedendum circle), respectively. The side surface of the tooth is divided into two regions, with the division line occurring at the location of the pitch circle. The region above the pitch circle is known as the face, and the region below is known as the flank.

Example Problem 7.1

A gear has the following inputs: number of teeth $N = 24$, diametral pitch $P = 4$, and pressure angle $\phi = 20$. Calculate the pitch diameter, radius of the base circle, and radius of the addendum circle. Plot points of the involute curve at five locations ranging from the base circle to the addendum circle.

Solution: The pitch circle diameter is calculated from Equation 7.9.

$$d = \frac{N}{P} = \frac{24}{4}$$

Answer: $d = 6$ in.
The radius of the base circle is calculated from Equation 7.10:

$$r_b = \frac{d}{2}\cos\phi = \frac{6}{2}\cos 20$$

Answer: $r_b = 2.819$ in.
The radius of the addendum circle is equal to the radius of the pitch circle plus the addendum distance. The addendum distance is determined from Equation 7.11 to be $1/P$:

$$r_a = \frac{d}{2} + \frac{1}{P} = \frac{6}{2} + \frac{1}{4}$$

Answer: $r_a = 3.25$ in.
Points on the involute curve can be calculated from equations developed in Section 7.2.2. For a radius equal to r_a, the corresponding value of β can be determined from Equation 7.4:

$$\beta = \tan^{-1}\left(\frac{\sqrt{r^2 - r_b^2}}{r_b}\right) = \tan^{-1}\left(\frac{\sqrt{3.25^2 - 2.819^2}}{2.819}\right) = 0.5208 \text{ rad}$$

Table 7.3

r (in.)	β (rad)	$\theta = \text{inv}\,\beta$ (°)	x (in.)	y (in.)
$r_a = 3.25$	0.5208	3.027	0.172	3.245
3.10	0.4290	1.628	0.088	3.099
3.00	0.3491	0.854	0.045	3.000
2.90	0.2368	0.259	0.013	2.900
$r_b = 2.819$	0.0000	0.000	0.000	2.819

The polar angle (or involute of β) is determined from Equation 7.6:

$$\text{inv}\,\beta = \tan\beta - \beta = \tan(0.5208) - 0.5208$$
$$\text{inv}\,\beta = \theta = 0.0528\,\text{rad} \quad (3.027°)$$

The polar coordinates (r, θ) for the involute curve at the addendum are then given as $(3.25, 3.027)$. For purposes of plotting, the polar coordinate values will be converted to Cartesian coordinates.

$$x_a = r_a \sin\theta = 3.25 \sin(3.027) = 0.1716$$
$$y_a = r_a \cos\theta = 3.25 \cos(3.027) = 3.2455$$

This process is repeated for other points to give the results shown in Table 7.3. The points are plotted, as shown in Figure 7.16, to give the involute profile.

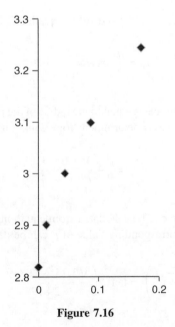

Figure 7.16

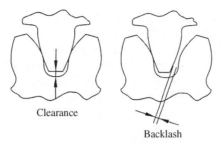

Clearance

Backlash

Figure 7.17   Clearance and backlash

7.3.4 Clearance and Backlash

During meshing of a gearset, it is undesirable to have the top land of one gear make contact with the bottom land of the mating gear. Clearance, as shown in Figure 7.17, is the space measured on the line of centers between the top land of one gear and the bottom land of the other.

Note that clearance is equal to the difference between dedendum and addendum. Referring to Equations 7.11 and 7.12, the clearance is given as $(0.2/P) + 0.002$ for a diametral pitch of 20 or above. Figure 7.18 shows how clearance changes with diametral pitch. It can be seen that for large values of diametral pitch, the constant term (shown as a dashed line at 0.002) plays a more significant role in clearance.

The distance, measured on pitch circle, between the tooth width of the driving gear and the space width of the driven gear is called backlash. In other words, backlash is the amount of motion a gear has when the gear it is meshed with is held stationary. Figure 7.17 shows exaggerated backlash. A gearset will function properly even if the width of the tooth of the driving gear is slightly less than the width of the space on the mating gear. However, if the rotational direction of such a drive is suddenly reversed, the tooth of the driving gear will

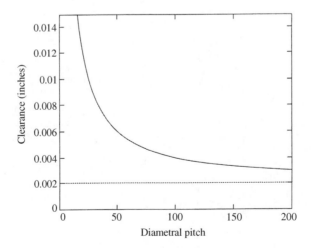

Figure 7.18   Clearance for different values of diametral pitch

temporarily lose contact with the mating gear tooth and "lash back" to strike the adjacent tooth. Therefore, backlash is not desired if the drive requires frequent reversal in direction.

The amount of allowable backlash will vary from case to case. Backlash is generally allocated according to diametral pitch; however, application can also play an important role in choosing backlash. As an example, more backlashes may be required for high-speed applications. In most cases, teeth of both mating gears are made thinner by an amount of one half the total backlash. However, for small pinions, all the backlash may need to be obtained by thinning the teeth of the mating gear.

Backlash also occurs due to increase of gear center distance. As the center distance increases by a value of ΔC (in inches), the change in backlash (in inches) will be

$$\Delta B = 2\Delta C \tan \phi \tag{7.13}$$

7.4 Tooth Contact

7.4.1 Involute Gear Tooth

Involute curves representing one side of N number of teeth can be generated by placing N equally spaced "knots" on the string before unwinding. As the string is unwound from the base circle, each knot will generate an involute curve, as shown in Figure 7.19a.

The base pitch is given as the circumference of the base circle divided by the number of teeth. Returning to the string analogy, the base pitch would be the distance between the knots, as represented in Figure 7.19a:

$$P_b = \frac{(2\pi)r_b}{N} \tag{7.14}$$

where P_b is the base pitch, r_b is the radius of the base circle, and N is the number of teeth. For the true sides of the gear teeth, the involute curve is trimmed at the addendum and dedendum circles. Figure 7.19b shows the completed gear teeth and the measure of base pitch.

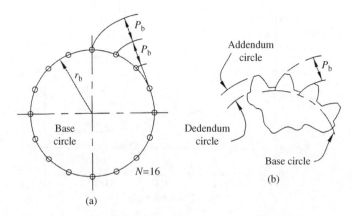

Figure 7.19   (a) Base pitch (P_b) using cord method. (b) Development of involute gear teeth

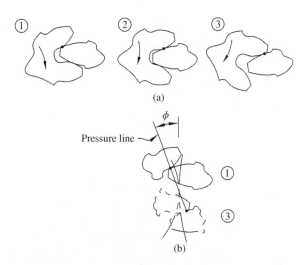

Figure 7.20 Contact point during gear motion ($P = 2$, $N_P = 12$, and $N_G = 30$). (a) Contact point for three different positions. (b) Pressure line orientation

7.4.2 Path of Contact

The contact point between two conjugate surfaces changes during the rotation causing the point of contact to travel along a path. Figure 7.20a shows two mating gears in three successive contact positions with the contact point shown. Initial contact will be located at the tip of the driven gear tooth and final contact will be at the tip of the driving gear tooth. The contact point will move along a straight pressure line located at an angle equal to the pressure angle, as shown in Figure 7.20b.

Because the path of contact is a straight line for involute gear teeth, it is fairly simple to determine the length of the path of contact. Figure 7.21a shows overlapping addendum circles for two mating gears. Contact between teeth will exist only in the region of the overlapping addendum circles. Placing the pressure line on the figure, the length of the path of contact will be the length of the pressure line within the region (represented as line AB). Therefore, only three parameters are needed to determine the path of contact: (i) pitch circles, (ii) addendum circles, and (iii) pressure angle.

The length of the path of contact can be determined graphically or analytically. Graphically determining the length requires drawing the addendum circles and pitch circles to scale and at the proper center distance. A line can then be drawn through the pitch point and at an angle equal to the pressure angle. This process results in a line similar to line APB shown in Figure 7.21a. The distance AB can then be measured to determine the length of contact.

The length of contact can also be determined analytically with use of trigonometry. The obtuse triangle shown in Figure 7.21b is drawn from the gear center (O_g), the pitch point (P), and intersection point of the gear addendum circle and the pressure line (A). Law of sines gives

$$\frac{(r_p)_g}{\sin \alpha_g} = \frac{(r_a)_g}{\sin(90 + \phi)} \tag{7.15}$$

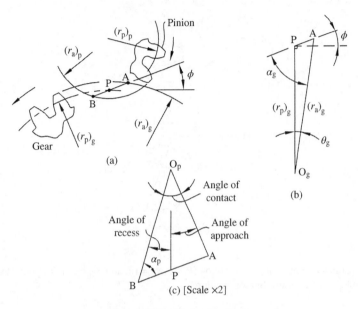

Figure 7.21   Path of contact

Making the trigonometric substitution of $\sin(90 + \phi) = \cos \phi$, the angle α_g can be determined as the following:

$$\alpha_g = \sin^{-1}\left(\frac{(r_p)_g \cos \phi}{(r_a)_g}\right) \tag{7.16}$$

The sum of the angles in the obtuse triangle must equal $180°$. Therefore, the angle θ_g can be determined by subtracting the two known angles from $180°$:

$$\theta_g = 180 - \alpha_g - (90 + \phi)$$
$$\theta_g = 90 - \alpha_g - \phi \tag{7.17}$$

The law of sines can be used again to give

$$\frac{AP}{\sin \theta_g} = \frac{(r_p)_g}{\sin \alpha_g} \tag{7.18}$$

which can be rearranged to solve for the distance AP.

$$AP = \frac{(r_p)_g \sin \theta_g}{\sin \alpha_g} \tag{7.19}$$

The length AP is only one portion of the total length of contact. The additional portion, as shown in Figure 7.21a, is the length BP. In general, length BP will be different from the length AP due to the difference in pitch diameter of the pinion and gear. Therefore, the process just completed for the gear must be repeated for the pinion using oblique triangle O_pPB shown in Figure 7.21c. Angle α_g is required and is determined using the law of sines. Once angle α_g is calculated, length BP is determined using the law of sines. The results are as follows:

$$\alpha_p = \sin^{-1}\left(\frac{(r_p)_p \cos \phi}{(r_a)_p}\right) \tag{7.20}$$

$$BP = \frac{(r_p)_p \sin \theta_p}{\sin \alpha_p} \tag{7.21}$$

The total length of contact is given by the summation of length AP and length BP.

$$L_c = AP + BP \tag{7.22}$$

Tracing an arc along the pitch circle from the time two teeth start contact until the time those same teeth loose contact yields the arc of contact. It is not important if this arc is measured on the pinion or gear since each will yield the same arc of contact.

Further separation of the arc of contact results in the arc of approach and arc of recess. The defining point of this separation is the pitch point. The arc of approach (Q_a) is the portion of the arc of contact from initial contact until the pitch point. The remaining portion of the arc of contact from pitch point until contact is lost is known as the arc of recess (Q_r).

7.4.3 Contact Ratio

Before one set of gear teeth loose contact, it is critical that the next set of teeth start contact. The minimum amount of that contact depends on many factors. The contact ratio is a measure of the number of teeth in contact. The contact ratio is determined by dividing the length of contact found from Equation 7.21 by the base pitch found from Equation 7.13.

$$m = \frac{L_c}{P_b} \tag{7.23}$$

The contact ratio is represented graphically in Figure 7.22.

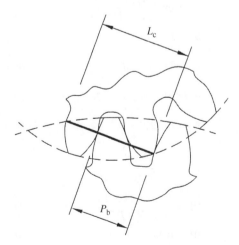

Figure 7.22   Contact ratio

In order to prevent intermittent contact, a contact ratio greater than 1 must exist. If intermittent contact does exist, the path of contact must be increased to a value above the base pitch. Increasing the path of contact can be accomplished by increasing the addendum, increasing the pitch diameters, or decreasing the pressure angle.

Example Problem 7.2

A pinion and gearset have a diametral pitch of 8 and a pressure angle of 20°. The pinion has 20 teeth and the gear has 42 teeth. Determine the length of contact and the contact ratio.

Solution: In order to calculate the length of contact, we need to have radius values for pitch and addendum circles. Then, the length of contact is divided into two parts: one for the pinion and one for the gear.

	Pinion	Gear
Pitch circle diameter from Equation 7.9	$d = \dfrac{N}{P} = \dfrac{20}{8} = 2.50$	$d = \dfrac{N}{P} = \dfrac{42}{8} = 5.25$
Pitch circle radius	$(r_p)_p = \dfrac{2.50}{2} = 1.25$	$(r_p)_g = \dfrac{5.25}{2} = 2.625$
Radius of addendum circle using Equation 7.11 for addendum distance	$(r_a)_p = (r_p)_p + \dfrac{1}{P} = 1.25 + \dfrac{1}{8}$ $(r_a)_p = 1.375$	$(r_a)_g = (r_p)_g + \dfrac{1}{P} = 2.625 + \dfrac{1}{8}$ $(r_a)_g = 2.75$
Equation 7.19 for pinion	$\alpha_p = \sin^{-1}\left[\dfrac{(r_p)_p \cos\phi}{(r_a)_p}\right]$	$\alpha_g = \sin^{-1}\left[\dfrac{(r_p)_g \cos\phi}{(r_a)_g}\right]$
Equation 7.15 for gear	$\alpha_p = \sin^{-1}\left[\dfrac{1.25 \cos 20}{1.375}\right]$	$\alpha_g = \sin^{-1}\left[\dfrac{2.625 \cos 20}{2.75}\right]$
	$\alpha_p = 58.6787$	$\alpha_g = 63.7638$
Equation 7.16	$\theta_p = 90 - \alpha_p - \phi$	$\theta_g = 90 - \alpha_g - \phi$
	$\theta_p = 90 - 58.6787 - 20$	$\theta_g = 90 - 63.7638 - 20$
	$\theta_p = 11.3213$	$\theta_g = 6.2362$
Equation 7.20 for pinion	$BP = \dfrac{(r_p)_p \sin\theta_p}{\sin\alpha_p}$	$AP = \dfrac{(r_p)_g \sin\theta_g}{\sin\alpha_g}$
Equation 7.18 for gear	$BP = \dfrac{1.25 \sin 11.3213}{\sin 58.6787}$	$AP = \dfrac{2.625 \sin 6.2362}{\sin 63.7638}$
	$BP = 0.2873$	$AP = 0.3179$

The total length of contact is then determined from Equation 7.21.

$$L_c = AP + BP = 0.3179 + 0.2873$$

Answer: $L_c = 0.6052$ in.

Contact ratio requires base pitch. Using the pinion, the base circle radius is calculated from Equation 7.10 to be 1.1746 inches. The base pitch is determined from Equation 7.13 to be

$$P_b = \frac{(2\pi)r_b}{N} = \frac{(2\pi)(1.1746)}{20} = 0.369$$

Note that the same base pitch is obtained using the gear. Contact ratio is calculated using Equation 7.22:

$$m = \frac{L_c}{P_b} = \frac{0.6052}{0.369}$$

Answer: $m = 1.64$

7.4.4 Interference

Another important factor to consider is interference. Interference is a condition when tooth profiles overlap each other and contact occurs between nonconjugate surfaces. Interference generally occurs due to using gears with too few teeth, but it also occurs when the center distance is shortened.

Conjugate surfaces exist only above the base circle. Therefore, contact below the base circle needs to be avoided. Interference is most likely to occur for gears with small pressure angle and low tooth numbers. The most basic method for checking interference would be graphical, although it can be checked analytically. Figure 7.23a shows an example of tooth contact with interference. A possible solution to the interference problem is to undercut the flank to create necessary clearance. Undercutting, as shown in Figure 7.23b, weakens the tooth. Undercutting can also result in increased noise and vibration.

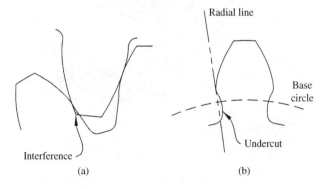

Figure 7.23   (a) Interference. (b) Undercutting

Interference becomes more intense as the tooth count on the pinion decreases. To avoid interference, the minimum number of teeth on the pinion is given by

$$N_P = \frac{2}{\sin^2\phi}$$

(7.24)

Perhaps the most obvious way to fix interference issues would be to shorten the path of contact by decreasing the addendum. Another method of eliminating interference issues would be to increase the pitch radius. An increase in pitch radius, while holding diametral pitch constant, will cause an increase in the number of teeth. Increasing the pressure angle will also help to eliminate interference issues because the base circle becomes smaller. The smaller base circle causes more of the tooth profile to be involute.

7.5   Analysis of Spur Gears

7.5.1   Basic Concepts of Spur Gears

Our discussion of gear analysis will start with the analysis of spur gears, which are the most common gear type in use. A typical set of spur gears is shown in Figure 7.24. Spur gears are easy to manufacture because the gear teeth are parallel to the axis of rotation. For two mating spur gears, the smaller is typically called the pinion, while the larger is called the gear. For typical applications in which the pinion is the driver, output speed decreases while torque increases. For the opposite case in which the gear is the driver, output speed increases while torque decreases. A gear segment with infinite radius is known as a rack. In a rack and pinion, the rotation of the pinion results in linear motion of the rack.

Spur gear teeth can either be external or internal. An external gear has teeth that protrude away from the axis of rotation, and an internal gear has teeth that protrude toward the axis of

Figure 7.24   Spur gears. Reproduced from Mabie and Reinholtz, Mechanisms and Dynamics of Machinery, 4th edition, John Wiley & Sons, © 1987

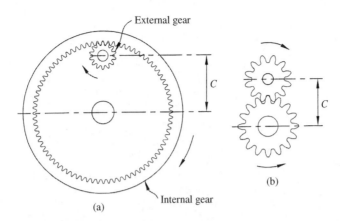

Figure 7.25 Mating gears and direction of rotation. (a) External and internal gears in contact. (b) External gears in contact

rotation. Examples of external and internal gears are shown in Figure 7.25. Figure 7.25a shows an external and internal gear in contact with center distance C. Note that a pinion and an internal gear rotate in the same direction. Figure 7.25b shows an example of two external gears in contact, which causes the rotation of the pinion and gear to be in opposite directions.

7.5.2 Speed Ratio of Spur Gears

Consider a gear and a pinion in contact, as shown in Figure 7.26a. As the pinion rotates through angle θ_P, the gear will rotate through angle θ_G in the opposite direction. The two angles of rotation will be determined by the arc lengths along the pitch circles as demonstrated in Figure 7.26b.

$$S_G = R_G\theta_G$$
$$S_P = R_P\theta_P \tag{7.25}$$

If no slipping occurs, then the arc length for the gear pitch circle (S_G) must equal the arc length of the pinion pitch circle (S_P). Therefore, setting the equations in 7.25 equal gives

$$R_G\theta_G = R_P\theta_P \tag{7.26}$$

Taking the time derivate of Equation 7.26, recognizing that the gear and pinion radius values are constant, gives the following.

$$R_G\frac{d\theta_G}{dt} = R_P\frac{d\theta_P}{dt} \Rightarrow R_G\omega_G = R_P\omega_P \tag{7.27}$$

Rearranging Equation 7.27 gives the angular velocity ratio:

$$\frac{\omega_G}{\omega_P} = \frac{R_P}{R_G} \tag{7.28}$$

Therefore, the angular velocity ratio is inversely proportional to the pitch circle radii (or pitch circle diameters).

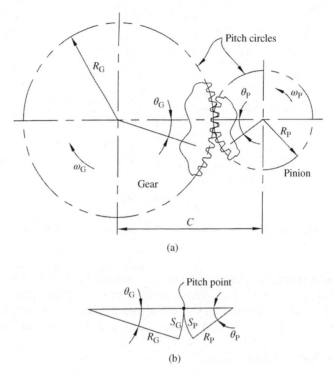

(a)

(b)

Figure 7.26 Spur gear velocity analysis

As defined earlier in Equation 7.9, the pitch diameter is equal to the number of teeth divided by the diametral pitch. Making that substitution gives

$$\frac{\omega_G}{\omega_P} = \frac{N_P/P}{N_G/P} \tag{7.29}$$

Since the diametral pitch (P) for mating gears must be equal, it can be seen that the angular velocity ratio is inversely proportional to the number of teeth:

$$\frac{\omega_G}{\omega_P} = \frac{N_P}{N_G} \tag{7.30}$$

Commonly, a specific velocity ratio is required, but the gear center distance is not specified. However, some cases specify a velocity ratio and a required gear center distance. Having a specified center distance will complicate the problem. Example Problem 7.3 will illustrate both scenarios.

Example Problem 7.3

A pair of spur gears is required to provide a velocity ratio of 3/5. The minimum number of teeth on the pinion to avoid interference will be 14. Design the gears if (a) center distance is not specified and (b) the required center distance is 5.5 inches.

Solution: For part (a), we must consider the velocity ratio. From Equation 7.30

$$\frac{\omega_G}{\omega_P} = \frac{N_P}{N_G}$$

$$\frac{3}{5} = \frac{N_P}{N_G}$$

For every three teeth on the pinion, the gear must have five teeth. To satisfy the limitation of a minimum of 14 teeth on the pinion, we can multiply each value by 5 to get

$$\left(\frac{5}{5}\right)\left(\frac{3}{5}\right) = \frac{N_P}{N_G}$$

$$\frac{15}{25} = \frac{N_P}{N_G}$$

Note that any diametral pitch could be used.

Answer: $\begin{aligned} N_P &= 15 \\ N_G &= 25 \end{aligned}$

Part (b) provided an additional constraint of center distance. The radius of the pinion plus the radius of the gear must equal the center distance:

$$R_P + R_G = C \Rightarrow d_P + d_G = 2C$$

$$d_P + d_G = 11 \tag{1}$$

The velocity ratio can be expressed using pitch diameters:

$$\frac{\omega_G}{\omega_P} = \frac{d_P}{d_G} \Rightarrow d_G = \left(\frac{\omega_P}{\omega_G}\right) d_P$$

$$d_G = \left(\frac{5}{3}\right) d_P \tag{2}$$

Equation (2) into Equation (1) gives

$$d_P + \left(\frac{5}{3}\right) d_P = 11$$

$$\left(\frac{8}{3}\right) d_P = 11$$

$$d_P = \frac{33}{8}$$

Substituting this value into Equation (1) gives

$$\frac{33}{8} + d_G = 11$$

$$d_G = \frac{55}{8}$$

The gear properties can be determined because the pitch diameter is equal to the number of teeth divided by the diametral pitch.

$$P = 8$$

Answer: $N_P = 33$

$$N_G = 55$$

7.5.3 Efficiency of Spur Gears

Customarily, gear efficiency is developed in terms of a coefficient of friction. Sliding friction between mating gear teeth will alter the force vector resulting in a minor loss of transmitted torque. The information in this chapter covering gear efficiency is presented as a summary of material covered in Buckingham's *Analytical Mechanics of Gears*. Interested readers should refer to this book for a more thorough discussion of efficiency.

For a spur gearset consisting of a pinion and a gear, the efficiency of the gearset is given by

$$e = 1 - \left[\frac{1 + (N_P/N_G)}{\beta_a + \beta_r}\right]\left(\frac{\mu_a}{2}\beta_a^2 + \frac{\mu_r}{2}\beta_r^2\right) \tag{7.31}$$

In Equation 7.31, N is the number of teeth, and the subscripts P and G indicate pinion and gear, respectively. The terms β_a and β_r represent the angle of approach and the angle of return, respectively. Both angles are defined in radians and can be seen in Figure 7.21c. The coefficient of friction during approach μ_a is different from the coefficient of friction during recess μ_r. These coefficients of friction can be determined from Figure 7.27.

Average sliding velocity is required for Figure 7.27, and the average sliding velocity for recess and approach are determined using Equations 7.32 and 7.33. Pitch line velocity used in the equations is the tangential velocity of either the pinion or the gear at a radial location equal to the pitch circle radius.

$$V_{sr} = \frac{V}{2}\left(1 + \frac{N_P}{N_G}\right)\beta_r \cos \phi \tag{7.32}$$

$$V_{sa} = \frac{V}{2}\left(1 + \frac{N_P}{N_G}\right)\beta_a \cos \phi \tag{7.33}$$

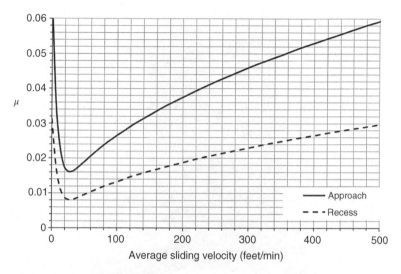

Figure 7.27 Average coefficient of friction for approach (μ_a) and recess (μ_r)

7.6 Analysis of Parallel Helical Gears

7.6.1 Parallel versus Crossed Helical Gears

Of all the types of gears available, helical gears (Figure 7.28) allow for the widest variety in shaft relationships. Two shafts at any angle can be connected by a pair of helical gears. Parallel helical gears are used for parallel shafts, but crossed helical gears can accommodate nearly any shaft angle.

Helical gears for parallel shafts, known as parallel helical gears, can carry larger loads than an equivalent spur gear pair. Another advantage of parallel helical gears is that they have a greater contact ratio than spur gears. Helical gears for nonparallel shafts, known as crossed helical gears, cannot carry very large loads due to point contact between mating teeth.

Figure 7.28 Helical gears (a) for parallel shafts and (b) for crossed shafts. Reproduced from Mabie and Reinholtz, Mechanisms and Dynamics of Machinery, 4th edition, John Wiley & Sons, © 1987

Figure 7.29 Herringbone gears. Reproduced from Mabie and Reinholtz, Mechanisms and Dynamics of Machinery, 4th edition, John Wiley & Sons, © 1987

7.6.2 Basic Concepts of Helical Gears

Teeth on helical gears are not straight like spur gear teeth. Helical gear teeth are cut at an angle to the gear face. Because of the angled teeth, tooth contact is more gradual compared with spur gear tooth contact. Gradual contact makes helical gears quieter and smoother than spur gears. Angled gear teeth are longer than straight gear teeth for the same face width. Because of the longer teeth, helical gears can carry higher loads than spur gears of the same size.

Forces on angled teeth result in an axial thrust force, which is not present for spur gears. Thrust bearings are needed to accommodate the axial thrust from helical gears. A double helical gear, also known as a herringbone gear, can be used to eliminate the axial thrust force. As seen in Figure 7.29, a herringbone gear consists of mirrored helical gears, which will cancel out thrust forces.

7.6.3 Terminology Specific to Helical Gears

The teeth of helical gears are oblique to the axis of rotation. The helix angle is the angle between a line parallel to the axis of rotation and a line tangent to the pitch helix.

$$\psi = \tan^{-1}\left(\frac{2\pi R}{L}\right) \tag{7.34}$$

In Equation 7.34, R is the pitch radius of the gear and L is the lead of the tooth. Helical gears can be right hand or left hand. A right-hand helical gear has teeth that twist clockwise away from an observer looking along the axis. The lead is the axial advance in one revolution. Equation 7.34 relates helix angle and lead by use of a right triangle with lead (L) as one leg and pitch circumference ($2\pi R$) as the other.

Unlike spur gears, the pitch diameter of a helical gear is not simply a direct linear relationship to the number of teeth. The helix angle will also affect the pitch diameter:

$$d = \frac{N}{P \cos \psi} \tag{7.35}$$

It can be seen that the pitch diameter will be minimum when the helix angle is zero, which would simply be a spur gear. Therefore, the center distance for helical gears will be larger compared to using equivalent spur gears.

7.6.4 Efficiency of Helical Gears

Efficiency analysis of helical gears is very similar to analysis for spur gears. The efficiency is determined from

$$e = 1 - \left[\frac{1 + (N_1/N_2)}{(Q_a + Q_r)\cos \psi} \right] \frac{\mu}{2} \left(Q_a^2 + Q_r^2 \right) \tag{7.36}$$

In Equation 7.36, N again represents the number of teeth, now with the subscript 1 indicating the driving gear and 2 indicating the driven gear. The arc of approach is Q_a for the driving gear and the arc of recess is Q_r. The coefficient of friction is μ and can be seen in Figure 7.30. The helix angle is ψ.

Equation 7.36 calculates efficiency for helical gears based on the assumption that the coefficient of friction is the same for approach and recess, which can be determined based on average sliding velocity. Average sliding velocity is calculated from Equation 7.37 using the pitch line velocity V in feet/min.

$$V_s = V \left(1 + \frac{N_1}{N_2} \right) \left(\frac{Q_a + Q_r}{4} \right) \tag{7.37}$$

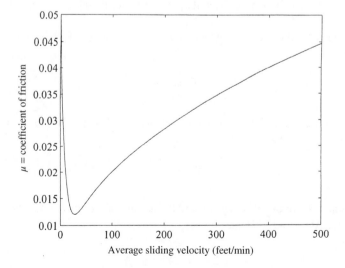

Figure 7.30 Coefficient of friction

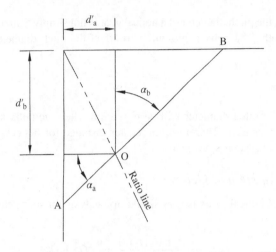

Figure 7.31 Graphical solution for crossed helical gears with 90° shaft angle

7.7 Analysis of Crossed Helical Gears

7.7.1 Graphical Solution for a Shaft Angle of 90°

Analysis of crossed helical gears makes use of equivalent diameter, which is the diameter of equivalent spur gears. For a desired speed ratio, a set of crossed helical gears for 90° shaft angle can be selected by first selecting equivalent spur gears. Let d'_a and d'_b be equivalent pitch diameters of spur gears a and b, respectively, to satisfy speed ratio requirements. Both equivalent diameters may be represented on a graph, as shown in Figure 7.31, to locate point O. A line drawn from the origin through point O gives a ratio line. For a line drawn diagonally through point O, distances AO and OB will represent the pitch diameter of helical gears a and b, respectively. Angles α_a and α_b will give tooth angles of helical gears a and b, respectively.

Pitch diameter of helical gear a, which is represented as the length of line AO, can be determined by

$$AO = \frac{d'_a}{\cos \alpha_a} \qquad (7.38)$$

Similarly, helical gear b will have a pitch diameter equal to the length of line BO and is determined by

$$BO = \frac{d'_b}{\cos \alpha_b} \qquad (7.39)$$

The center distance C for the helical gears will be the sum of the above pitch diameters divided by 2. Therefore, $AO + BO = 2C$ giving

$$\frac{d'_a}{\cos \alpha_a} + \frac{d'_b}{\cos \alpha_b} = 2C \qquad (7.40)$$

It can also be seen from Figure 7.31 that $\alpha_a + \alpha_b = 90$. Making the substitution of $\alpha_b = 90 - \alpha_a$ in Equation 7.40 and noting that $\cos(90 - \alpha_a) = \sin \alpha_a$ gives the following:

$$\frac{d'_a}{\cos \alpha_a} + \frac{d'_b}{\sin \alpha_a} = 2C \qquad (7.41)$$

Multiplying by $\sin \alpha_a$ and noting that $\sin \alpha_a / \cos \alpha_a = \tan \alpha_a$

$$d'_a \tan \alpha_a + d'_b = 2C \sin \alpha_a \qquad (7.42)$$

Equation 7.42 is a nonlinear equation with one unknown value of the tooth angle of helical gear a. Once the equation is solved, the tooth angle of helical gear b can be determined from

$$\alpha_b = 90 - \alpha_a \qquad (7.43)$$

Pitch diameters of the helical gears, represented by lengths AO and BO, can then be determined from Equations 7.38 and 7.39.

Example Problem 7.4

Select a set of crossed helical gears for a 90° shaft angle, which has a speed ratio of 3 : 1 and a center distance of 4.5 inches.

Solution: As an initial trial, we will try a diametral pitch of 10 and 14 teeth on gear a. For a speed ratio of 3 : 1, gear b must have $3(14) = 42$ teeth. Equivalent pitch diameters are calculated first:

$$d'_a = \frac{N_a}{P} = \frac{14}{10} = 1.4$$

$$d'_b = \frac{N_b}{P} = \frac{42}{10} = 4.2$$

Equation 7.42 can then be used to calculate the tooth angle for gear a:

$$d'_a \tan \alpha_a + d'_b = 2C \sin \alpha_a$$
$$1.4 \tan \alpha_a + 4.2 = 2(4.5)\sin \alpha_a$$

This equation can be solved numerically by the Newton–Raphson method, which requires an initial guess for α_a. This initial guess could be based on a graphical layout such as Figure 7.31. To show the effectiveness of the Newton–Raphson method, we will use an initial guess value of 45° (0.7854 rad).

$$f(\alpha_a) = 1.4 \tan \alpha_a + 4.2 - 9 \sin \alpha_a$$
$$f'(\alpha_a) = \frac{1.4}{\cos^2 \alpha_a} - 9 \cos \alpha_a$$

Table 7.4

N_a	N_b	α_a	α_b	d_a	d_b
10	30	22.2610°	67.7390°	1.0805	7.9192
12	36	28.1206°	61.8794°	1.3606	7.6380
14	42	35.1966°	54.8034°	1.7132	7.2868
16	48	45.6680°	44.3320°	2.2896	6.7104

For the trial value of 0.7854 rad,

$$f(\alpha_a^1) = 1.4\tan(0.7854) + 4.2 - 9\sin(0.7854) = -0.764$$
$$f'(\alpha_a^1) = \frac{1.4}{\cos^2(0.7854)} - 9\cos(0.7854) = -3.564$$

The next value is determined as

$$\alpha_a^{(2)} = \alpha_a^1 - \frac{f(\alpha_a^1)}{f'(\alpha_a^1)}$$

$$\alpha_a^{(2)} = 0.7854 - \frac{-0.764}{-3.564} = 0.5710$$

Repeating the process gives

$$\alpha_a^{(3)} = 0.6130 \text{ rad}$$

$$\alpha_a^{(4)} = 0.6143 \text{ rad}$$

$$\alpha_a^{(5)} = 0.6143 \text{ rad}$$

This gives $\alpha_a = 35.1966°$ and $\alpha_b = 90° - 35.1966° = 54.8034°$. The diametral pitches of the two helical gears are determined from Equations 7.38 and 7.39.

$$AO = d_a = \frac{d_a'}{\cos\alpha_a} = \frac{1.4}{\cos 35.1966} = 1.7132$$

$$BO = d_b = \frac{d_b'}{\cos\alpha_b} = \frac{4.2}{\cos 54.8034} = 7.2868$$

This gives one possible set of helical gears. Use of computer software can give several possible solutions. Maintaining a diametral pitch of 10 and changing the number of teeth of gear a gives possible solutions shown in Table 7.4. Figure 7.32 shows the graphical representation for N_a of 14 teeth.

7.7.2 Graphical Solution for Other Shaft Angles

Similar analysis can be performed for crossed helical gears with shafts that are not perpendicular to each other. Figure 7.33 shows the procedure for graphical solution. Two lines are drawn

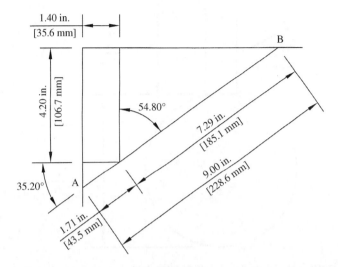

Figure 7.32

at an angle λ, which is equal to the angle between shafts. A ratio line is drawn based on equivalent pitch diameters (d'_a and d'_b) of spur gears. Line AB can then be drawn at a length equal to twice the required center distance. As before, distances AO and AB represent the pitch diameters of gears a and b, respectively. The helix angles of gears a and b are given by α_a and α_b.

7.7.3 Versatility of Helical Gears for Nonparallel Shafts

It has already been mentioned that helical gears are highly versatile due to the ability to use for parallel or nonparallel shaft arrangements. However, the fact that pitch diameter changes with

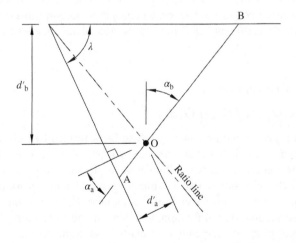

Figure 7.33 Graphical solution for general crossed helical gears

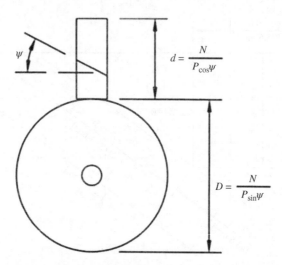

$$d = \frac{N}{P \cos \psi}$$

$$D = \frac{N}{P \sin \psi}$$

Figure 7.34   Gears with different pitch diameters for a gear ratio of $1:1$

helix angle will allow for more unique versatility. For example, helical gears with different pitch diameters can have a gear ratio of $1:1$. Consider a set of crossed helical gears with a shaft angle of $90°$, as shown in Figure 7.34. The pitch diameter of each gear is determined using the equations shown. For a gear ratio of $1:1$, both gears must have the same number of teeth (N) and the ratio of pitch diameters will be

$$\frac{d}{D} = \frac{N}{P \cos \psi} \frac{P \sin \psi}{N} = \frac{\sin \psi}{\cos \psi} \qquad (7.44)$$

Therefore, the two gears will have the same pitch diameter only if the helix angle is $45°$. For any other helix angle, the gears will have different pitch diameters but maintain a ratio of $1:1$. Similarly, the helix angle can be adjusted to modify an existing gearbox to a different ratio. For example, a gearbox uses two gears at a set center distance to accomplish a specific ratio. A new set of helical gears with a different helix angle can be used at the same center distance to obtain a different ratio.

7.8 Analysis of Bevel Gears

7.8.1 Basic Concepts of Bevel Gears

Much like how spur gears have imaginary pitch surfaces formed by right circular cylinders, bevel gears have imaginary pitch surfaces formed by frustums of right circular cones. In other words, a pair of bevel gears acts as two rolling cones in the same sense as a pair of spur gears acts as two rolling cylinders. Similar to the terminology used for spur gears, mating bevel gears consist of a pinion (member with fewer teeth) and a gear. Bevel gear pairs are the simplest available for an intersecting axes of rotation, and they are most commonly used for $90°$ shaft angles. Use of bevel gears for perpendicular shafts is so common that if no shaft angle is specified, $90°$ shaft angles can be assumed. Figure 7.35 shows a typical bevel gear.

Figure 7.35 Straight bevel gears. Reproduced from Mabie and Reinholtz, Mechanisms and Dynamics of Machinery, 4th edition, John Wiley & Sons, © 1987

7.8.2 Terminology Specific to Bevel Gears

Figure 7.36 shows terminology for bevel gears. The pitch apex represents the common apex point for two rolling cones in contact. Cone distance is measured with respect to the apex. Teeth for bevel gears are specified by diametral pitch, just as spur gear teeth are specified. Diametral pitch for bevel gears, however, varies with cone distance due to change in center distance.

7.8.3 Speed Ratio and Direction of Rotation

For a point in pure rotation about a fixed point, tangential velocity increases as the distance from the axis of rotation increases. Applying this concept to velocity of a point on the surface of a rotating cone means that the velocity will increase as the distance from the apex increases. The velocity at the apex is zero due to the fact that the radius is zero. Therefore, two rolling

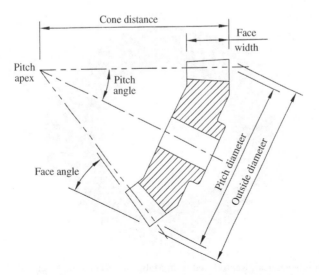

Figure 7.36 Bevel gear geometry and terminology

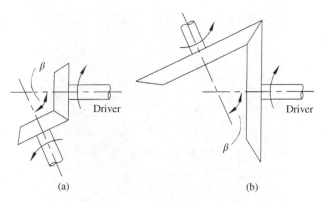

(a)   (b)

Figure 7.37   Bevel gear direction of rotation

cones in contact must have the apices located at the same point in order to have proper matching of velocities.

For two intersecting shaft centerlines that are not perpendicular, bevel gear pairs could be developed to span an acute angle or an obtuse angle, as shown in Figure 7.37a and b, respectively. For a specified rotation direction of the driver, the follower for the acute angle and the follower for the obtuse angle will have opposite directions. Therefore, the desired rotational direction of the follower controls whether the bevel gear pair should be placed on the acute or obtuse angle between shafts. Due to the fact that the follower sizes are different for nonperpendicular shaft angles, angular speeds of the follower for the acute angle and the follower for the obtuse angle will not be the same. The difference in angular speeds would not be the case, however, for bevel gears for perpendicular shafts. For perpendicular shafts, direction of follower rotation can be reversed without affecting speed ratio by simply shifting either the driver or the follower by 180°.

7.8.4 Other Types of Bevel Gears

This section has discussed straight bevels, which are the simplest form of bevel gears. Other forms of bevel gears exist. Spiral bevel gears, such as the one shown in Figure 7.38, allow for

Figure 7.38   Spiral bevel gear. Reproduced from Mabie and Reinholtz, Mechanisms and Dynamics of Machinery, 4th edition, John Wiley & Sons, © 1987

Figure 7.39 Hypoid bevel gear. Reproduced from Mabie and Reinholtz, Mechanisms and Dynamics of Machinery, 4th edition, John Wiley & Sons, © 1987

smoother and quieter operation compared with straight bevel gears. They have curved teeth causing gradual contact. Spiral bevels will have two or more teeth in contact at all times, which results in smoother operation.

Hypoid gears developed out of necessity for the automotive industry. A method was required to give quiet rear-wheel drive but lower the drive shaft below the centerline of the rear axle. Figure 7.39 shows the arrangement of hypoid bevel gears.

7.9 Analysis of Worm Gearing

7.9.1 Basic Concepts of Worm Gearing

A major application for worm gearing is one that requires a large speed reduction with a single pair of gears. Worm gearing, therefore, has an advantage of a more compact design. The compact size comes at a cost of lower efficiency and large axial forces. Tooth engagement for worm gearing occurs without shock, resulting in quiet operation.

A worm gear set consists of the driver, or worm, and the follower, or worm gear, as seen in Figure 7.40. The worm has the same basic characteristics of a screw, and it has helical teeth called threads. Worms can be single-, double-, triple-, or even quadruple-threaded.

Worm gear teeth are not involutes over the entire face. Therefore, proper conjugate action depends heavily on center distance. Tolerance on center distance for a worm gear pair is much smaller compared to spur or helical gears. Small changes in center distance, as small as 0.005 inches, for a worm gear pair can cause noticeable changes in speed ratio. In comparison, a pair of helical gears can have center distance changes 10 times that amount with essentially no change in running qualities.

7.9.2 Terminology Specific to Worm Gearing

Two basic terms used to describe a worm are pitch and lead, as shown in Figure 7.41. Even though these terms are basic, they are often confused. The pitch is the distance from one thread

Figure 7.40 Worm and worm gear. Reproduced from Mabie and Reinholtz, Mechanisms and Dynamics of Machinery, 4th edition, John Wiley & Sons, © 1987

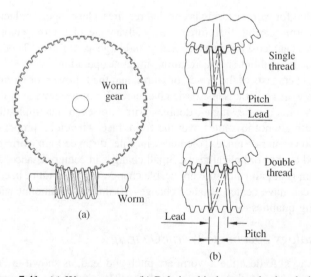

Figure 7.41 (a) Worm gearset. (b) Relationship between lead and pitch

to the next thread if measured parallel to the axis of the screw. Lead is the distance advanced in a single revolution of the worm. The lead of a worm is found by multiplying the pitch by the number of threads (single thread, double thread, etc.). A single-threaded worm will have a lead equal to the pitch, whereas the lead is twice the pitch for a double-threaded worm and thrice the pitch for a triple-threaded worm.

7.9.3 Speed Ratios of Worm Gearing

Speed ratios for worm gearing (typically ranging from $10:1$ to $200:1$) are high compared to spur or helical gears. In fact, speed ratio ranges for worm gears roughly begin where ranges for other types of gears end. Velocity ratio of a worm and worm gear is found by dividing the number of teeth on the gear by the number of threads on the worm. As an example, if the worm is single-threaded, the speed ratio of the gearset is 1 over the number of teeth on the worm gear.

7.9.4 Efficiency of Worm Gearing

Although efficiency for spur gearing is very high and essentially constant over a wide range of speeds, this is not the case for worm gearing. In worm gearing, the efficiency varies nonlinearly with lead angle (Figure 7.42), with the highest value of efficiency occurring with a lead angle of 45°. Lead angle needs to be high (customarily ranging from 25° to 45°) for worm gearing applications to avoid unreasonable losses:

$$e = \frac{\tan \lambda (1 - \mu \tan \lambda)}{\mu + \tan \lambda} \tag{7.45}$$

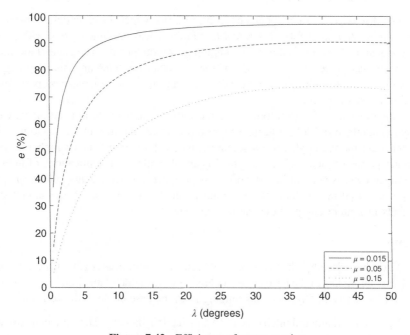

Figure 7.42 Efficiency of worm gearing

Figure 7.42 shows plots of efficiency (e) versus lead angle (λ) for three different values of coefficient of friction. Efficiency does not drastically change for lead angles above 30°, especially for low values of friction.

7.9.5 Self-locking Condition

Unlike spur and helical gears, which can be driven from either shaft, worm gears have an interesting property. The worm can easily turn the worm gear, but the gear may not easily turn the worm. A worm gearset is called self-locking when the worm gear cannot drive the worm, and a pair in which the worm gear can drive the worm is called overhauling. Self-locking exists when the friction angle exceeds the lead angle, which causes the friction between worm and worm gear to hold the worm in place. For a thread angle just small enough to become self-locking, the efficiency is given as

$$e = \frac{1}{2}\left(1 - \tan^2 \lambda\right) \tag{7.46}$$

The maximum efficiency will exist when λ equals zero, giving a maximum efficiency of self-locking worm gears of 0.5.

The condition used to define the condition of self-locking is based on the coefficient of friction, and therefore the coefficient should be assumed low to be conservative. The boundary between gear pairs being self-locking or overhauling is very hard to determine, due to the fact that the coefficient of friction is very sensitive to lubrication. Self-locking can be a very desirable property in certain situations, such as hoists, because it is necessary to have the load hold its position when power is shut off.

7.10   Review and Summary

This chapter discussed the fundamental concepts of gears, which are essential for discussion of gear trains presented in Chapter 8. Critical terminology was presented along with properties of involute gear teeth, which is the most common gear tooth profile. The general mechanics of tooth contact was presented. The study of tooth contact provides an understanding of the limitations of gears and helps to ensure the design of a gearset that will mesh properly and offer a smooth transfer of motion.

The chapter focused largely on spur gears, due to their simplicity. Spur gears are widely used and easy to manufacture. Helical gears were also presented and offer more versatility. Though helical gears are more complex to manufacture, they can operate at much higher speeds and have higher power capacity. Crossed helical gears also allow for applications of nonparallel shaft arrangements. A general introduction was provided for bevel gears and worm gearing. Both types are used in applications with nonparallel shafts. The use of nonparallel shafting often results in a more compact design and save space.

Problems

P7.1   A gear has 18 teeth, a diametral pitch of 8, and a pressure angle of 20°. Calculate the pitch diameter, radius of the base circle, and radius of the addendum circle. Plot points of the involute curve at five locations ranging from the base circle to the addendum circle.

P7.2   A gear has a diametral pitch of 8 and a 4 inch pitch diameter. Determine the (a) number of teeth, (b) addendum, and (c) outside diameter of the gear.

P7.3   A gear has an outside diameter of 4.8 inches and has 46 teeth. Determine the diametral pitch.

Answer: $P = 10$

P7.4   A gear has 32 teeth, a diametral pitch of 4, and a 14.5° pressure angle. Calculate the pitch diameter, radius of the base circle, and radius of the addendum circle. Plot points of the involute curve at five locations ranging from the base circle to the addendum circle.

P7.5   Two gears with a module of 2.31 mm have a center distance of 4 inches and a velocity ratio of 4.5 : 1. Determine the number of teeth on each gear. (*Note*: Diametral pitch is related to module by $mP = 25.4$).

Answer: $N_p = 16$
$N_g = 72$

P7.6   A pinion and gearset have a diametral pitch of 12 and a pressure angle of 14.5°. The pinion has 16 teeth and the gear has 30 teeth. Determine the length of contact and the contact ratio.

P7.7   Two gears, each having 16 teeth, are going to be used together. Each has a pressure angle of 20° and a diametral pitch of 10. Determine the contact ratio.

P7.8   A pinion has 18 teeth and meshes with a gear to give a center distance of 4.5 inches. What is the velocity ratio if the diametral pitch of each gear is 12?

Answer: 5 : 1

P7.9   Solve Problem P7.8 for a diametral pitch of 8.

P7.10 A pinion and gear, each with a diametral pitch of 10, are meshed together to give a velocity ratio of 6.5 : 1. The pinion has 16 teeth. What is the center distance?

Answer: 12 inches

P7.11 A two-gear mesh needs to be designed to give a velocity ratio of 7 : 1. The shaft centers are 8 inches apart. Determine the gear parameters for the design. Indicate diametral pitch and the number of teeth on both gears.

P7.12 A pinion spur gear having 14 teeth drives a spur gear to give a speed ratio of 9 : 1. Both have a diametral pitch of 10 and a pressure angle of 20°. The center distance is 7 inches. What is the contact ratio?

P7.13 Select a set of crossed helical gears for a 90° shaft angle, which has a speed ratio of 2 : 1 and a center distance of 3 inches.

P7.14 Select a set of crossed helical gears for a 90° shaft angle, which has a speed ratio of 4 : 1 and a center distance of 6 inches.

P7.15 A machine is currently driven with the use of a belt drive. The small pulley is 3 inches in diameter and the large pulley is 15 inches in diameter. The shaft centers are 10 inches apart. It is desired to replace the belt drive system with a gearset while maintaining the velocity ratio and shaft center distance. Design a suitable gearset.

Further Reading

Buckingham, E. (1949) *Analytical Mechanics of Gears*, McGraw-Hill, New York.

Budynas, R.G. and Nisbett, J.K. (2011) *Shigley's Mechanical Engineering Design*, McGraw-Hill, New York.

Colbourne, J.R. (1987) *The Geometry of Involute Gears*, Springer, New York.

Dudley, D.W. (1962) *Gear Handbook: The Design, Manufacture, and Application of Gears*, McGraw-Hill, New York.

8

Gear Trains

8.1 Introduction

Chapter 7 focused on concepts of the interaction and performance of a single pair of gears. Commonly, gears are used for a reduction in speed. Though this speed reduction can be accomplished with a single pair of gears, the applications would be very limited. For a large reduction in speed, for example, you would need to use one very large gear and one very small gear. The use of extremely different gear sizes is not practical and is not a good use of space. Another complication with using a single pair of gears would be difficulty in obtaining a specific speed reduction ratio (ratio of the output speed to the input speed). In order to get some specific speed reduction ratio, one may need to specify nonstandard gear sizes, which will greatly increase manufacturing costs.

Because of complications such as those mentioned above, speed reduction is often accomplished using a set of gears in a train. A gear train, such as the one shown in Figure 8.1, is a combination or series of several pairs of gears. Gear trains can utilize the different types of gears discussed in Chapter 7 and can have numerous arrangements. Two major classifications of gear trains are ordinary and planetary. Ordinary gear trains, which are further divided into simple and compound, consist of gears that have axes fixed relative to the frame. Planetary gear trains will have at least one gear with an axis that moves relative to the frame.

Gear trains allow for a positive drive (no slipping) mechanism to generate a constant ratio modification of speed. Other types of mechanisms, such as belt drives, can modify speed while allowing for slipping. Though slipping is good for dissipation of shock loading, it is very undesirable when exact speed and position relationships must exist between the input and output shafts. Belt drives will offer speed modification at lower costs in some applications that allow for some amount of slipping. Chain drives can also serve as an alternative to gear trains.

When designing a gear train, there will not be just one possible solution. There is not an equation that will result in the best possible gear arrangement. The process of gear train design is iterative. A satisfactory solution must consider all design requirements as well as limitations. Though this chapter focuses only on the kinematics of gear trains, full design must also consider forces produced in gear trains. The gears must be designed so that the gear teeth can withstand the forces. Because gears function due to contact between mating teeth, gear materials must be chosen so that the gears do not have excessive wear. Designing gears for

Machine Analysis with Computer Applications for Mechanical Engineers, First Edition. James Doane.
© 2016 John Wiley & Sons, Ltd. Published 2016 by John Wiley & Sons, Ltd.
Companion Website: www.wiley.com/go/doane0215

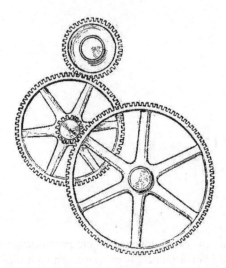

Figure 8.1 Example of a compound gear train. Source: Wikimedia [http://commons.wikimedia.org/ wiki/File:Transmission_of_motion_by_compund_gear_train_(Army_Service_Corps_Training, _Mechanical_Transport,_1911).jpg]

strength and wear considerations is beyond the scope of this textbook. Machine design textbooks can be referred to for more information.

8.2　Simple Gear Trains

In a simple gear train, each gear is mounted on its own shaft. In other words, all gears in a simple gear train are arranged in series. Figure 8.2 shows a general setup for a simple gear train with three gears.

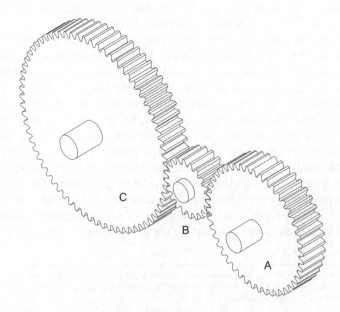

Figure 8.2　General simple gear train

For the simple gear train shown in Figure 8.2, we will define gear A as the driver and gear C as the output. The speed of gear B can be determined using velocity ratio equations developed in Chapter 7. The speed ratio between the first two gears in the train will be inversely proportional to the number of teeth:

$$\frac{\omega_B}{\omega_A} = -\frac{N_A}{N_B} \Rightarrow \omega_B = -\omega_A \frac{N_A}{N_B} \tag{8.1}$$

The negative sign is there because the rotation direction changes, meaning that gear B will rotate in the opposite direction of gear A.

Similarly, the speed of gear C can be determined from the mating of gears B and C:

$$\frac{\omega_C}{\omega_B} = -\frac{N_B}{N_C} \Rightarrow \omega_C = -\omega_B \frac{N_B}{N_C} \tag{8.2}$$

Substituting Equation 8.1 into Equation 8.2 gives

$$\omega_C = \omega_A \left(\frac{N_A}{N_B}\right)\left(\frac{N_B}{N_C}\right)$$

$$\omega_C = \omega_A \frac{N_A}{N_C} \tag{8.3}$$

It can be seen from Equation 8.3 that the output speed ω_C depends only on the input speed ω_A and the number of teeth on the input and output gears. The properties of the middle gear (gear B) drops out of the equation. In any simple gear train, the speed ratio between the input and output depends only on the number of teeth on the input and output gears. The intermediate gears, which are called idlers, affect the rotation direction of the output gear but not the speed. Each idler added to the train will reverse the rotational direction of the output gear. The idlers also change the center distance between the input and output gears. Therefore, if a larger center distance is required, idler gears can be used. The general speed ratio of the input and output gears in a simple gear train will be

$$\frac{\omega_{in}}{\omega_{out}} = \pm\frac{N_{out}}{N_{in}} \tag{8.4}$$

The sign can be determined by close inspection of the system. A sketch of the gear train with arrows showing rotation directions is an easy and direct approach to determine gear rotational directions.

It was discussed in Chapter 7 that a minimum number of teeth on the pinion exists in order to avoid interference problems. Large speed ratios will require a gear that is very large compared to the pinion. Therefore, there are practical limits on how high the speed ratio can be for a simple gear train. Typically, the speed ratio limit for a single set of gears is limited to around 10. If higher speed ratios are required, it is necessary to use compound gear trains.

When designing a simple gear train, the fundamental factor to be considered is speed ratio. Some applications require an exact speed ratio, while others may have large tolerance. Another factor to consider is rotation direction. Direction of rotation changes between two mating gears. Therefore, it must be known whether the input and output rotation directions must be the same or opposite. Typically, it is desired to achieve the required speed ratio with the minimum number of gears possible. Using more gears increases cost and adds to the overall size of the system. It is also important to use standard gear sizes. Though special nonstandard gear sizes may provide an easier solution to the required speed ratio, standard gears are less expensive and easier to replace.

8.3   Compound Gear Trains

8.3.1   Speed Ratio Calculations

Unlike simple gear trains, compound gear trains contain at least one shaft carrying more than one gear. Compound gear trains allow for a more compact design and are commonly used for speed reducers such as that shown in Figure 8.3.

Figure 8.4 shows a general arrangement for a compound gear train consisting of four gears. Notice that gears B and C share a common shaft.

The procedure for compound gear train analysis is similar to that of simple gear trains. Consider the compound gear train shown in Figure 8.4 with gear A acting as the driver and gear D as the output. The rotation speed of gear B can be determined from

$$\omega_B = -\omega_A \frac{N_A}{N_B} \tag{8.5}$$

Figure 8.3   Triple-reduction speed reducer. Reproduced from Mabie and Reinholtz, Mechanisms and Dynamics of Machinery, 4th edition, John Wiley & Sons, © 1987

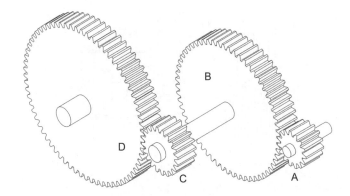

Figure 8.4 General compound gear train

Because gear C is on the same shaft as gear B, the rotation speed of those two gears must be equal: $\omega_C = \omega_B$. The rotation of gear D is determined using the mesh between gears C and D.

$$\omega_D = -\omega_C \frac{N_C}{N_D}$$

$$\omega_D = \omega_A \left(\frac{N_A}{N_B}\right) \left(\frac{N_C}{N_D}\right) \tag{8.6}$$

Now, we can see a very important difference between simple gear trains and compound gear trains. It can be seen in Equation 8.6 that terms for intermediate gears do not cancel out, unlike simple gear trains. The rotation speed of the output depends on the number of teeth of all gears in the system. Therefore, the speed ratios for compound gear trains can be larger than the speed ratios using simple gear trains. Large speed reductions can be obtained with small gears, allowing for more practical and compact designs.

It is important to be able to determine gear ratio for compound gear trains with any number of gears. Careful inspection of Equation 8.6 will show that the terms in the numerator all apply to driving gears and terms in the denominator all apply to followers. Therefore, the total gear ratio can be determined by dividing the product of the number of teeth on drivers by the product of the number of teeth on the followers.

Example Problem 8.1

The hoist shown in Figure 8.5 is driven by a four-stage gear reducer. The input gear A has 12 teeth and rotates at 1800 rpm. Gears B and D have 32 teeth, gear C has 12 teeth, gear E has 16 teeth, gear F has 54 teeth, gear G has 14 teeth, and gear H has 56 teeth. What is the rotation speed of the hoist drum?

Solution: Expanding Equation 8.6 to include all the gearsets will give

$$\omega_H = \omega_A \left(\frac{N_A}{N_B}\right) \left(\frac{N_C}{N_D}\right) \left(\frac{N_E}{N_F}\right) \left(\frac{N_G}{N_H}\right)$$

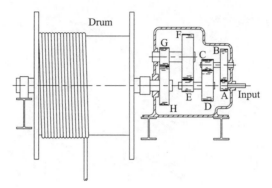

Figure 8.5

Substituting the values gives

$$\omega_H = 1800 \left(\frac{12}{32}\right)\left(\frac{12}{32}\right)\left(\frac{16}{54}\right)\left(\frac{14}{56}\right)$$

Answer: $\omega_H = 18.75$ rpm

8.3.2 *Design of Compound Gear Trains*

Given a particular arrangement of gears and an input rotational speed, it is relatively simple to perform analysis of compound gear trains to determine the output rotational speed. Engineers, however, are often required to perform a design of a compound gear train for a specific performance. Designing a compound gear train for a specific speed ratio is more complicated, especially if an exact speed ratio is required. Due to the versatility of gear trains, there will not be one unique solution for a specific speed ratio. The designer has a lot of freedom in the design process. At first attempt, it may appear that the design process is largely based on trial and error, but the process becomes more streamlined with experience. Keep in mind that the design process outlined below is for the kinematic design only. Although outside the scope of this text, additional design work would be required to ensure safe stress levels for the application.

Let us consider a general example to illustrate the overall thought process for designing a compound gear train. Once a specific speed ratio is determined, the first step is to estimate the number of gears required in the train. To minimize cost, it is desirable to use the minimum number of gears necessary. As a general example, let us say that the required speed ratio is 310 : 1. We will limit the speed ratio for each pair of gears to 10 : 1. From Equation 8.6, we can see that the speed ratio of each pair is multiplied together to get the final speed ratio. Therefore, if we assume that each set of gears will provide the same speed ratio (which is a good starting point), two sets of gears would each need to have a speed ratio of the square root of 310 $\left(\sqrt{310} = 17.60\right)$. Because that is higher than our cutoff value of 10, we will add another set of gears. For three sets, the speed ratio of each will be the cube root of 310 $\left(\sqrt[3]{310} = 6.77\right)$. Based on this, we can conclude that three sets of gears (six total gears) will likely be a good design for the required speed ratio.

A gearset consisting of 14 teeth on the pinion and 95 teeth on the gear would give a speed ratio of $95/14 = 6.786$. Three sets of that gear ratio would yield a total speed ratio of $6.786^3 = 312.5$, which is a little higher than the required ratio. At this point, we would need to determine if that ratio is good enough or do we need to make modifications. Several combinations of gearsets can be analyzed until the optimal design is determined. The process is explained further in the following two examples. Each example approaches the design process a little differently.

Example Problem 8.2

A compound gearset is required to give a speed ratio of exactly $608 : 1$. The output gear must rotate in the same direction as the input gear.

Solution: The first step is to determine the number of gearsets required. The cube root gives an approximate ratio per set that is less than 10, so we will begin with three sets of gears:

$$\sqrt[3]{608} = 8.47$$

If all three sets of gears produced a speed ratio of 8.47, our problem would be complete. If the pinion has 14 teeth, the gear would need $14 \times 18.47 = 118.6$ teeth to produce a ratio of 8.47. This is not possible because the gear must have an integer number of teeth. Other possibilities are for pinions of 15 and 16 teeth and gears of 12.07 and 135.55 teeth, respectively. Therefore, we will use 8.47 only as a guideline for speed ratios.

It is noted that $8 \times 8 \times 9.5 = 608$. So, if two sets of gears provide speed ratios of $8 : 1$ and one gearset provides a speed ratio of $9.5 : 1$, we will get a total speed ratio of $608 : 1$. A gearset of 16 teeth on the pinion and 128 teeth on the gear will give a speed ratio of 8. A gearset of 14 teeth on the pinion and 133 teeth on the gear will give a speed ratio of 9.5. Therefore, that will be a possible solution. A sketch of that solution is shown in Figure 8.6. Notice that an idler must be used to cause the output gear to rotate in the same direction as the input gear.

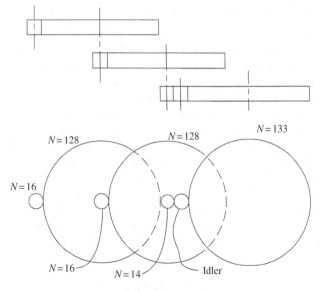

Figure 8.6

Example Problem 8.3

A compound gearset is required to give a speed ratio of exactly $400:1$. The smallest gear must have at least 16 teeth and no gear can be larger than 96 teeth.

Solution: Because the minimum and maximum gear sizes are listed, we can determine the maximum possible reduction from one set of gears. The maximum velocity ratio will be

$$\text{VR}_{\text{max}} = \frac{96}{16} = \frac{6}{1}$$

Two sets of gears would give a velocity ratio of $(6)(6) = 36$, three sets of gears would give (6) $(6)(6) = 216$, and four sets of gears would give a velocity ratio of $(6)(6)(6)(6) = 1296$. Four gearsets will be required because it is the smallest number of gears to give larger than the $400:1$ ratio required. If we use a 6/1 ratio for the first three gearsets, we can determine the required velocity ratio of the fourth set.

$$(\text{VR}_{1-3})(\text{VR}_4) = \text{VR}_{\text{total}}$$

$$(216)(\text{VR}_4) = 400$$

$$\text{VR}_4 = \frac{400}{216} = \frac{50}{27}$$

The total velocity ratio can be expressed by ratios of each gearset.

$$\text{VR} = \left(\frac{96}{16}\right)\left(\frac{96}{16}\right)\left(\frac{96}{16}\right)\left(\frac{50}{27}\right) = \frac{400}{1}$$

8.4   Reverted Compound Gear Trains

A reverted compound gear train is a special case of the compound train arrangement that results in the input shaft and output shaft being coaxial. The fact that the axes of the input and output shafts are in line with each other adds restrictions to the design of a reverted compound gear train. Gear dimensions must be chosen strategically to ensure proper gear ratio and proper alignment of the input and output shafts. A major advantage of reverted gear trains is that they have a compact design. Figure 8.7 shows a reverted gear train using helical gears. The arrangement has the motor flange mounted directly on the gearbox.

Consider the reverted gear train arrangement shown in Figure 8.8 consisting of four gears. Gear A is the input gear and gear D is the output gear. The arrangement is a reverted compound gearset because the shafts for gears A and D are on the same axis. Just like the standard compound gear arrangement, gears B and C share a common shaft.

The overall process of design will be similar to that of a general compound gear train. However, in order for the input and output shaft to be coaxial, the center distance for both pairs of mating gears must be the same. That center distance can be expressed in terms of pitch radii:

$$d_{\text{c}} = R_{\text{A}} + R_{\text{B}} = R_{\text{C}} + R_{\text{D}} \tag{8.7}$$

Figure 8.7 Application of a reverted geartrain

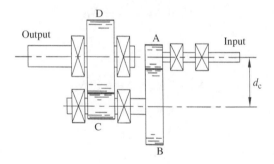

Figure 8.8 Reverted compound gear train

Multiplying by 2 to get in terms of pitch diameter, recalling that pitch diameter is equal to the number of teeth divided by the diametral pitch, gives

$$2d_c = D_A + D_B = D_C + D_D$$

$$2d_c = \frac{N_A}{P_A} + \frac{N_B}{P_B} = \frac{N_C}{P_C} + \frac{N_D}{P_D} \tag{8.8}$$

Assuming all gears have the same diametral pitch $P_A = P_B = P_C = P_D = P$, the equation can be simplified:

$$(2d_c)P = N_A + N_B = N_C + N_D \tag{8.9}$$

Therefore, in a reverted compound gear train, the number of teeth on the input side (gears A and B) must equal the number of teeth on the output side (gears C and D). The design of reverted compound trains is similar to that of compound trains with the additional constraint of Equation 8.9.

Example Problem 8.4

A set of reverted gears similar to those in Figure 8.8 is required for a reduction of 70 : 1. Design the reverted gearset such that the minimum number of teeth on any gear is 14.

Solution: The gear ratio will be divided into two reductions. Ideally each reduction would be $\sqrt{70} = 8.37$. However, it is more convenient to work with whole numbers. Working with two whole numbers that are close in value and give a product of 70, we will divide the ratio into two sets 1 : 7 and 1 : 10 giving a total of 1 : 70.

$$\frac{N_A}{N_B} = \frac{1}{7} \Rightarrow N_B = 7 N_A$$

$$\frac{N_C}{N_D} = \frac{1}{10} \Rightarrow N_D = 10 N_C \tag{1}$$

From Equation 8.9

$$N_A + N_B = N_C + N_D \tag{2}$$

Substituting Equation 1 into Equation 2 gives

$$N_A + 7 N_A = N_C + 10 N_C$$

$$8 N_A = 11 N_C$$

$$\frac{N_A}{N_C} = \frac{11}{8} \tag{3}$$

Equation 3 tells us that gear A has 11 teeth and gear C has eight teeth. The minimum number of teeth from the problem statement is 14. Multiplying the ratio in Equation 3 by 2 gives
   Answer: $N_A = 22$ $N_C = 16$
   Plugging the values obtained into Equation 1,

$$N_B = 7 N_A = 7(22) = 154$$
$$N_D = 10 N_C = 10(16) = 160$$

   Answer: $N_B = 154$ $N_D = 160$

8.5   Gear Trains with Different Types of Gears

Gear trains only utilizing spur gears will have limitations of output shaft position relative to the input shaft. Other gear types can be used to expand the arrangement options. Figure 8.9, for example, shows a drive with the input and output shafts arranged at a 90° angle. The angle is accomplished by combining helical and bevel gears. Speed ratio calculation procedures above focused on spur gears, but the methods apply to other gear types with the exception of worm gears. The following example will illustrate speed reducers utilizing different types of gears.

Figure 8.9 Geared motor unit arranged for a right-angle drive

Example Problem 8.5

The gear train shown in Figure 8.10 consists of a right angle helical gearset (gears A and B) and a bevel gearset (gears C and D). The input gear A has 30 teeth, gear B has 42 teeth, gear C has 36 teeth, and the output gear D has 50 teeth. What is the rotational speed of gear D if gear A rotates at 350 rpm?

Solution: The rotation speed of gear B is determined based on the ratio of teeth on gears A and B.

$$\omega_B = \omega_A \frac{N_A}{N_B} = 350\left(\frac{30}{42}\right) = 250 \text{ rpm}$$

Gears B and C are on a common shaft, so gear C will have the same rotation speed as gear B. The speed of gear D is determined from the ratio of teeth on gears C and D.

$$\omega_D = \omega_C \frac{N_C}{N_D} = 250\left(\frac{36}{50}\right)$$

Answer: $\omega_D = 180$ rpm

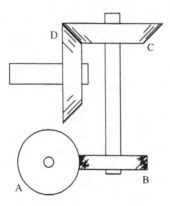

Figure 8.10

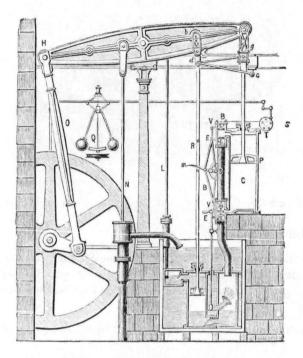

Figure 8.11 Watt's steam engine

8.6 Planetary Gear Trains

8.6.1 Investigation of a Historical Application

As an introduction to the benefits of planetary gear trains, let us briefly examine a historical application. Figure 8.11 shows a drawing of Watt's steam engine, which is the oldest surviving engine that converted steam power to rotary motion. The engine was installed in a brewery in 1785 to replace a team of horses. A museum in Sydney now has the engine and provides demonstrations using actual steam as the power source. Videos of the engine in motion can be found on the Internet to aid in your understanding of its functionality.

As mentioned in Chapter 1, Watt did not invent the steam engine but made improvements over previous designs. On the right side of the figure is a cylinder that contains a piston. Steam moves the piston, which then moves the rocker arm at the top. The opposite end of the rocker arm moves the connecting rod that turns a set of gears called the sun and planet gears. The gears, with the aid of a flywheel to maintain momentum, turn the drive wheel to drive equipment.

Though Watt's steam engine contained several ingenious inventions, our focus will be on the sun and planet gears. Figure 8.12 shows a general arrangement similar to the sun and planet gears used on Watt's steam engine. The sun gear is attached to the shaft of the drive wheel. The planet gear, which gets its name because it rotates around the sun gear like a planet, is rigidly fixed to the connecting rod that leads to the rocker arm. The sun and planet gears are of the same size and are held in contact by a swinging link, which is not keyed to the shaft allowing it to rotate freely.

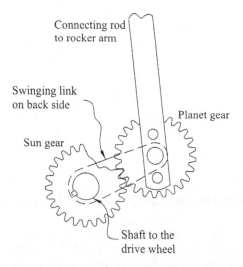

Connecting rod
to rocker arm

Swinging link
on back side

Sun gear

Planet gear

Shaft to the
drive wheel

Figure 8.12   Planetary gearset similar to that used in Watt's steam engine

At first glance, the sun and planet gears may seem unnecessarily complex. Why not just use a solid link and eliminate the gears? Though it may not be obvious, using this simple planetary gearset offers a useful advantage. The planet gear has two motions that are cumulative. As the planet gear moves one complete turn around the sun, it also makes one complete revolution relative to the sun. Because of this, the sun gear will complete two revolutions every time the planet gear makes one complete rotation around the sun gear. Watt's creative sun and planet gearset shown in Figure 8.12 is a planetary gearset in its most basic form. The sections to follow will demonstrate how to analyze the basic planetary gearset and build into more complex planetary gearsets.

8.6.2 Basic Planetary Gear Train

In a typical pair of gears, as shown in Figure 8.13a, the two gears move and a stationary ground link (or arm) connects the two gear centers (this forms a three-link mechanism). We can take an inversion of that three-link mechanism and hold one of the gears stationary. The arm that connects the two gears is now a moving link. The inversion, which is shown in Figure 8.13b, shows the most basic form of a planetary gearset where a planet gear rotates around the fixed sun gear.

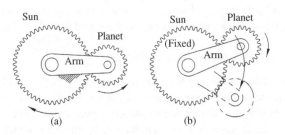

Sun   Planet   Sun   Planet

Planet   (Fixed)

Arm   Arm

(a)   (b)

Figure 8.13   (a) Standard gearset with fixed arm. (b) Basic planetary gear train with fixed sun gear

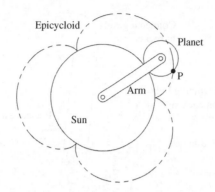

Figure 8.14   Epicycloid path of a point on a planet gear

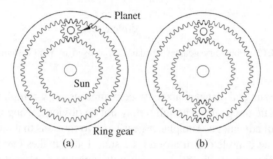

(a)   (b)

Figure 8.15   (a) Addition of an internal ring gear. (b) Additional planet

The path of a point on the planet gear can be traced as the planet revolves around the stationary sun. Figure 8.14 shows a sun that has a radius three times that of the planet. The planet gear rolls without slipping around the outer circumference of the sun gear. As the planet gear revolves around the sun gear, point P will trace the path shown as the phantom line. The path is known as an epicycloid. Therefore, planetary gear trains are also known as epicyclic gear trains.

Once the basic planetary gearset is fully understood, we can develop more complex planetary gearsets. One common addition is an internal ring gear, as shown in Figure 8.15a. It is also common to have additional planets, as shown in Figure 8.15b.

Figure 8.16 shows an example of a planetary gear train containing a ring gear and multiple planet gears.

There are many advantages of using a planetary gear train. One is that you can obtain large speed ratios with very few gears, which gives them weight and space advantages. Planetary gear trains offer in-line arrangements of gears that are compact.

8.6.3   Speed Ratio of Planetary Gear Trains

Now that we have an understanding of the basic structure and function of planetary gearsets, we can look at procedures to determine speed ratios. Compared with other types of gear trains, it is often more difficult to visualize the motion of planetary gear trains. This difficulty is primarily due to the fact that planet gears are rotating relative to the arm, which is also in motion. Therefore, speed ratio calculations require using velocity difference equations.

Figure 8.16 Planetary gearset

There are numerous methods for analyzing planetary gear trains, and a couple will be presented in this section. The first approach is a mathematical approach based on the idea of equal arcs. Consider the planetary gear train shown in Figure 8.17a consisting of a sun, arm, planet, and fixed ring gear. To determine the motion of the system, we will rotate the sun gear

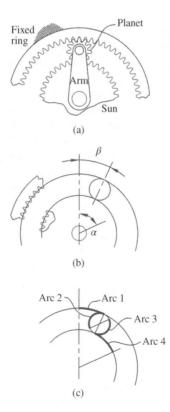

Figure 8.17 Equal arc approach. (a) Planetary gear train with fixed ring gear. (b) Gears in position when sun is rotated an angle α. (c) Equal arcs

an arbitrary angle α, as shown in Figure 8.17b. This rotation will result in the arm rotating some angle β. The angular rotation of the arm will be determined using the equal arc approach. The planet gear will rotate about its own central axis as it rolls without slipping along both the ring and sun gears. Rolling along the ring gear will result in arc 1 and arc 2 shown in Figure 8.17c being of the same length. If the ring gear has a radius r_r, the planet gear has a radius r_p, and the planet gear rotates an angle ϕ about its own axis, then the equal arc lengths give the following:

$$r_r\beta = r_p\phi \tag{8.10}$$

Similarly, arc 3 shown in Figure 8.17c will be of the same length as arc 4. If the sun gear has a radius r_s, then the equal arc equation gives

$$r_s(\alpha - \beta) = r_p\phi \tag{8.11}$$

From Equations 8.10 and 8.11

$$r_r\beta = r_s(\alpha - \beta)$$

$$\beta = \left(\frac{r_s}{r_r + r_s}\right)\alpha \tag{8.12}$$

Equation 8.12 will now provide the rotation angle α of the arm for a given rotation β of the sun gear. This equation is only applicable to the case presented in Figure 8.17 and is not a general relationship for all planetary gearsets. The speed ratio will either be β/α or α/β depending on which element is the input.

Next, we will discuss the tabular approach to solving planetary gear trains. The tabular method is more general than the method of equal arcs. The general tabular analysis of planetary gear trains can be broken into two steps, which will be the same regardless of the gear arrangement. A nice advantage of the tabular method is that the two steps are the same no matter which gear is the driver and which gear is the follower. To illustrate the two-step procedure, let us start with a basic planetary gearset, as shown in Figure 8.18. For this particular set the sun gear is fixed, the arm A is the input (driver), and the planet gear is the output

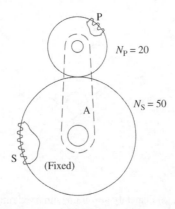

Figure 8.18 Illustrative example for the tabular approach for planetary gear trains

Table 8.1   Speed ratio solution

	Arm (driver)	Sun (fixed)	Planet (follower)
Step 1	+1	+1	+1
Step 2	0	−1	$\dfrac{50}{20} = +2.5$
Total	+1	0	+3.5

(follower). It will be seen that this two-step method will work regardless of which gear is the driver and which is the follower.

Step 1 in our procedure is to treat the entire gear train as if it were one solid unit and rotate the complete system one revolution clockwise (clockwise rotation will be treated as the positive direction). This rotation will cause every member to rotate one revolution in the positive direction, which is tabulated in Table 8.1. Step 2 involves holding the arm stationary and rotating the fixed gear one revolution in the counterclockwise (negative) direction to return it to its original fixed position. Step 2 will result in no rotation of the driver and −1 rotation of the fixed gear, as shown in Table 8.1. Because the arm is now held stationary, the number of rotations of the follower can now be determined using simple gear ratios. For the set shown in Figure 8.18, the planet will rotate 50/20 = 2.5 times in the clockwise direction for one rotation of the sun in the counterclockwise direction. All rotations for step 2 are tabulated in Table 8.1. The last step is to add the rotations from both steps as shown in Table 8.1. The final ratio is now determined from the totals. For one rotation of the driver arm, the planet follower will rotate 3.5 times in the same direction.

The relatively simple process illustrated above can be used for more complex arrangements of planetary gearsets. The following examples will demonstrate the process.

Example Problem 8.6

The planetary gearset shown in Figure 8.19 has a fixed sun gear with 70 teeth. The planet gear has 35 teeth and the follower has 20 teeth. Determine the gear ratio.

Solution: Table 8.2 shows the final results. Step 1 is the result of all gears rotating as a unit clockwise one rotation. The fixed sun gear is rotated one full revolution counterclockwise while the arm is fixed in step 2. That rotation would cause the planet with 35 teeth to rotate two revolutions clockwise. That planet serves as an idler. The follower will rotate 2.5 rotations counterclockwise. If the arm is fixed, the gears act as a simple gear train. Therefore, the idler does not affect the rotation speed of the follower.

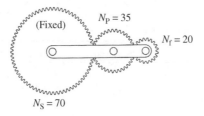

Figure 8.19

Table 8.2   Speed ratio for Example Problem 8.6

	Arm (driver)	Sun (fixed)	Planet idler	Follower
Step 1	+1	+1	+1	+1
Step 2	0	−1	$\dfrac{70}{35} = +2$	$-\dfrac{70}{20} = -3.5$
Total	+1	0	+3	−2.5

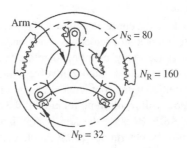

Figure 8.20

Example Problem 8.7

The planetary gearset shown in Figure 8.20 has a fixed ring gear (internal gear) with 160 teeth. The three planet gears, each having 32 teeth, are connected by the arm. Determine the rotational speed and direction of the sun (80 teeth) if the arm rotates clockwise at 200 rpm.

***Solution*:**   The solution process is the same as previously described and illustrated in Table 8.3.

    The planet gears act as idlers, so the calculation of the planet gear ratio is unnecessary. For one complete positive revolution of the driver arm, the sun gear will turn three revolutions in the same direction.

$$\omega_s = 3\omega_A = 3(200 \text{ rpm})$$

Answer: $\omega_s = 600 \text{ rpm}$ (cw)

Table 8.3   Speed ratio for Example Problem 8.7

	Arm (driver)	Ring gear (fixed)	Planet gears	Sun gear (follower)
Step 1	+1	+1	+1	+1
Step 2	0	−1	$-\dfrac{160}{32} = -5$	$+\dfrac{160}{80} = +2$
Total	+1	0	−4	+3

8.7 Differentials

Discussion of gear trains in this chapter focused primarily on the idea of speed reduction. Though that is a very common application of gear trains, many other applications also exist. This section will briefly discuss one application of differentials. Differentials developed due to a major problem in automotive design when you have two wheels that are on a common shaft. In straight line motion, both wheels turn at the same speed. In a turn, however, the outside wheel must cover more distance than the inside wheel requiring it to turn faster. The nondriven wheels can be allowed to spin independently, so this speed difference is not an issue. The driven wheels, however, are linked together, so a solution to this problem is necessary. One easy solution would be to separate the driven wheels and only drive one wheel. Though this solves the problem, one wheel drive is not very practical. A much better solution is a clever gearbox (called a differential) that will drive each wheel, on two separate axles, with equal speed until it enters a curve. In a curve, the gearbox needs to slow down one axle and speed up the other.

To develop an understanding of how a differential functions, let us consider the very basic rack and pinion system shown in Figure 8.21. Consider the rack and pinion set shown in Figure 8.21a. If the gear is moved to the right, but not allowed to rotate, both racks will move at the same speed with the gear. The gear acts as a rigid connection between the two racks. This is similar to the function of the differential in straight line motion transferring torque equally to each axle.

A very different motion occurs if one rack is held fixed, such as that shown in Figure 8.21b. Now the gear rotates as it moves to the right causing the bottom rack to move at twice the rate. This arrangement demonstrates the general function of a differential in a curve. A differential, such as the one shown in Figure 8.22, is the same concept but in a different arrangement.

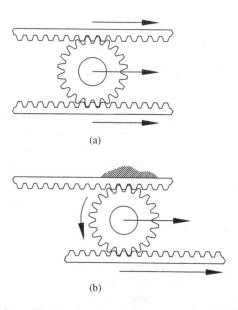

(a)

(b)

Figure 8.21 Rack and pinion to illustrate function of differential (a) in straight motion and (b) in a curve

Figure 8.22   Automotive differential. Reproduced from Mabie and Reinholtz, Mechanisms and Dynamics of Machinery, 4th edition, John Wiley & Sons, © 1987

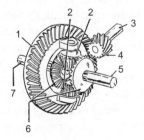

Figure 8.23   Differential gearset

Figure 8.23 shows a general schematic of a differential gearset. Shaft 3 is the drive shaft and shafts 5 and 7 are the axles. Gear 4 is the drive gear, and it is driving gear 1.

8.8   Computer Methods for Gear Train Design

It is probably clear now that gear train design can be a tedious trial and error process. The process is not typically complex, but can require a lot of time and effort to develop a system for optimal performance. Computer software can aid in the design process, especially when optimization is required. Many software packages could be used, but many of the problems can be solved with the use of spreadsheets. Planetary gear train analysis, as an example, involves simple tables that could easily be modeled in a spreadsheet allowing for "what if" style analysis or optimization.

8.9   Review and Summary

Gear trains are extremely versatile and can serve many functions. It would be impossible to explore all possible gear train arrangements in this chapter. Therefore, several arrangements of gear trains were discussed in this chapter in order to build a foundational understanding of gear train analysis and design. Several books have been devoted to topics associated with gears and gear trains. Interested readers may refer to Section "Further Reading" for examples.

As a general design note, gear trains used for high-power transmission can benefit from the use of hardened teeth. It is possible to reduce the space required for a gear train by half, simply by changing from low hardness steel to hardened teeth. Surface durability is improved that leads to less pitting, scoring, and other forms of abrasive wear. Though there are many benefits of using hardened teeth, it does add cost for heat treating and machining.

Problems

P8.1 The simple gear train shown in Figure 8.24 consists of five gears arranged as shown. The driving gear rotates clockwise at 50 rpm. (a) What is the rotation speed and direction of gear C? (b) What is the rotation speed and direction of the output gear? (c) Would the output gear motion be the same if gears A and B were removed and gear C meshed directly with the input gear?

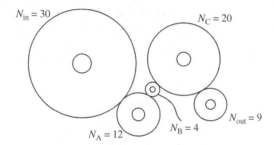

Figure 8.24

P8.2 Gear A in Figure 8.24 rotates at a speed of 80 rpm and the output gear rotates clockwise. What is the rotation direction of gear A? What is rotation speed of the input gear?

P8.3 A simple gear train is required for a velocity ratio of 8 : 1. The output gear must rotate in the same direction as the input gear. All gears have a diametral pitch of 2 and must have a minimum of 12 teeth each. Design the gear train if the center distance between the input and output must be 36 in. Provide a sketch of the layout.

P8.4 Solve Problem P8.3 if the diametral pitch is 4, the velocity ratio is 6 : 1, and the center distance is 18 in.

P8.5 The winch shown in Figure 8.25 has an 8″ diameter drum attached to gear A, which has 24 teeth. Gear A turns gear B, which has 52 teeth. The 5″ diameter drum attached to gear B lifts a load. A rope wrapped around the upper drum is pulled with a velocity v_1. Determine the ratio of v_2 to v_1.

P8.6 The gear train shown in Figure 8.26 consists of six gears in a compound train. The number of teeth on each gear is given in parenthesis. Input gear A rotates at a speed of 1400 rpm. Determine the rotation speed of output gear F.

P8.7 The compound gearset shown in Figure 8.27 consists of three sets of gears. Gear A rotates at 2400 rpm and gear F rotates at 20 rpm. Determine the number of teeth on gear D.

Answer: $N_D = 64$

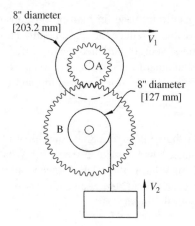

Figure 8.25

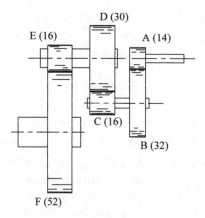

Figure 8.26

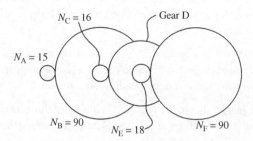

Figure 8.27

P8.8   The compound gear train shown in Figure 8.28 consists of three external gears (A–C) and one internal gear (D). Gears B and C are on a common shaft. Gear A drives the system by rotating at 1400 rpm clockwise. Determine the rotation speed and direction of gear D.

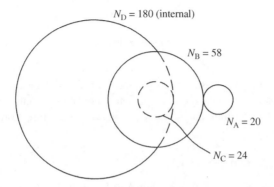

$N_D = 180$ (internal)

$N_B = 58$

$N_A = 20$

$N_C = 24$

Figure 8.28

P8.9   A reverted compound gear train similar to that shown in Figure 8.8 has 20 teeth on gear A and 70 teeth on gear B, both having a diametral pitch of 10. Gear C has 14 teeth and gear D has 40 teeth. What is the diametral pitch of gears C and D? What is the velocity ratio of the total gear train?

Answer: $\begin{aligned} P_C &= P_D = 6 \\ VR &= 10:1 \end{aligned}$

P8.10   A reverted compound gear train similar to that shown in Figure 8.8 has 20 teeth on gear A and 60 teeth on gear B, both having a diametral pitch of 10. Gear C has 16 teeth. How many teeth will gear D have if (a) gears C and D have a diametral pitch of 8 and (b) gears C and D have a diametral pitch of 6?

P8.11   The planetary gearset shown in Figure 8.29 has a sun gear with 40 teeth, an idler planet gear with 20 teeth, and the follower gear with 30 teeth. The arm rotates at a constant speed of 100 rpm clockwise. Determine the rotation speed and direction of the follower.

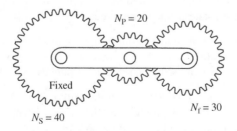

$N_P = 20$

Fixed

$N_S = 40$

$N_f = 30$

Figure 8.29

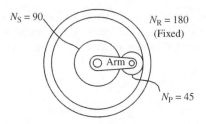

$N_S = 90$ $N_R = 180$ (Fixed)

Arm

$N_P = 45$

Figure 8.30

P8.12 In the planetary gearset shown in Figure 8.30, the sun gear (90 teeth) is the input rotating at 1800 rpm clockwise. The planetary gear has 45 teeth, and the fixed ring gear has 180 teeth. Determine the output rotational speed and direction of the arm using the tabular method.

Answer: $\omega_{\mathrm{arm}} = 600\,\mathrm{rpm}$ (cw)

P8.13 The compound planetary gearset shown in Figure 8.31 has a fixed sun gear with 40 teeth. The outer ring gear has 90 teeth and is rotating 3.4 rad/s clockwise. Each intermediate gear A has 20 teeth, and each planet gear B has 30 teeth. Determine the rotational velocity and direction of the arm.

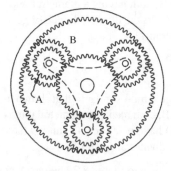

Figure 8.31

P8.14 Design a simple gear train using diametral pitch of 12 resulting in a gear ratio of 18 : 1. Sketch the gear train indicating number of teeth and pitch diameter. The input and output gears must rotate in the same direction.

P8.15 Design a compound gear train resulting in a speed ratio of 20 : 1. The smallest gear available has 14 teeth and the largest available has 90 teeth. Sketch the gear train indicating the number of teeth. The diametral pitches available are 6, 8, and 10.

P8.16 A compound gear speed reducer has an input of 1500 rpm and a required output of 35 rpm. No gear in the train should have less than 14 teeth or more than 120 teeth. Design the gear train and indicate the number of teeth on each gear. All gears must have the same diametral pitch.

P8.17   Design a compound gear train resulting in a speed ratio of 900 : 1. The smallest gear available has 15 teeth and the largest available gear has 90 teeth. Sketch the gear train indicating the number of teeth.

P8.18   A compound gear speed reducer has an input of 800 rpm and a required output of 40 rpm. No gear in the train should have less than 16 teeth or more than 80 teeth. Design the gear train and indicate the number of teeth on each gear.

P8.19   A set of reverted gears must be designed when the input speed is 1800 rpm. Design the reverted gearset such that the minimum number of teeth on any gear is 18 if the output speed must be between 102 and 103 rpm.

P8.20   A reverted gear train must be designed to give a velocity ratio of 30 : 1 using only gears with a diametral pitch of 12. The minimum number of teeth is 14 and the maximum is 100.

Further Reading

Buckingham, E. (1949) *Analytical Mechanics of Gears*, McGraw-Hill, New York.
Budynas, R.G. and Nisbett, J.K. (2011) *Shigley's Mechanical Engineering Design*, McGraw-Hill, New York.
Dudley, D.W. (1962) *Gear Handbook: The Design, Manufacture, and Application of Gears*, McGraw-Hill, New York.

9

Cams

9.1 Introduction

This chapter will focus on a different type of mechanism. It will present concepts for analysis and design of cam follower mechanisms. A cam is a mechanism that causes a specified nonuniform motion of a follower by direct contact. In most common applications, the cam is mounted to a shaft and moves in pure rotation. Other designs, however, have a stationary cam and the follower moves across the cam. Cams are used in numerous types of machines, but are especially common in automatic machines. Washing machines, for example, use cams to automatically activate the different cycles. Cams are critical components in internal combustion engines where they actuate the inlet and exhaust valves. Other examples of cam applications would include exercise equipment, sewing machines, toys, and material-handling equipment.

One major advantage of cam mechanisms is their versatility. A basic rotational input of the cam can generate various forms of rather complex motion of the follower (a cam follower mechanism is essentially a very versatile function generator). Another advantage is that the design procedure for cams is straightforward compared with linkages, although a poor understanding of proper dynamic design can lead to an unacceptable cam follower design. While there are obvious advantages of cams compared with linkages, there are also disadvantages. A cam follower mechanism is typically more expensive and more difficult to manufacture. Contact stresses are typically high, causing component wear and increased maintenance. Cams are typically milled and ground to correct surface error or distortion. In some cases, the cam will be polished after the grinding process to get a smoother finish. All of these finishing processes add cost to the overall mechanism.

The cam shape is determined by the desired motion of the follower. Therefore, the design process actually begins with the follower and its motion. The follower motion can be a basic rise–fall motion, oscillating motion, or numerous forms of complex motion requiring specific timing. Dwells, or periods of time where the follower is stationary while input cam motion continues, are also very common and straightforward in cam mechanisms. As demonstrated in Chapter 5, dwells can be very difficult to obtain in other types of mechanisms such as linkages.

Early design of cams focused only on the motion and not acceleration, which is acceptable if the cam operates at low speeds. It will be seen throughout this chapter that acceleration

---

Machine Analysis with Computer Applications for Mechanical Engineers, First Edition. James Doane.
© 2016 John Wiley & Sons, Ltd. Published 2016 by John Wiley & Sons, Ltd.
Companion Website: www.wiley.com/go/doane0215

becomes a very important parameter in cam design, especially in high-speed machinery. Compared with linkage mechanisms, however, control of acceleration is typically easier in cam mechanisms. This chapter will cover different categories of common mathematical functions used in cam design, and many of them are developed primarily based on desired properties of acceleration.

9.2 Types of Cams and Followers

9.2.1 Common Cam Configurations

Cams can be designed in many different physical forms, as introduced in Chapter 1. The two major categories of cams are radial cams and axial cams. In radial cams, the follower moves in a radial direction, and in axial cams the follower moves in a direction parallel to the axis of cam rotation. Perhaps the most common radial cam is the plate cam. A plate cam is a flat disk with the desired cam shape (determined from required follower motion) developed along its circumference. Another common cam is a drum cam, otherwise known as a cylindrical cam. A drum cam is a cylinder with a track, which provides the path for the follower, machined around its circumference.

Cams can also be used for indexing, which is simply converting constant rotation of the input to intermittent motion of the output. Geneva mechanisms were previously discussed in Chapter 4 as one possible method of indexing. One thing to notice about Geneva mechanisms would be that the input and output shafts are parallel. Indexing can be achieved for a perpendicular shaft arrangement by using a barrel cam such as the one shown in Figure 9.1.

9.2.2 Follower Types

The first way to classify follower types would be based on the surface contact or configurations for the follower. Some common examples are illustrated in Figure 9.2. A knife edge follower, as shown in Figure 9.2a, is one in which the follower tapers to a point. Knife edge followers are easier to analyze, but are rarely used because the contact between the follower and the cam is a point that is not very practical due to high wear. A roller follower uses a roller, or circular disk,

Figure 9.1 Cylindrical cam indexing drive. Reproduced from Tuttle, Mechanisms for Engineering Design, John Wiley & Sons, © 1967

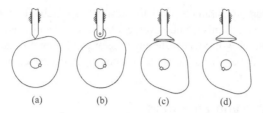

Figure 9.2 Follower configurations. (a) Knife edge. (b) Roller. (c) Flat-faced. (d) Mushroom

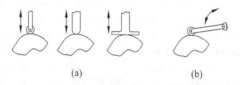

Figure 9.3 Follower motion. (a) Translating follower examples. (b) Rotating follower example

which rotates about its center as the contact between the follower and the cam. Roller followers, as shown in Figure 9.2b, are typically preferred in applications with higher speed because of reduced wear and heat. Flat-faced followers, as shown in Figure 9.2c, have a flat plane as the contact surface between the follower and the cam. Though flat followers can be used with steep cams, misalignment issues can cause very high surface stresses. To compensate for misalignment issues, the flat follower can have a slight modification to create a spherical surface or mushroom follower, as shown in Figure 9.2d.

Regardless of the cam type chosen, the cam shape will be determined by the required motion of the follower. The motion of the follower can be translation or rotation. Examples of translating followers are shown in Figure 9.3a, and an example of a rotating (or swinging) follower is shown in Figure 9.3b. Translating and rotating followers both are commonly used, and the choice is determined by the required output motion.

The follower must be constrained in some fashion so that the follower always remains in contact with the cam. Contact between the cam and the follower can be maintained by different methods. Figure 9.4a shows a roller follower that rides in a groove in the cam. The follower is constrained by the geometry of the cam, and this method of joint closure is known as a form-closed cam joint.

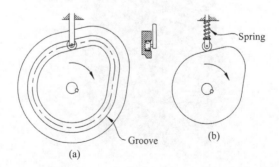

Figure 9.4 (a) Form closed. (b) Force closed

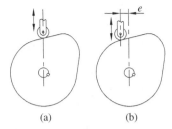

Figure 9.5   (a) In-line follower arrangement. (b) Offset follower arrangement

Figure 9.4b shows a roller follower in contact with the outside profile of the cam. The contact is maintained with the use of an external force generated by the preloaded compression spring. The use of external force to maintain joint closure is known as a force-closed cam joint.

Figure 9.5 illustrates the difference between two follower arrangements in relation to the cam center axis. In Figure 9.5a, the line of follower motion passes through the axis of cam rotation. Such an arrangement is known as an in-line follower arrangement. Figure 9.5b shows the follower in an offset arrangement. The line of follower action does not pass through the axis of rotation, but is instead offset by a distance e.

9.3   Basic Concepts of Cam Geometry and Cam Profiles

9.3.1   Follower Displacement

Regardless of the cam and follower types chosen, the starting point of the cam design process is determining the required motion of the follower. The required motion must be converted to a specific mathematical expression, which is then used to develop the geometry of the cam. The position of the follower at a given time within the cycle measured from a reference starting position (zero position of the follower) is known as follower displacement. One cycle of follower displacement can be plotted on a displacement diagram, which shows displacement on a rectangular coordinate plot. This displacement diagram gives a quick visual representation of the follower displacement. Displacement is generally plotted along with velocity, acceleration, and jerk to get a complete picture of cam performance.

Figure 9.6 shows two very common classifications of follower displacement. The double-dwell displacement diagram shown in Figure 9.6a has four phases: The follower will rise into a high dwell, remain at the high dwell for a defined period of time, fall to the original position, and then remain in a low dwell. All four steps occur in 360° of cam rotation. Figure 9.6b shows a single-dwell displacement diagram consisting of three steps: The follower will rise to a specific location, immediately fall back to the starting position, and then remain in a low dwell for a defined period. Again, the complete cycle occurs over 360° of cam rotation.

9.3.2   SVAJ Diagrams

The most straightforward way to examine cam motion is through the use of *SVAJ* diagrams. Consider a plate cam profile that has been "unwrapped" and shown linearly. The plot of this flattened cam profile would represent the follower displacement function s. After taking the first three time derivatives, we could plot velocity v, acceleration a, and jerk j along with displacement. This plot, usually plotted on a common axis representing either time or cam

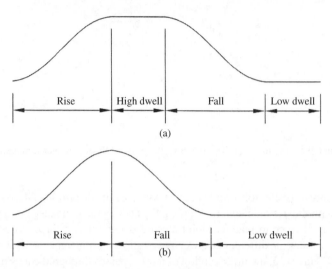

(a)

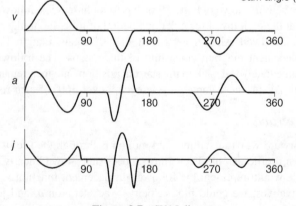

(b)

Figure 9.6 Common classifications of follower displacement. (a) Double dwell. (b) Single dwell

angle, is known as the *SVAJ* diagram. An example of a general *SVAJ* diagram is shown in Figure 9.7.

Concepts of velocity and acceleration have been defined previously. Jerk, on the other hand, is potentially a new term and needs some explanation. Jerk, which is also referred to as pulse, is the time rate of change of acceleration.

$$j(t) = \frac{da}{dt} = \frac{d^3s}{dt^3} \qquad (9.1)$$

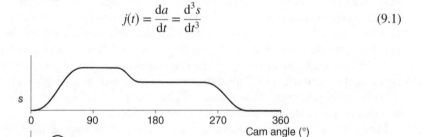

Figure 9.7 *SVAJ* diagram

We know from Newton's law that inertial force is proportional to acceleration. These inertial forces are balanced by elastic deformation in the mechanism. Jerk measures the rate at which elastic deformation occurs. Therefore, energy available for vibration will increase as jerk increases. Jerk magnitude should be kept small, particularly in high-speed applications, to minimize vibration.

9.3.3 General Rules of Cam Design

A common mistake is to consider the displacement function to be the primary concern when designing a cam mechanism. A poor choice of displacement function, however, can lead to serious problems caused by infinite accelerations or discontinuities in the acceleration function. It is more important to focus on the acceleration curve to ensure that it is continuous throughout the cam motion.

Good cam designs should have finite jerk over the entire 360°. A continuous acceleration function ensures that the jerk function will be finite, but not necessarily continuous. Discontinuities in the acceleration, however, will cause infinite spikes in the jerk function. The functions introduced in Section 9.4 will focus on functions that are continuous through first and second derivatives.

9.4 Common Cam Functions

9.4.1 Introduction to Cam Functions

Follower displacement is given by a mathematical expression $s = f(\theta)$, where θ is the angle of rotation of the cam and s is the cam displacement. Cam rotation speed ω is most commonly constant, giving the relationship $\theta = \omega t$. Requirements of the cam function are not typically defined for every position but rather defined at certain key locations.

One of the most common applications of cams is to rise or fall from one constant location to another. The cam functions derived in this section will all be in a very convenient normalized form. The functions will be in the form $y(\beta)$ and will increase from $y = 0$ to $y = 1$ over a region of $0 \leq \beta \leq 1$. These normalized cam functions can then be easily transformed into any required changes in cam displacement over any required range of cam rotation. The transformation process will be explained in Section 9.5. It will be seen that the transformation process can also be applied to the derivatives to give velocity, acceleration, and jerk functions.

Though numerous displacement functions exist, the two major categories for cam displacement functions discussed in this chapter are trigonometric functions and simple polynomial functions (more advanced function types are discussed in Chapter 13). Trigonometric functions have the appealing feature that they are continuously differentiable. Three major examples of trigonometric cam functions would be simple harmonic functions, cycloidal functions, and double harmonic functions.

Polynomial functions are very useful in cam design. One important note is that polynomial functions decrease in order with every differentiation. Therefore, the original order of the function must be selected at a high-enough order to ensure proper acceleration and jerk. Polynomial cam displacement functions have the basic form

$$y(\beta) = k\beta^n \tag{9.2}$$

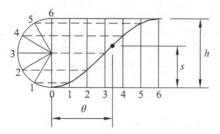

Figure 9.8   Graphical development of simple harmonic motion

where k is a constant and n is an integer value. Constant velocity occurs when $n = 1$, constant acceleration exists for $n = 2$, and constant jerk occurs when $n = 3$.

9.4.2 Simple Harmonic Function

Let us start by examining the simple harmonic function. This function is a good starting point because it is relatively simple to layout and analyze. It will later be seen that the simple harmonic function is not ideal, but it will give moderate performance at slower speeds. The general geometric construction of a simple harmonic function can be examined in Figure 9.8. A half circle of a diameter equal to the total rise (h) of the harmonic function is drawn and divided into equal parts. A horizontal line is drawn of the desired length (required cam rotation) and divided into the same number of equal parts. Projection lines can then be drawn from the half circle to generate the simple harmonic function. The path will give the function of follower displacement (s) versus cam rotation (θ).

A normalized form of the simple harmonic displacement function is given by

$$y(\beta) = \frac{1}{2}[1 - \cos(\pi\beta)] \tag{9.3}$$

We are interested in the plot of the function for $0 \leq \beta \leq 1$, which is shown in Figure 9.9a. The extreme values would be $y = 0$ when $\beta = 0$ and $y = 1$ when $\beta = 1$.

The first three time derivatives of Equation 9.3 are given in Equation 9.4:

$$\frac{dy}{d\beta} = \frac{\pi}{2}\sin(\pi\beta)$$

$$\frac{d^2y}{d\beta^2} = \frac{\pi^2}{2}\cos(\pi\beta) \tag{9.4}$$

$$\frac{d^3y}{d\beta^3} = -\frac{\pi^3}{2}\sin(\pi\beta)$$

We are again interested in the plots for $0 \leq \beta \leq 1$. The plot of the first derivative, as shown in Figure 9.9b, equals zero at the extremes and reaches a maximum value of $\pi/2$ at $\beta = 0.5$. The second derivative reaches a maximum magnitude at the extreme ends, which is shown in Figure 9.9c. Substituting $\beta = 0$ into the second derivative function gives the maximum positive

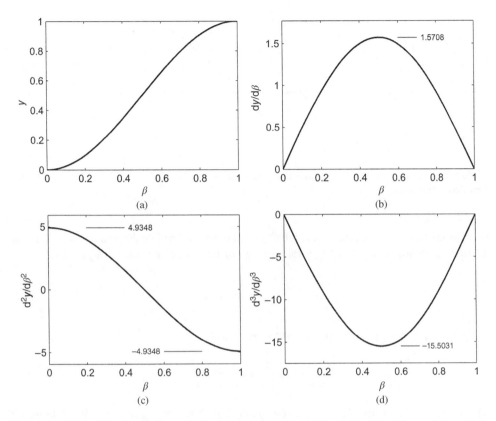

Figure 9.9   Plots of (a) simple harmonic displacement function, (b) first derivative, (c) second derivative, and (d) third derivative

value of $\pi^2/2$. The maximum negative value of $-\pi^2/2$ is found by substituting $\beta = 1$ into the second derivative function. Figure 9.9d shows the plot of the third derivative function, which obtains the maximum value of $-\pi^3/2$ at $\beta = 0.5$ and is equal to zero at the extreme end values.

  An obvious advantage of harmonic functions would be that the derivative always results in another harmonic function (with a 90° phase shift). A less obvious disadvantage comes from the fact that the second derivative is nonzero at the boundaries. The nonzero boundaries will often cause the overall acceleration function to be discontinuous. A primary example would be dwell mechanisms, which will be discussed later in this chapter.

9.4.3 The Cycloidal Function

An improvement to the simple harmonic function would be to set the acceleration boundary conditions to zero. A common example is a cycloidal function. The cycloidal function is developed to eliminate the nonzero boundaries of the second derivative, which exists for the simple harmonic function. Therefore, the derivation of the cycloidal function requires two integrations of the desired second derivative function. The general shape of the second derivative, which is illustrated in Figure 9.10a, needs to be a full cycle sine wave (requires

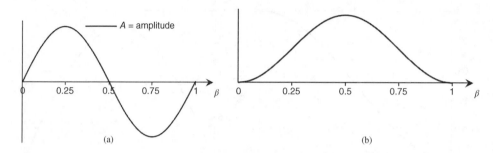

Figure 9.10 Derivation of the cycloidal function. (a) Desired second derivative function. (b) Desired first derivative function

a multiple of 2π). The amplitude A is left as an unknown and will be determined later from the desired amplitude of $y(\beta)$. The general equation of the second derivative is given as

$$\frac{d^2y}{d\beta^2} = A\sin(2\pi\beta) \tag{9.5}$$

Integration gives the general first derivative expression:

$$\frac{dy}{d\beta} = -\frac{A}{2\pi}\cos(2\pi\beta) + C_1 \tag{9.6}$$

The desired first derivative function is shown in Figure 9.10b. Integration constant is determined from boundary conditions:

$$\left(\frac{dy}{d\beta}\right)_{\beta=0} = 0 \Rightarrow C_1 = \frac{A}{2\pi} \tag{9.7}$$

Substituting the constant into Equation 9.6 gives the final form of the first derivative:

$$\frac{dy}{d\beta} = -\frac{A}{2\pi}\cos(2\pi\beta) + \frac{A}{2\pi} \tag{9.8}$$

Second integration gives

$$y = -\frac{A}{4\pi^2}\sin(2\pi\beta) + \frac{A}{2\pi}\beta + C_2 \tag{9.9}$$

The integration constant C_2 and acceleration amplitude A can now be determined from boundary conditions:

$$y_{\beta=0} = 0 \Rightarrow C_2 = 0$$
$$y_{\beta=1} = 1 \Rightarrow A = 2\pi \tag{9.10}$$

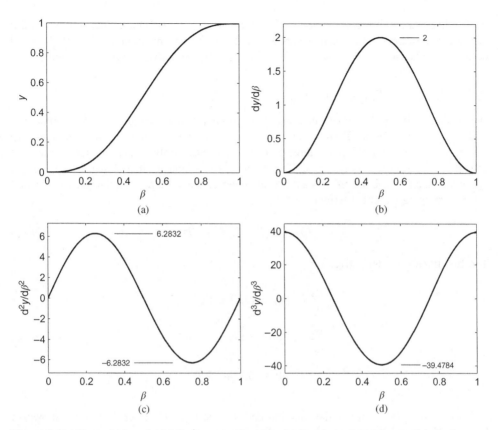

Figure 9.11 Plots of (a) cycloidal displacement function, (b) first derivative, (c) second derivative, and (d) third derivative

The final expression for the cycloidal displacement is

$$y(\beta) = \beta - \frac{1}{2\pi}\sin(2\pi\beta) \tag{9.11}$$

The plot of the displacement can be seen in Figure 9.11a. The first three time derivatives are

$$\frac{dy}{d\beta} = 1 - \cos(2\pi\beta)$$

$$\frac{d^2y}{d\beta^2} = 2\pi\sin(2\pi\beta) \tag{9.12}$$

$$\frac{d^3y}{d\beta^3} = 4\pi^2\cos(2\pi\beta)$$

The first derivative becomes maximum when $\cos(2\pi\beta) = -1$. For that to occur, $2\pi\beta = \pi$, which gives the maximum value of 2 located at $\beta = 0.5$. The second derivative will be maximum when $\sin(2\pi\beta) = 1$, which will occur when $2\pi\beta = \pi/2$. Therefore, the maximum second derivative value will be 2π and it will occur at $\beta = 0.25$. Similarly, for the maximum location of the third derivative, $\cos(2\pi\beta) = 1$, which occurs when $2\pi\beta = 0$. The maximum value of the third derivative will be $4\pi^2$ at a location $\beta = 0$. All plots for the derivatives are shown in Figure 9.11.

9.4.4 The 3-4-5 Polynomial Function

We now move into polynomial functions. A polynomial's order will decrease by a value of 1 with every differentiation. Therefore, it is important to start with a higher order polynomial to ensure proper behavior of the first three derivatives. The first polynomial function that we will consider is a fifth-order polynomial. The derivation is similar to that shown for the simple harmonic function, and the result is a function known as the 3-4-5 polynomial function (named for its remaining exponent terms).

$$y(\beta) = 10\beta^3 - 15\beta^4 + 6\beta^5 \tag{9.13}$$

The first three time derivatives are

$$\frac{dy}{d\beta} = 30\beta^2 - 60\beta^3 + 30\beta^4$$

$$\frac{d^2y}{d\beta^2} = 60\beta - 180\beta^2 + 120\beta^3 \tag{9.14}$$

$$\frac{d^3y}{d\beta^3} = 60 - 360\beta + 360\beta^2$$

Plots for the 3-4-5 polynomial function and its first three time derivatives are shown in Figure 9.12. It can be seen that the acceleration is not only continuous but it also equals zero at both ends. The jerk function, however, is nonzero on the boundaries. Therefore, the jerk function would not be continuous across the full 360° in dwell mechanisms.

9.4.5 The 4-5-6-7 Polynomial Function

We can achieve a continuous jerk function that is zero on both boundaries by increasing the order of the polynomial to seventh order. The result, shown in Equation 9.15, is known as a 4-5-6-7 polynomial function (named for its remaining exponent terms):

$$y(\beta) = 35\beta^4 - 84\beta^5 + 70\beta^6 - 20\beta^7 \tag{9.15}$$

The first three time derivatives are

$$\frac{dy}{d\beta} = 140\beta^3 - 420\beta^4 + 420\beta^5 - 140\beta^6$$

$$\frac{d^2y}{d\beta^2} = 420\beta^2 - 1680\beta^3 + 2100\beta^4 - 840\beta^5 \tag{9.16}$$

$$\frac{d^3y}{d\beta^3} = 840\beta - 5040\beta^2 + 8400\beta^3 - 4200\beta^4$$

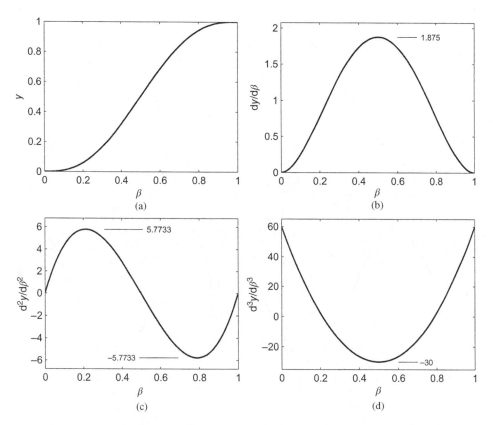

Figure 9.12 Plots of (a) 3-4-5 polynomial displacement function, (b) first derivative, (c) second derivative, and (d) third derivative

The 4-5-6-7 polynomial function, along with the first three derivatives, can be seen in Figure 9.13. Maximum values are shown on the individual plots. The 4-5-6-7 polynomial will lead to higher peak acceleration and jerk compared with the 3-4-5 polynomial.

9.4.6 The Double Harmonic Function

The functions described thus far are common functions for double-dwell mechanisms. The double harmonic function, however, is common in single-dwell mechanisms that perform a rise–fall–dwell motion. The overall benefits of using this function for single-dwell mechanisms will be discussed further in Section 9.7. The double harmonic function is developed as a combination of two harmonic functions:

$$y(\beta) = \frac{3}{8} - \frac{1}{2}\cos(\pi\beta) + \frac{1}{8}\cos(2\pi\beta) \tag{9.17}$$

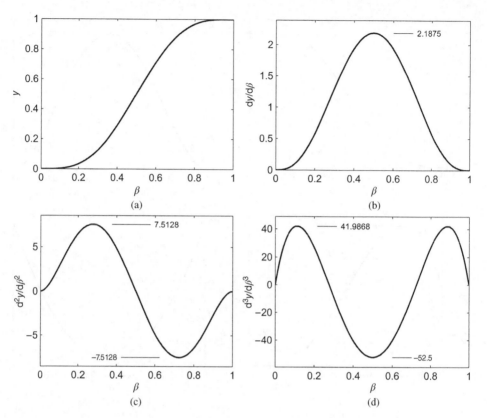

Figure 9.13 Plots of (a) 4-5-6-7 polynomial displacement function, (b) first derivative, (c) second derivative, and (d) third derivative

It can be seen that the second harmonic function has one-fourth the amplitude and twice the frequency of the first harmonic function. The first three time derivatives give

$$\frac{dy}{d\beta} = \frac{\pi}{2}\sin(\pi\beta) - \frac{\pi}{4}\sin(2\pi\beta)$$

$$\frac{d^2y}{d\beta^2} = \frac{\pi^2}{2}\cos(\pi\beta) - \frac{\pi^2}{2}\cos(2\pi\beta) \qquad (9.18)$$

$$\frac{d^3y}{d\beta^3} = -\frac{\pi^3}{2}\sin(\pi\beta) + \pi^3\sin(2\pi\beta)$$

Plots for the double harmonic functions and the derivatives are shown in Figure 9.14.

9.4.7 Comparison of Cam Functions

The functions defined in previous sections are only some of the possible functions used in cam design. Each function has its own unique advantages and disadvantages. The simple

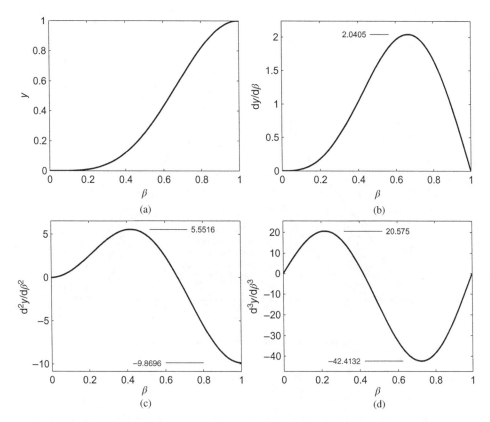

Figure 9.14 Plots of (a) double harmonic displacement function, (b) first derivative, (c) second derivative, and (d) third derivative

harmonic function, for example, has low peak acceleration values but will have infinite jerk values at dwell locations. The cycloidal function offers lower vibration and noise. Zero acceleration at the ends also allows cycloidal functions to be used for dwells. The 4-5-6-7 polynomial function also provides smooth operation and low vibration but has high peak acceleration amplitudes.

Velocity considerations are important because the stored kinetic energy will be proportional to the square of the velocity. Figure 9.15 shows a comparison of the first derivative plots.

High acceleration will lead to high dynamic forces. Therefore, it is common to pick functions based on peak acceleration values. Figure 9.16 shows a comparison of second derivative plots for the functions discussed in this chapter. It can be seen that the 4-5-6-7 polynomial function has a higher peak acceleration value compared with the other function types. Double harmonic functions have small acceleration amplitude at the beginning of the cycle, which nearly eliminates shock and vibration at the start of the rise function.

Figure 9.17 shows a comparison plot of third derivative magnitudes for the different function types discussed in this chapter.

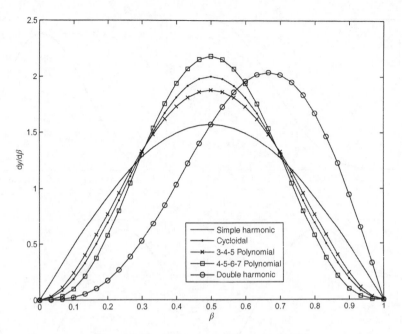

Figure 9.15 Comparison of plots for $dy/d\beta$

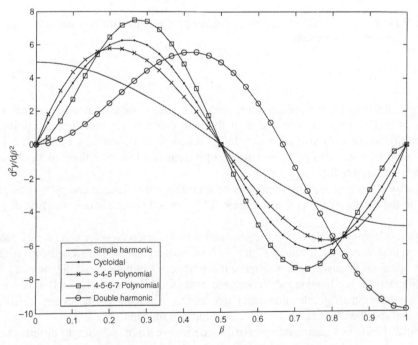

Figure 9.16 Comparison of plots for $d^2y/d\beta^2$

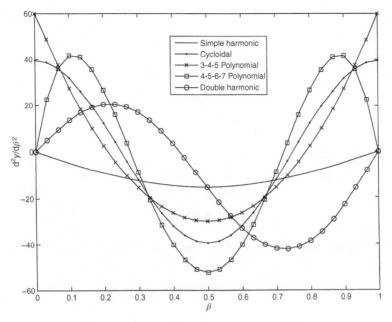

Figure 9.17   Comparison of plots for $\mathrm{d}^3 y/\mathrm{d}\beta^3$

9.5   Using Cam Functions for Specific Applications

In order to use any of the cam functions in the previous section, developed in the form $y = f(\beta)$, they need to be modified to give the form $s = f(\theta)$. This modification is accomplished by making the following substitutions:

$$\beta = \frac{\theta - \theta_1}{\theta_2 - \theta_1}$$

$$y = \frac{s - s_1}{s_2 - s_1}$$

(9.19)

It is important to realize that the above values are dimensionless, which in trigonometric equations represent radian values. The actual process is best represented with examples.

Example Problem 9.1

A cam function is required to produce a rise from 0 to 2.5 inches for a cam rotation from 0 to 80° using a simple harmonic function. Determine and plot the displacement function.

Solution: The simple harmonic function was given in Equation 9.3:

$$y(\beta) = \frac{1}{2}[1 - \cos(\pi\beta)]$$

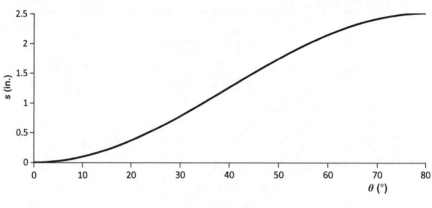

Figure 9.18

Making the substitutions from Equation 9.19 gives the displacement function:

$$\frac{s - s_1}{s_2 - s_1} = \frac{1}{2}\left[1 - \cos\left(\pi\frac{\theta - \theta_1}{\theta_2 - \theta_1}\right)\right]$$

The problem statement gives values of

$$s_1 = 0 \quad s_2 = 2.5$$
$$\theta_1 = 0 \quad \theta_2 = 80$$

Substituting the given values into the displacement function,

$$s = \frac{2.5}{2}\left[1 - \cos\left(\pi\frac{\theta}{80}\right)\right]$$

The cosine terms are in radians. Converting to degrees gives the final displacement function.

Answer: $s = 1.25\left[1 - \cos\left(\dfrac{180\theta}{80}\right)\right]$

Figure 9.18 shows the displacement function.

Example Problem 9.2

A cam function is required to produce a fall from 4.25 to 1 inches for a cam rotation from 20 to 110° using a 3-4-5 polynomial function. Determine and plot the displacement function.

Solution: The 3-4-5 polynomial function was given in Equation 9.13:

$$y(\beta) = 10\beta^3 - 15\beta^4 + 6\beta^5$$

Making the substitutions from Equation 9.19 gives the displacement function:

$$\frac{s - s_1}{s_2 - s_1} = 10\left(\frac{\theta - \theta_1}{\theta_2 - \theta_1}\right)^3 - 15\left(\frac{\theta - \theta_1}{\theta_2 - \theta_1}\right)^4 + 6\left(\frac{\theta - \theta_1}{\theta_2 - \theta_1}\right)^5$$

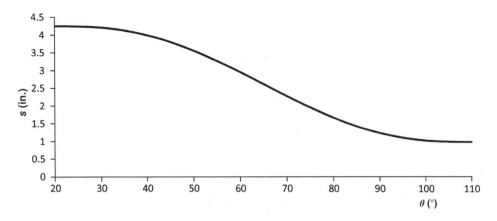

Figure 9.19

The problem statement gives values of

$$s_1 = 4.25 \quad s_2 = 1$$
$$\theta_1 = 20 \quad \theta_2 = 110$$

Substituting the given values into the displacement function,

$$\frac{s - 4.25}{1 - 4.25} = 10\left(\frac{\theta - 20}{110 - 20}\right)^3 - 15\left(\frac{\theta - 20}{110 - 20}\right)^4 + 6\left(\frac{\theta - 20}{110 - 20}\right)^5$$

The following is the final displacement function:

Answer: $s = -32.5\left(\dfrac{\theta - 20}{90}\right)^3 + 48.75\left(\dfrac{\theta - 20}{90}\right)^4 - 19.5\left(\dfrac{\theta - 20}{90}\right)^5 + 4.25$

Figure 9.19 shows the plot of the displacement function.

Time derivatives are required in order to calculate velocity, acceleration, and jerk. Modification to the previously developed time derivatives gives the following:

$$\frac{ds}{d\theta} = \left(\frac{s_2 - s_1}{\theta_2 - \theta_1}\right)\frac{dy}{d\beta}$$

$$\frac{d^2s}{d\theta^2} = \left(\frac{s_2 - s_1}{(\theta_2 - \theta_1)^2}\right)\frac{d^2y}{d\beta^2} \qquad (9.20)$$

$$\frac{d^3s}{d\theta^3} = \left(\frac{s_2 - s_1}{(\theta_2 - \theta_1)^3}\right)\frac{d^3y}{d\beta^3}$$

Again, use of the equations is best illustrated with examples.

Example Problem 9.3

Determine and plot the velocity and acceleration functions for the cam displacement in Example Problem 9.1.

Solution: The first derivative of the simple harmonic function was given in Equation 9.4:

$$\frac{dy}{d\beta} = \frac{\pi}{2}\sin(\pi\beta)$$

The velocity is determined from Equation 9.20:

$$\frac{ds}{d\theta} = \left(\frac{s_2 - s_1}{\theta_2 - \theta_1}\right)\frac{dy}{d\beta}$$

Combining the two above equations and making the substitutions from Equation 9.19 gives the velocity function:

$$\frac{ds}{d\theta} = \left(\frac{s_2 - s_1}{\theta_2 - \theta_1}\right)\left(\frac{\pi}{2}\right)\sin\left(\pi\frac{\theta - \theta_1}{\theta_2 - \theta_1}\right)$$

The problem statement from Example Problem 9.1 gives values of

$$s_1 = 0 \quad s_2 = 2.5$$
$$\theta_1 = 0 \quad \theta_2 = 80$$

Substituting the given values into the velocity function and converting to degrees gives

$$v = \frac{ds}{d\theta} = \left(\frac{2.5}{80}\right)\left(\frac{\pi}{2}\right)\sin\left(\frac{180\theta}{80}\right)$$

The following is the final velocity function:

Answer: $v = 0.0491\sin\left(\dfrac{180\theta}{80}\right)$

Figure 9.20 shows the plot of the velocity.

The second derivative of the simple harmonic function was given in Equation 9.4:

$$\frac{d^2y}{d\beta^2} = \frac{\pi^2}{2}\cos(\pi\beta)$$

and the acceleration is determined from Equation 9.20:

$$\frac{d^2s}{d\theta^2} = \left(\frac{s_2 - s_1}{(\theta_2 - \theta_1)^2}\right)\frac{d^2y}{d\beta^2}$$

Combining the two above equations and making the substitutions from Equation 9.19 gives the acceleration function:

$$\frac{d^2s}{d\theta^2} = \left(\frac{s_2 - s_1}{(\theta_2 - \theta_1)^2}\right)\left(\frac{\pi^2}{2}\right)\cos\left(\pi\frac{\theta - \theta_1}{\theta_2 - \theta_1}\right)$$

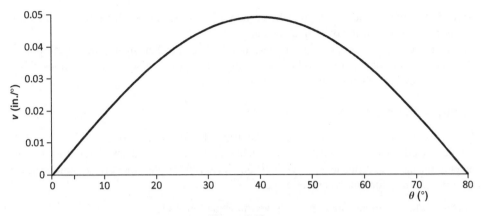

Figure 9.20

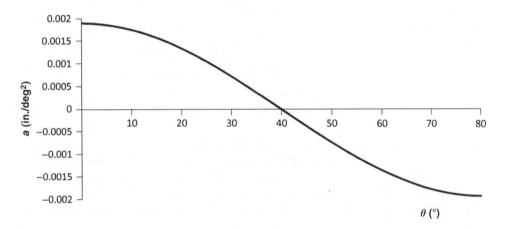

Figure 9.21

Substituting the given values

$$a = \frac{d^2s}{d\theta^2} = \left(\frac{2.5}{(80)^2}\right)\left(\frac{\pi^2}{2}\right)\cos\left(\frac{180\theta}{80}\right)$$

The following is the final acceleration function:

Answer: $a = 0.0019 \cos\left(\frac{180\theta}{80}\right)$

Figure 9.21 shows the acceleration plot.

9.6 Application of Cam Functions for Double-Dwell Mechanisms

Figure 9.6a showed a typical arrangement of a double-dwell cam mechanism. It can be seen that the function is a piecewise function over the entire 360° interval. The sections are rise–dwell–fall–dwell. The follower position remains stationary during the dwells causing the velocity, acceleration, and jerk to be zero. As previously stated, the acceleration for a simple harmonic function is not continuous when used with dwells. The acceleration is a half period cosine function, which is nonzero at both ends.

Developing *SVAJ* diagrams for a complete double-dwell mechanism requires developing the cam functions for both the rise and fall following the procedures discussed in Sections 9.4 and 9.5. The following example illustrates the complete process.

Example Problem 9.4

A cam is required for a double-dwell mechanism. The follower must rise 0.5 inches in 1.5 seconds, dwell for 0.75 seconds, fall 0.5 inches in 2 seconds, and dwell for 0.75 seconds. Develop the cam functions using cycloidal functions. Plot displacement (as functions of cam angle θ and time) for one complete cycle.

Solution: The cycloidal function was given in Equation 9.11:

$$y(\beta) = \beta - \frac{1}{2\pi}\sin(2\pi\beta)$$

Using Equation 9.19,

$$\frac{s - s_1}{s_2 - s_1} = \frac{\theta - \theta_1}{\theta_2 - \theta_1} - \frac{1}{2\pi}\sin\left(2\pi\frac{\theta - \theta_1}{\theta_2 - \theta_1}\right)$$

$$s = (s_2 - s_1)\left[\frac{\theta - \theta_1}{\theta_2 - \theta_1} - \frac{1}{2\pi}\sin\left(2\pi\frac{\theta - \theta_1}{\theta_2 - \theta_1}\right)\right] + s_1 \tag{1}$$

This equation will be used to develop both the rise and fall portions of the total cycle. The data provided were in terms of time not angle. The total time for one complete cycle is the sum of the time required for the rise, fall, and both dwells giving a total time of 5 seconds per cycle.

We can now consider the rise portion of the cycle. Knowing that it takes 5 seconds to complete 360°, the rise in 1.5 seconds will be completed in 108°. Therefore, the parameters for the rise function will be

$$s_1 = 0 \quad s_2 = 0.5$$
$$\theta_1 = 0 \quad \theta_2 = 108$$

Substitution into Equation 1 gives

$$s = (0.5 - 0)\left[\frac{\theta - 0}{108 - 0} - \frac{1}{2\pi}\sin\left(2\pi\frac{\theta - 0}{108 - 0}\right)\right] + 0$$

$$s = \frac{\theta}{216} - \frac{1}{4\pi}\sin\left(2\pi\frac{\theta}{108}\right) \tag{2}$$

The displacement function will maintain a value of 0.5 inches through the dwell for a period of 0.75 seconds. Therefore, the fall function will begin at 2.25 seconds, which occurs at an angle of 162°. The fall lasts for 2 seconds, which means the fall ends at an angle of 306. The parameters for the fall function are

$$s_1 = 0.5 \quad s_2 = 0$$
$$\theta_1 = 162 \quad \theta_2 = 306$$

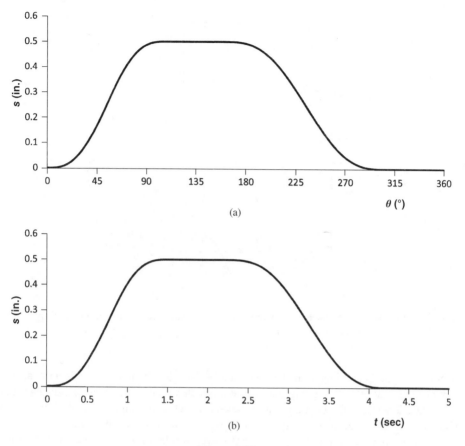

Figure 9.22

Substituting the parameters into Equation 1 gives

$$s = (0 - 0.5)\left[\frac{\theta - 162}{306 - 162} - \frac{1}{2\pi}\sin\left(2\pi\frac{\theta - 162}{306 - 162}\right)\right] + 0.5$$

$$s = -\frac{\theta - 162}{288} + \frac{1}{4\pi}\sin\left(2\pi\frac{\theta - 162}{144}\right) + 0.5 \tag{3}$$

The remainder of the cycle will be at a location of 0 for the low dwell. The entire cycle is plotted in Figure 9.22a versus angle and Figure 9.22b versus time.

9.7 Application of Cam Functions for Single-Dwell Mechanisms

Another common classification of motion is a single dwell. Figure 9.23a shows a general displacement diagram for a rise–fall displacement where the rise time equals to the fall time. The resulting acceleration functions are shown for simple harmonic displacement and cycloidal displacements in Figure 9.23b and c. The simple harmonic displacement results in nonzero

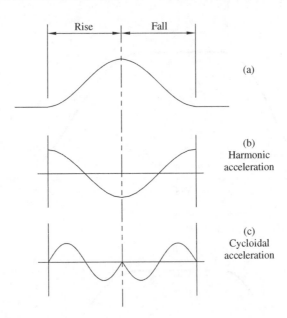

Figure 9.23   (a) Rise–fall displacement diagram. (b) Acceleration for simple harmonic displacement. (c) Acceleration for cycloidal displacement

acceleration at the beginning of the rise and end of the fall. The nonzero ends will cause an abrupt change in acceleration going into a dwell. Acceleration for the cycloidal displacement has zero acceleration at the beginning and the end, which is an improvement. However, the double point at the peak displacement location will result in increased vibration.

One way to improve performance of a single-dwell mechanism is to use a combination of cycloidal and simple harmonic displacement functions. Figure 9.24a shows the same general rise–fall displacement diagram utilizing the combination, and Figure 9.24b shows the resulting acceleration. Notice now that using the cycloidal allows the acceleration to be zero at the terminal locations, thereby allowing continuous acceleration in and out of the dwells. Utilizing the simple harmonic function in the middle eliminates the double point at the peak displacement. In addition, the resulting velocity will also be continuous through the complete cycle.

To better understand the combined cycloidal harmonic function, let us examine the rise portion. The velocity function must be continuous throughout the rise function. From Section 9.4, it was determined that the peak velocity value for a simple harmonic is 1.5708 and the peak velocity for a cycloidal function is 2. Therefore, the following ratio must exist for continuous velocity:

$$\frac{2}{1.5708}\frac{h_1}{h_2} = \frac{\phi_1}{\phi_2} \tag{9.21}$$

If we define $\phi_2 = k\phi_1$ and $h_1 + h_2 = h$,

$$\frac{2}{1.5708}\left(\frac{h_1}{h - h_1}\right) = \frac{\phi_1}{k\phi_1}$$

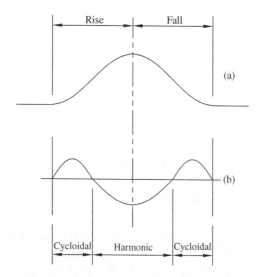

Figure 9.24 (a) Rise–fall displacement diagram. (b) Acceleration for combined cycloidal and harmonic displacement

$$2h_1k = 1.5708(h - h_1)$$

$$(2k + 1.5708)h_1 = 1.5708h$$

$$h_1 = \left(\frac{\pi}{4k + \pi}\right)h \tag{9.22}$$

Then, from $h_1 + h_2 = h$

$$h_2 = \left(\frac{4k}{4k + \pi}\right)h \tag{9.23}$$

It is important to note that the displacement diagram in Figure 9.24a is symmetrical, meaning that it has the rise and fall occurring over equal time periods. Unequal rise and fall times, or nonsymmetrical displacement, will result in a discontinuity in the acceleration at the location of peak displacement. It is important to design the system to attempt to minimize that discontinuity.

Example Problem 9.5

Develop a cam displacement function for a symmetric single-dwell mechanism with a rise of 2 inches occurring in 144°, a fall occurring in 144°, and a dwell for 72°. The cam rotates at 80 rpm. Use a combination cycloidal and simple harmonic displacement function (60° cycloidal and 84° harmonic). Plot velocity and acceleration.

Solution:

$$\phi_2 = k\phi_1$$

$$84° = k(60°) \Rightarrow k = 1.4$$

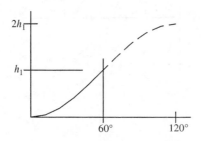

Figure 9.25

The rise values for each section from Equation 9.23 are

$$h_1 = \left(\frac{\pi}{4k + \pi}\right)h = \left(\frac{\pi}{4(1.4) + \pi}\right)(2) = 0.7188 \text{ in.}$$

$$h_2 = \left(\frac{4k}{4k + \pi}\right)h = \left(\frac{4(1.4)}{4(1.4) + \pi}\right)(2) = 1.2812 \text{ in.}$$

First, we will develop the cycloidal function that will be used for the first 60° of the rise:

$$y(\beta) = \beta - \frac{1}{2\pi}\sin(2\pi\beta)$$

$$\frac{s - s_1}{s_2 - s_1} = \frac{\theta - \theta_1}{\theta_2 - \theta_1} - \frac{1}{2\pi}\sin\left(2\pi\frac{\theta - \theta_1}{\theta_2 - \theta_1}\right) \qquad (1)$$

Figure 9.25 shows the desired plot. Only the first half of the cycloidal function will be used to reach a height of h_1 in 60°. The conditions for the entire plot are given in Equation 2:

$$\begin{aligned} s_1 &= 0 \quad s_2 = 2h_1 \\ \theta_1 &= 0 \quad \theta_2 = 2\phi_1 \end{aligned} \qquad (2)$$

Substituting Equation 2 into Equation 1 gives

$$\frac{s}{2h_1} = \frac{\theta}{2\phi_1} - \frac{1}{2\pi}\sin\left(2\pi\frac{\theta}{2\phi_1}\right)$$

$$s = 0.7188\left(\frac{\theta}{60} - \frac{1}{\pi}\sin\left(2\pi\frac{\theta}{120}\right)\right)$$

From 60° to 144°, the displacement function will use a simple harmonic:

$$y(\beta) = \frac{1}{2}[1 - \cos(\pi\beta)]$$

$$\frac{s - s_1}{s_2 - s_1} = \frac{1}{2}\left[1 - \cos\left(\pi\frac{\theta - \theta_1}{\theta_2 - \theta_1}\right)\right] \qquad (3)$$

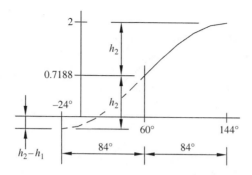

Figure 9.26

Figure 9.26 shows the desired plot. The upper half of the plot starts from 0.7188 in 84° of cam rotation. The dashed line shows the lower portion of the curve in order to develop the following conditions:

$$s_1 = -(h_2 - h_1) \quad s_2 = 2$$
$$\theta_1 = -24 \quad \theta_2 = 144 \tag{4}$$

Substituting Equation 4 into Equation 3 gives

$$\frac{s + 0.5624}{2 + 0.5624} = \frac{1}{2}\left[1 - \cos\left(\pi\frac{\theta + 24}{144 + 24}\right)\right]$$

$$s = 1.2812\left[1 - \cos\left(\pi\frac{\theta + 24}{168}\right)\right] - 0.5624$$

The total displacement function will be

$$Answer:\ s = \begin{cases} 0.7188\left(\dfrac{\theta}{60} - \dfrac{1}{\pi}\sin\left(2\pi\dfrac{\theta}{120}\right)\right) & 0 < \theta < 60 \\[2mm] 1.2812\left[1 - \cos\left(\pi\dfrac{\theta + 24}{168}\right)\right] - 0.5624 & 60 < \theta < 144 \end{cases}$$

Differentiating the displacement function will give velocity and acceleration functions, which are plotted in Figure 9.27.

Double harmonic functions developed in Section 9.4.6 are also useful in single-dwell mechanisms. The following example illustrates the use of double harmonic functions and compares that with the combined functions.

Example Problem 9.6

Develop the single-dwell system from Example Problem 9.5 using double harmonic functions for the rise and fall. Compare the acceleration characteristics of combined functions with double harmonic functions.

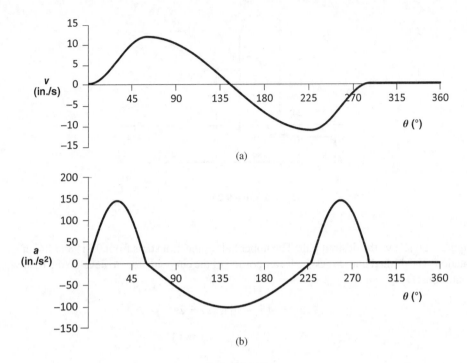

Figure 9.27

Solution: The general form of the double harmonic function is

$$y(\beta) = \frac{3}{8} - \frac{1}{2}\cos(\pi\beta) + \frac{1}{8}\cos(2\pi\beta)$$

For the rise portion,

$$s_1 = 0 \quad s_2 = 2$$
$$\theta_1 = 0 \quad \theta_2 = 144$$

$$\frac{s - s_1}{s_2 - s_1} = \frac{3}{8} - \frac{1}{2}\cos\left(\pi\frac{\theta - \theta_1}{\theta_2 - \theta_1}\right) + \frac{1}{8}\cos\left(2\pi\frac{\theta - \theta_1}{\theta_2 - \theta_1}\right)$$

$$\frac{s - 0}{2 - 0} = \frac{3}{8} - \frac{1}{2}\cos\left(\pi\frac{\theta - 0}{144 - 0}\right) + \frac{1}{8}\cos\left(2\pi\frac{\theta - 0}{144 - 0}\right)$$

Answer: $s_{\text{rise}}(\theta) = \frac{3}{4} - \cos\left(\pi\frac{\theta}{144}\right) + \frac{1}{4}\cos\left(2\pi\frac{\theta}{144}\right)$

For the fall portion,

$$s_1 = 2 \quad s_2 = 0$$
$$\theta_1 = 144 \quad \theta_2 = 288$$

$$\frac{s - 2}{0 - 2} = \frac{3}{8} - \frac{1}{2}\cos\left(\pi\frac{\theta - 144}{288 - 144}\right) + \frac{1}{8}\cos\left(2\pi\frac{\theta - 144}{288 - 144}\right)$$

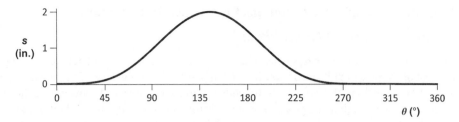

Figure 9.28

Answer: $s_{\text{fall}}(\theta) = -\frac{3}{4} + \cos\left(\pi\frac{\theta-144}{144}\right) - \frac{1}{4}\cos\left(2\pi\frac{\theta-144}{144}\right) + 2$

Figure 9.28 shows the displacement plot.

The general acceleration equation for a double harmonic is developed from the second derivative equation:

$$\frac{d^2s}{d\theta^2} = \left(\frac{s_2 - s_1}{(\theta_2 - \theta_1)^2}\right)\left(\frac{\pi^2}{2}\cos\left(\pi\frac{\theta-\theta_1}{\theta_2-\theta_1}\right) - \frac{\pi^2}{2}\cos\left(2\pi\frac{\theta-\theta_1}{\theta_2-\theta_1}\right)\right)$$

For the rise portion, the acceleration will be

$$\frac{d^2s}{d\theta^2} = (0.001)\left(\frac{\pi^2}{2}\cos\left(\pi\frac{\theta}{144}\right) - \frac{\pi^2}{2}\cos\left(2\pi\frac{\theta}{144}\right)\right)$$

For the fall portion, the acceleration will be

$$\frac{d^2s}{d\theta^2} = (-0.001)\left(\frac{\pi^2}{2}\cos\left(\pi\frac{\theta-144}{144}\right) - \frac{\pi^2}{2}\cos\left(2\pi\frac{\theta-144}{144}\right)\right)$$

Figure 9.29 shows the acceleration plot along with the acceleration for the combination function. The double harmonic function generated a smoother acceleration plot but with higher peak magnitude.

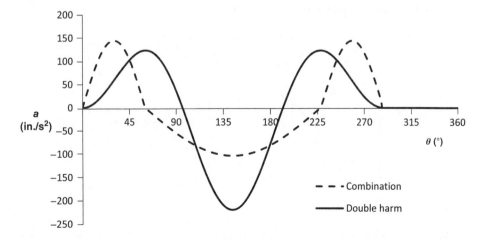

Figure 9.29

9.8   Application of Cam Functions for Critical Path Motion

9.8.1   Basic Concept of Critical Path Motion

Previous methods focused on cases where the extreme position was critical, but the path to get there was not. As an example, let us consider a double-dwell cam function. The critical points are the extreme positions of the dwells. A function is defined to move the follower from the low dwell position to the high dwell position over a specific time interval. The method of moving between dwells is not critical, but the extreme positions are. Critical path motion is different. In critical path motion, the path itself, or in some cases a derivative of the path, is specifically defined over a region of the cycle (or the complete cycle). Therefore, critical path motion is similar to the concept of function generators. Common types of critical path motion are cam functions to generate constant velocity or constant acceleration.

9.8.2   Cam Functions for Constant Acceleration

First we will consider critical path cam functions for constant or uniform acceleration. The process for graphically constructing a constant acceleration cam displacement is shown in Figure 9.30a. The resulting curve is parabolic, so uniformly accelerated motion is commonly referred to as parabolic motion. A general example of displacement, velocity, and acceleration is shown in Figure 9.30b. Due to the abrupt changes in acceleration, constant acceleration functions will have infinite spikes in the jerk function.

9.8.3   Cam Functions for Constant Velocity

Another common requirement for cam motion is to have constant velocity over a period of the cycle. Cam functions for constant velocity will result in infinite spikes in acceleration, which gives an unacceptable cam. Therefore, it is common to alter the uniform velocity motion by adding short curved sections at the extreme ends. There are different methods for adding the curved portions. The first method discussed will utilize short sections of uniform acceleration (or deceleration). Figure 9.31a shows a cam displacement function for constant velocity with the ends modified using uniform acceleration. The modified ends can be constructed using methods discussed in Section 9.8.2. The total rise occurs from $0 < t < t_3$. The actual section of constant velocity occurs in the cropped region $t_1 < t < t_2$.

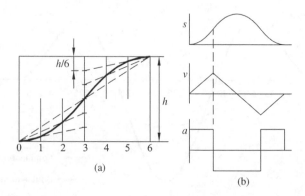

(a)

(b)

Figure 9.30   (a) Graphical construction of a constant acceleration cam displacement. (b) Displacement velocity and acceleration plots

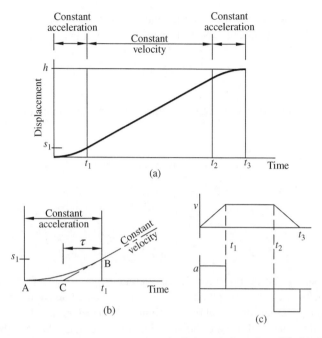

Figure 9.31 (a) Modified constant velocity motion. (b) Enlarged view of modified end. (c) Velocity and acceleration plots

Figure 9.31b shows an enlarged view of the modified end. The uniform acceleration curve must now be defined. The curve shown from point A to point B is developed as a uniform acceleration curve. In the region from $0 < t < t_1$, displacement will be defined by

$$s = \frac{1}{2}at^2 \tag{9.24}$$

and the velocity will be

$$v = at \tag{9.25}$$

At the end of the constant acceleration curve, the displacement can be determined from

$$s_1 = \frac{1}{2}at_1^2 \tag{9.26}$$

and the velocity will be

$$v_1 = at_1 \tag{9.27}$$

The constant velocity line can be extended to intersect the horizontal axis at point C, which is located some time τ from point B. Moving along the constant velocity line, the displacement at point B would be $V\tau$, where V is the constant velocity (or slope of the line). That displacement must match the displacement obtained in Equation 9.26 giving

$$\frac{1}{2}at_1^2 = V\tau \tag{9.28}$$

Also, the velocity obtained in Equation 9.27 must equal the constant velocity V. Making the substitution of $v_1 = V$ in Equation 9.28 gives

$$\frac{1}{2}at_1^2 = v_1\tau$$

$$\frac{1}{2}at_1^2 = (at_1)\tau$$

$$\tau = \frac{1}{2}t_1 \tag{9.29}$$

Similar calculation can be done for the deceleration phase. Therefore, in order to have continuous velocity, the constant velocity portion must lie on a line beginning at the midpoint of the acceleration phase and ending at the midpoint of the deceleration phase. The velocity and acceleration plots for the total rise are shown in Figure 9.31c. It can be seen that both velocity and acceleration are continuous. However, the jerk function will have infinite spikes. The performance of the uniform velocity cam function can be improved by using higher order polynomials at the extreme ends to replace the second-order constant acceleration functions.

9.9 Cam Geometry

9.9.1 Basic Concepts

Now that the appropriate cam functions have been determined, we can move into discussions on the actual cam profile. The cam profile is the geometry or contour of the cam surface. There are some important factors to help determine the overall size of the cam. In general, it is desired to keep the cam as small as possible and still maintain the performance. Larger cams obviously take up more space and are more expensive, but there are other reasons for trying to minimize the overall size. For large cams, the outer perimeter increases, which increases the follower path per cycle. To compensate for that increase, the cam must operate at a higher speed. Because the speed is increased, any small manufacturing error (deviation from theoretical displacement diagram) will cause larger acceleration error. Larger cams also have a larger unbalanced mass, which leads to increased vibration.

There are, however, limitations on how small the cam can become. Smaller cams tend to have sharper curves and less efficient force transmission. To determine cam size and performance, we must define certain terms related to cam geometry. The following sections will discuss a couple of important geometry terms for cams.

9.9.2 Base Circle

As illustrated in Figure 9.32, the smallest possible circle on the cam profile drawn from the center of the rotational axis is known as the base circle. The base circle is the zero position for the follower or its lowest position. The cam profile, or working surface of the cam, is developed

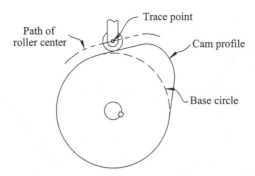

Figure 9.32 Cam geometry terminology

radially off the base circle. In other words, the cam displacement (determined from follower displacement) is measured in reference to the base circle. For roller followers, such as the follower in Figure 9.32, the roller center will follow a path offset from the cam profile. The roller center will be the trace point.

9.9.3 Pressure Angle

The cam pressure angle, as shown in Figure 9.33a, is the angle measured between the line normal to contact and the line of motion of the follower. Essentially, it is a measure of the steepness of the profile. The pressure angle changes as the cam rotates and will be zero when the follower is in contact at the base circle. Force is transmitted to the follower along the normal line. It can be seen that if the pressure angle were equal to zero, then all the force is transferred to motion of the follower. As the pressure angle increases, less of the force is transferred to motion of the follower and the remaining force acts in the tangent direction as slip. Figure 9.33b shows that for excessive pressure angles, the side thrust will become large, which can cause the follower to jam. Therefore, it is typically desired to keep the pressure

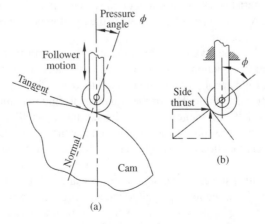

Figure 9.33 (a) Pressure angle. (b) Side thrust

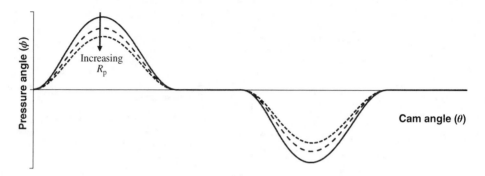

Figure 9.34 Pressure angle for different values of prime circle radius

angle below 30° to improve the performance of the cam follower mechanism. That limitation is simply a guideline and is somewhat conservative. With caution, larger pressure angles can be used if necessary.

Pressure angle will decrease as the roller diameter increases. For an in-line roller follower, the pressure angle ϕ is shown in Figure 9.34 for different values of prime circle radius R_p. Notice that the general shape of the curve is similar to a velocity curve and that the peak value decrease as the prime circle radius increases.

The radius of the prime circle is generally limited based on space availability. If the pressure angle cannot be sufficiently reduced by increasing the prime circle radius, an offset follower may be required. The pressure angle can also be reduced by allowing more time for the rise and fall.

9.10 Determining Cam Size

9.10.1 General Ideas

Due to infinite choices of base circle size, in general there would be infinite possibilities for the cams to be generated for any given displacement diagram. Obviously, making a very large cam would be a waste of material and space. Limitations also exist on how small a cam could be and still function properly. A common constraint to aid in the development of overall cam size is the maximum allowed pressure angle.

It may seem that finding the maximum pressure angle for a complete cam cycle would be a trivial task. Mathematical equations can be developed to calculate pressure angle for any angular location of the follower. These equations could theoretically be used to size the cam for a specific maximum pressure angle. Use of the equations to find maximum pressure angle is often not simple, especially for more complex displacement functions. To illustrate these difficulties, let us examine the most basic rise function, which is a straight line. Though a straight line displacement function is unacceptable, it will help show some important aspects of pressure angle.

Figure 9.35a shows the displacement diagram for a linear rise function with a slope C. The horizontal axis is drawn with the length equal to the arc length along the prime circle for the cam rotation β. The pressure angle would be the angle of the line normal to the displacement curve as shown. In theory, the pressure angle would remain constant throughout the rise.

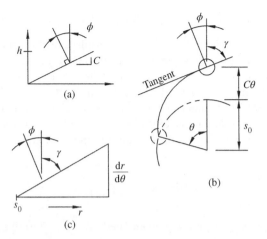

Figure 9.35 (a) Displacement diagram for linear rise. (b) Cam profile for linear rise. (c) Relationship between $\tan\phi$ and $\tan\gamma$

As shown in Figure 9.35b, the radial position to the center of the roller will change with the follower function of cam angle, where s_0 is the radial location at the zero position and C is the slope of the linear displacement function:

$$r(\theta) = s_0 + C\theta \tag{9.30}$$

We will define an angle γ that is an angle between the radius line and the tangent line. From Figure 9.35c, it can be seen that

$$\tan\gamma = \frac{r}{dr/d\theta} = \frac{s_0 + C\theta}{C} \tag{9.31}$$

The basic right triangle shown in Figure 9.35c gives the relationship needed to determine the pressure angle. The tangent of the pressure angle will simply be the inverse of the expression in Equation 9.31. Therefore, the pressure angle can be expressed as a function of cam rotation angle:

$$\tan\phi = \frac{C}{s_0 + C\theta}$$

$$\phi = \tan^{-1}\left(\frac{C}{s_0 + C\theta}\right) \tag{9.32}$$

Therefore, the pressure angle will not be constant, even with a linear rise displacement function. It will actually increase as the cam angle increases. This distortion of the displacement diagram when plotted along a radial disk cam will occur for all categories of displacement functions.

9.10.2 Maximum Pressure Angle for Simple Harmonic Functions

The previous section showed how distortion of the displacement diagram can cause complications in determining the maximum pressure angle. Though this distortion complicates the calculations, the maximum pressure angle can be mathematically determined for basic functions such as the simple harmonic function for in-line followers. The math gets more complex with advanced cam functions or with the use of an offset follower.

For a disk cam with an in-line roller follower, the following expression can be developed. Though the derivation is not provided here, one can refer to Rothbart (1956) for details:

$$\tan \phi_m = \frac{(d^2s/dt^2)_p}{\omega(ds/dt)_p} \tag{9.33}$$

In Equation 9.33, ϕ_m is the maximum pressure angle and the subscript p indicates the pitch point (location of the maximum pressure angle) value. Once the angular location is determined, the radius of the pitch circle (circle passing through roller center when it is at θ_p) is determined from

$$\rho_p = \frac{(ds/dt)_p^2}{(d^2s/dt^2)_p} \tag{9.34}$$

Let us consider simple harmonic cam function to illustrate the use of the equation. For a rise of h over a cam rotation of β, the simple harmonic function is defined as

$$s(\theta) = \frac{h}{2}\left(1 - \cos\left(\frac{\pi\theta}{\beta}\right)\right) \tag{9.35}$$

The first two time derivatives are needed for Equation 9.35. Noting that $\theta = \omega t$, the time derivatives are

$$\frac{ds}{dt} = \frac{h}{2}\left(\frac{\pi\omega}{\beta}\right)\sin\left(\frac{\pi\theta}{\beta}\right)$$

$$\frac{d^2s}{dt^2} = \frac{h}{2}\left(\frac{\pi\omega}{\beta}\right)^2\cos\left(\frac{\pi\theta}{\beta}\right) \tag{9.36}$$

Substituting Equation 9.36 into Equation 9.33 gives

$$\tan \phi_m = \frac{h/2(\pi\omega/\beta)^2 \cos(\pi\theta_p/\beta)}{\omega(h/2)(\pi\omega/\beta) \sin(\pi\theta_p/\beta)} = \frac{\pi}{\beta}\cot\left(\frac{\pi\theta_p}{\beta}\right) \tag{9.37}$$

For a desired maximum pressure angle (such as 30°), Equation 9.37 could be solved for the angular location of the follower θ_p that results in the maximum pressure angle. The pitch circle radius is determined by substituting Equation 9.36 into Equation 9.34:

$$\rho_p = \frac{\left(h/2(\pi\omega/\beta)\sin(\pi\theta_p/\beta)\right)^2}{h/2(\pi\omega/\beta)^2\cos(\pi\theta_p/\beta)} = \frac{h\sin^2(\pi\theta_p/\beta)}{2\cos(\pi\theta_p/\beta)} \tag{9.38}$$

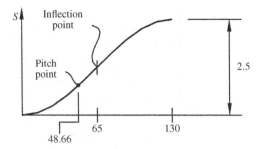

Figure 9.36

Example Problem 9.7

A cam will have a rise of 2.5 inches in 130° using a simple harmonic function. Find the exact pitch point location and pitch circle radius if the maximum allowed pressure angle is 30°.

Solution: Equation 9.37 can now be used:

$$\tan \phi_m = \frac{\pi}{\beta} \cot \left(\frac{\pi \theta_p}{\beta} \right)$$

$$\tan (30) = \frac{\pi}{2.27} \cot \left(\frac{\pi \theta_p}{2.27} \right)$$

$$\theta_p = 0.849 \text{ rad}$$

Answer: $\theta_p = 48.66°$

It is interesting to note that the location of the maximum pressure angle again illustrates the distortions discussed in Section 9.10.1. The displacement diagram is shown in Figure 9.36. The maximum pressure angle would be expected to occur at the location of maximum slope, which is the inflection point at $\beta/2$. The actual maximum pressure angle location is located at 48.66°, which is distorted in the direction of smallest radial distance.

From Equation 9.38,

$$\rho_p = \frac{h \sin^2 \left(\pi \theta_p / \beta \right)}{2 \cos \left(\pi \theta_p / \beta \right)} = \frac{2.5 \sin^2 (0.849\pi/2.27)}{2 \cos (0.849\pi/2.27)} = 2.77$$

As shown in Figure 9.37, the roller center will be at a radial position of 2.77 inches when the cam rotation angle is 48.66°. If the cam profile is designed to have this position, the maximum pressure angle for the rise will be 30°.

Answer: $\rho_p = 2.77$ in.

9.10.3 Maximum Pressure Angle for More Complex Cam Functions

For more complex cam functions, Equation 9.34 will typically generate an equation that cannot be solved analytically (due to more complicated time derivatives). Numerical methods could be utilized to solve the expression for maximum pressure angle location. A more practical approach is to use an approach based on approximate maximum pressure angle.

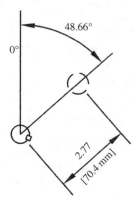

Figure 9.37

Defining f as a cam factor, the prime circle radius can be determined as

$$R_\text{p} = \frac{fh}{\beta} - s_{\beta/2} \qquad (9.39)$$

where β is the cam rotation required for a change in height h. The cam factor depends on the function type.

$$f = 2 \cot \phi_\text{max} \quad \begin{cases} \text{Double harmonic} \\ \quad \text{Cycloidal} \\ \quad \text{Parabolic} \end{cases} \qquad (9.40)$$

$$f = \frac{\pi}{2} \cot \phi_\text{max} \quad \{\text{Simple harmonic}$$

Example Problem 9.8 will provide an example of estimating the required cam size for a given pressure angle maximum.

It is difficult to use analytical expressions to determine maximum pressure angle. The nomogram shown in Figure 9.38 can be used to determine the maximum pressure angle. The parameter R_0 is the minimum pitch radius, L is the total displacement, and β is the cam rotation angle for the displacement. The nomogram shows a specific example for a cycloidal cam for a displacement of 0.75 inches that takes place in a cam rotation of 45° and a maximum allowed pressure angle of 30°. The line is drawn connecting $\beta = 45°$ and $\alpha_\text{max} = 30°$. The line crosses the axis at a point for a cycloidal cam to give $L/R_0 = 0.26$. Knowing that $L = 0.75$, the minimum pitch radius is calculated to be $R_0 = 0.75/0.26 = 2.88$ in.

9.11 Design of Cam Profiles

9.11.1 Graphical Methods for Plate Cams with In-Line Followers

Cam profile construction is a fairly basic graphical procedure. Once the appropriate cam function has been chosen, it is time to draw the actual cam profile to provide the desired follower motion. For any given cam displacement, the cam profile will change based on the type and size of the cam and follower. Therefore, the first step is to determine the type of cam

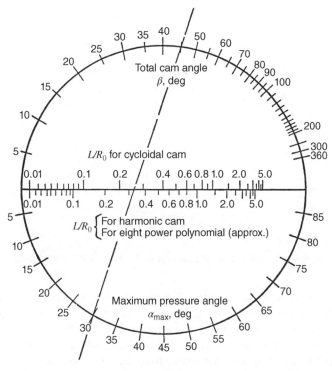

Figure 9.38 Nomogram to determine maximum pressure angle in a disk cam with a radial roller follower. Reproduced from Mabie and Reinholtz, Mechanisms and Dynamics of Machinery, 4th edition, John Wiley & Sons, © 1987

and follower to be used. Once the cam profile is developed, changes may need to be made to improve performance.

Consider the double-dwell displacement diagram shown in Figure 9.39 that was constructed using simple harmonic functions for the rise and fall.

We will first construct the cam profile for an in-line knife edge follower. The steps for the process are illustrated in Figure 9.40. The first step is drawing the base circle at radius R_B. The follower is then drawn in the home position, which relates to 0° in the displacement diagram in Figure 9.39.

Displacement measurement can now be transferred from the diagram in Figure 9.39. It is very important to notice the direction of rotation. The cam shown in Figure 9.40 has a

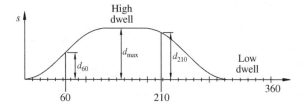

Figure 9.39 Displacement diagram measurements

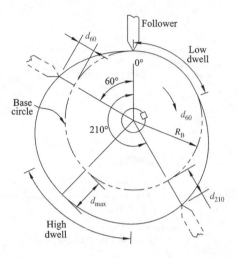

Figure 9.40   Development of cam profile for an in-line knife edge follower

clockwise rotation. As dimensions are transferred to the cam, we will hold the cam stationary and rotate the follower in the opposite direction of cam rotation. To illustrate the process, we will consider the displacement at 60° of cam rotation. The cam displacement d_{60} is measured from the diagram. A radial line is then drawn in Figure 9.40 at a location of 60° counterclockwise (opposite direction of cam rotation) from the home position. The dimension d_{60} is then measured radially from the base circle to locate the point of the follower at the 60° position. The process of transferring dimensions is continued around the base circle to draw the knife edge follower in numerous locations. The cam profile can then be drawn as a smooth curve connecting the points of the various follower positions.

A very similar process is used to develop the cam profile for an in-line roller follower. In addition to the base circle, a prime circle must be drawn. As shown in Figure 9.41, the prime

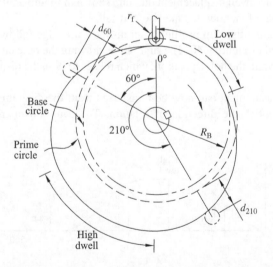

Figure 9.41   Development of cam profile for an in-line roller follower

circle is centered on the cam shaft and passes through the center of the roller follower. Once both circles are drawn, the roller follower is drawn in the home position with the roller tangent to the base circle. The dimensions from the displacement diagram can then be transferred to locations along the base circle. The dimension d_{60} is once again shown as an example. The roller follower can be drawn in various locations throughout the full 360°. The cam profile can now be drawn with a slight change to that of the knife edge follower. The cam profile will be a smooth curve tangent to the follower in each position.

Example Problem 9.8

A disk cam with an in-line roller follower is required for a rise of 0.75 inches in 130°, dwell for 40°, fall back to 0 in 150°, and then dwell the remaining 40°. The cam rotates at 250 rpm clockwise and the roller diameter is 0.3 inches. Both the rise and fall are 3-4-5 polynomial functions. Develop the displacement functions for the rise and fall. Design and construct the cam profile without exceeding a maximum pressure of 25°.

Solution: The 3-4-5 polynomial function $y(\beta) = 10\beta^3 - 15\beta^4 + 6\beta^5$ with substitutions gives

$$\frac{s - s_1}{s_2 - s_1} = 10\left(\frac{\theta - \theta_1}{\theta_2 - \theta_1}\right)^3 - 15\left(\frac{\theta - \theta_1}{\theta_2 - \theta_1}\right)^4 + 6\left(\frac{\theta - \theta_1}{\theta_2 - \theta_1}\right)^5$$

For the rise portion,

$$s_1 = 0 \quad s_2 = 0.75$$
$$\theta_1 = 0 \quad \theta_2 = 130$$

Substituting the given values into the displacement function,

$$\frac{s - 0}{0.75 - 0} = 10\left(\frac{\theta - 0}{130 - 0}\right)^3 - 15\left(\frac{\theta - 0}{130 - 0}\right)^4 + 6\left(\frac{\theta - 0}{130 - 0}\right)^5$$

The final displacement function for the rise is

$$s_{\text{rise}} = 7.5\left(\frac{\theta}{130}\right)^3 - 11.25\left(\frac{\theta}{130}\right)^4 + 4.5\left(\frac{\theta}{130}\right)^5$$

For the fall portion,

$$s_1 = 0.75 \quad s_2 = 0$$
$$\theta_1 = 170 \quad \theta_2 = 320$$

$$\frac{s - 0.75}{0 - 0.75} = 10\left(\frac{\theta - 170}{320 - 170}\right)^3 - 15\left(\frac{\theta - 170}{320 - 170}\right)^4 + 6\left(\frac{\theta - 170}{320 - 170}\right)^5$$

The final displacement function for the fall is

$$s_{\text{fall}} = -7.5\left(\frac{\theta - 170}{150}\right)^3 + 11.25\left(\frac{\theta - 170}{150}\right)^4 - 4.5\left(\frac{\theta - 170}{150}\right)^5 + 0.75$$

Converting the rotation speed gives

$$\omega = 250\frac{\text{rev}}{\text{min}}\left(\frac{2\pi\,\text{rad}}{\text{rev}}\right)\left(\frac{1\,\text{min}}{60\,\text{s}}\right) = 26.18\frac{\text{rad}}{\text{s}}$$

Next we can determine the cam size to satisfy the pressure angle requirements. For a parabolic cam profile, from Section 9.10.3, the cam factor is

$$f = 2\cot\phi_{\text{max}} = 2\cot(25°) = 4.3$$

The prime circle radius is determined using Equation 9.39:

$$R_{\text{p}} = \frac{fh}{\beta} - s_{\beta/2}$$

Both the rise and fall portions need to be considered to determine which one will cause a larger prime circle radius. Both the rise and fall will give the same displacement at $\beta/2$ (which will be half the rise distance). Therefore, the prime circle radius will be largest for the displacement that occurs over the smallest rotation angle β. For our example that would be the rise portion:

$$\beta_{\text{rise}} = 130°$$

$$s_{\theta=\beta/2} = 7.5\left(\frac{65}{130}\right)^3 - 11.25\left(\frac{65}{130}\right)^4 + 4.5\left(\frac{65}{130}\right)^5 = 0.375$$

$$R_{\text{p}} = \frac{4.3(0.75\,\text{in.})}{2.27\,\text{rad}} - 0.375\,\text{in.} = 1.05\,\text{in.}$$

Figure 9.42 shows the prime circle with roller positions (based on displacement diagram) at $10°$ increments taken counterclockwise (opposite direction of cam rotation).

The complete cam profile is shown in Figure 9.43 with the roller in the location of approximate maximum pressure angle. From the figure, the pressure angle is approximately $23°$ at that location, which satisfies the requirements.

9.11.2 Graphical Methods for Offset Followers

The cam profile for an offset follower can also be developed graphically. The general process is similar to that for the in-line roller follower and is illustrated in Figure 9.44. Again, the prime circle is centered on the cam shaft and passes through the center of the roller follower. The roller follower is drawn in the home position tangent to the base circle and offset the distance e from the centerline. As before, the dimensions are transferred from the displacement diagram while maintaining the offset. The process is repeated for the full $360°$ of cam rotation, and the profile can then be drawn connecting the positions.

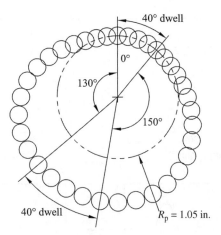

Figure 9.42

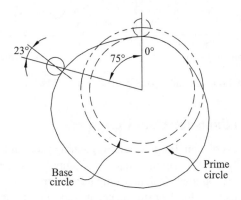

Figure 9.43

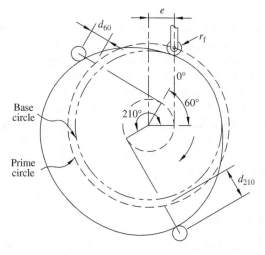

Figure 9.44 Development of cam profile for an offset roller follower

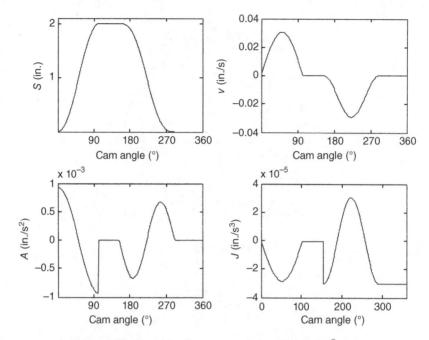

Figure 9.45 Example plots generated from MATLAB® code

9.12   Computer Methods for Cam Design

Though the general process of cam design is relatively straightforward, the design process can become tedious, especially if optimization is required. Computer methods can greatly aid in the process of finding a more optimal design. Computer programs can be developed to use different mathematical functions for the cam and can include all aspects of the cam kinematics. Cam geometry can easily be manipulated to examine the effects on the kinematics. A properly developed computer program can later be expanded to include the cam kinetics, which will be discussed in Chapter 13.

Because the displacement diagram is built from mathematical functions, it can be easily generated with the use of spreadsheets or math software such as MathCAD or MATLAB®. Software packages can also be used to differentiate the displacement functions to give velocity, acceleration, and jerk. Consider an application of software to automatically generate *SVAJ* diagrams for a double-dwell mechanism. User inputs would be required for total rise height as well as time is required for rise, fall, and dwell. For more versatility, the user should also input the function type for the rise and fall. Figure 9.45 shows an example set of plots generated from a MATLAB® code.

Once computer code has been developed to generate *SVAJ* diagrams, it can be continued to determine cam size and generate the cam profile. Coordinate information can be plotted as a polar plot to show the cam profile.

9.13   Review and Summary

This chapter discussed the general concepts of cam design. Some common cam functions were developed and used for various types of follower motion. It is important to remember that the cam functions discussed in this chapter do not provide an exhaustive list of available cam functions. The reader is encouraged to explore other sources for a more complete list. New cam function types are still being developed to provide unique benefits, such as reduced vibration. Advanced concepts of cam analysis and design will be examined in Chapter 13.

Problems

P9.1 The eccentric circular cam shown in Figure 9.46 rotates clockwise with $r = 4$ inches and $e = 2.5$ inches. Calculate and plot the follower displacement for the full cycle at $20°$ increments.

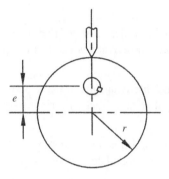

Figure 9.46

P9.2 The cam in Problem P9.1 rotates at a speed of 90 rpm. Plot the velocity, acceleration, and jerk for one complete cycle labeling maximum values.

P9.3 The acceleration plot shown in Figure 9.47 is for one complete cycle. Sketch the velocity and displacement diagram. Label maximum values in terms of a_{max}.

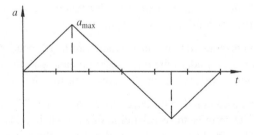

Figure 9.47

P9.4 The velocity diagram shown in Figure 9.48 is a portion of a proposed cam design. The displacement function is zero at time equal to zero and 5.33 inches at a time of 2 seconds. Sketch the displacement and acceleration functions and label maximum values.

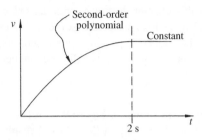

Figure 9.48

P9.5 The fundamental law of cam design states that the displacement function must have continuous first and second derivatives across the entire 360° interval. Discuss why this is important.

P9.6 Discuss advantages and disadvantages of using a simple harmonic function. Would you recommend using that cam function for a low-speed double-dwell mechanism? Explain your answer.

P9.7 Derive the 3-4-5 polynomial function Equation 9.13. Start with the general form of $y(\beta) = a_0 + \sum_{n=1}^{5} a_n \beta^n$ and calculate the constants.

P9.8 Derive the 4-5-6-7 polynomial function Equation 9.15. Start with the general form of $y(\beta) = a_0 + \sum_{n=1}^{7} a_n \beta^n$ and calculate the constants.

P9.9 A cam function is required to produce a rise from 1–3.25 inches for a cam rotation from 0–65° using a simple harmonic function. Determine and plot the displacement function.

Answer: $s = \dfrac{2.25}{2} \left[1 - \cos\left(\dfrac{180\theta}{65} \right) \right] + 1$

P9.10 A cam function is required to produce a fall from 5 to 2 inches for a cam rotation from 85 to 135°. Determine and plot the velocity function using (a) a 3-4-5 polynomial function and (b) a 4-5-6-7 polynomial function.

P9.11 A cam function is required to produce a rise from 1 to 3 inches for a cam rotation from 15 to 75° using a cycloidal function. Determine and plot the displacement function, velocity function, acceleration function, and jerk function.

P9.12 A cam function is required to produce a rise from 0 to 2.75 inches for a cam rotation from 20 to 55°. Determine the acceleration functions using (a) a cycloidal function and (b) a 3-4-5 polynomial function. Plot and compare the acceleration functions.

P9.13 A cam function is required to produce a fall from 2.75 to 0 inches for a cam rotation from 200 to 360° using a 4-5-6-7 polynomial function. Determine and plot the displacement function, velocity function, and acceleration function.

P9.14 A cam function is required to produce a fall from 4 to 2 inches for a cam rotation from 150 to 250° using a 4-5-6-7 polynomial function. Determine and plot the displacement function, velocity function, and acceleration function.

P9.15   A cam is required to rise from 0 to 1.25 inches in 70°, dwell for 38°, fall from 1.25 to 0 inches in 65°, then dwell at 0 inches for 187°. The cam rotation speed is 120 rpm. Develop rise and fall functions using cycloidal functions. Draw *SVAJ* diagrams for one complete cycle.

P9.16   Solve Problem P9.15 using a 3-4-5 polynomial for the rise function and a 4-5-6-7 polynomial for the fall function.

P9.17   A designer wants to develop a double-dwell cam mechanism to give the displacement shown in Figure 9.49. He proposes to use a third-order polynomial in the form $y(\beta) = k_1\beta^2 + k_2\beta^3$ for the rise and fall functions. Would this design yield continuous velocity for the full 360° cycle? Would this design yield continuous acceleration for the full 360° cycle? Sketch the velocity and acceleration diagrams labeling maximum values.

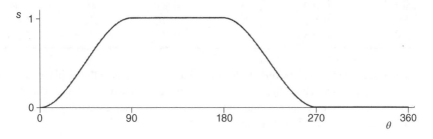

Figure 9.49

P9.18   A cam is required to rise from 0 to 4 inches in 2.5 seconds, dwell for 1.75 seconds, fall from 4 to 0 inches in 2 seconds, and then dwell for 0.5 seconds. Develop rise and fall functions using 4-5-6-7 polynomial functions. Draw *SVAJ* diagrams for one complete cycle.

P9.19   A cam is required to rise from 0 to 1.5 inches in 80°, dwell for 70°, fall from 1.5 to 0.5 inches in 45°, dwell for 50°, fall to 0 inches in 30°, and then dwell for 85°. The cam rotation speed is 200 rpm. Develop rise and fall functions using 4-5-6-7 polynomial functions for all rise portions and simple harmonic functions for all fall portions. Draw *SVAJ* diagrams for one complete cycle. Mark any discontinuities in the functions.

P9.20   A cam is required for a single-dwell mechanism with a rise of 1.5 inches occurring in 170°, a fall occurring in 170°, and a dwell for 20°. Develop the cam function using a combination of cycloidal and simple harmonic displacement functions. Plot velocity and acceleration labeling maximum values.

P9.21   A cam is required to rise from 0 to 0.75 inches in 80°, fall from 0.75 to 0 inches in 95°, and then dwell at 0 inches for 185°. The cam rotation speed is 120 rpm. Develop rise and fall functions using cycloidal functions. Draw *SVAJ* diagrams for one complete cycle. Is this a good design? Explain your answer.

P9.22   Solve Problem P9.21 using double harmonic functions.

P9.23   Develop a cam function to uniformly accelerate from 0 to 4 in./s in 0.2 seconds, maintain the constant velocity of 4 in./s for 1.5 seconds, and then uniformly decelerate back to zero velocity in 0.2 seconds. Plot the displacement, velocity, and acceleration.

P9.24 A disk cam with an in-line roller follower is required to provide the displacement values given in Table 9.1. The roller diameter is 0.5 inches. The cam must a have a base circle with a diameter of 5 inches and turns clockwise. Construct the cam profile graphically using CAD software.

Table 9.1 Displacement data for Problems P9.24 and P9.25

Cam angle (°)	Follower displacement (in.)	Cam angle (°)	Follower displacement (in.)	Cam angle (°)	Follower displacement (in.)
0	0.000	120	0.498	240	0.486
20	0.002	140	0.500	260	0.402
40	0.046	160	0.500	280	0.250
60	0.169	180	0.500	300	0.098
80	0.332	200	0.500	320	0.015
100	0.455	220	0.500	340	0.000

P9.25 A disk cam with an offset roller follower is required to provide the displacement values given in Table 9.1. The roller diameter is 0.3 inches. The cam must a have a base circle with a diameter of 6 inches and turns counterclockwise. Construct the cam profile graphically using CAD software.

P9.26 A disk cam with an in-line knife edge follower is required to provide the displacement for Example Problem 9.4. The cam must a have a base circle with a diameter of 3 inches and turns clockwise. Construct the cam profile graphically.

P9.27 A disk cam with an in-line roller follower is required for a rise of 1.2 inches in 110°, dwell for 70°, fall back to zero in 160°, and then dwell the remaining 20°. The cam rotates at 110 rpm and the roller diameter is 1 inch. Both the rise and fall are simple harmonic functions. Design and construct the cam profile to be as small as possible without exceeding a maximum pressure of 30°.

P9.28 An in-line roller cam is required to operate the slider–crank mechanism shown in Figure 9.50. The cam should move the slider to provide 20° rotation of the crank. The motion required is to dwell in the position shown for 0.5 seconds, move to vertical position in 0.75 seconds, dwell in vertical position for 0.6 seconds, and return to start

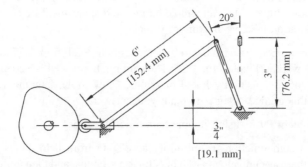

Figure 9.50

position in 0.75 seconds. The follower is held in contact with the cam by a spring (not shown). Design the cam using 4-5-6-7 polynomial functions for the rise and fall. The maximum pressure angle is limited to 30°. Determine rotation speed, base circle radius, and roller diameter. Plot the crank angle versus cam angle for one cycle.

P9.29 You have been asked to design a cam for the following critical path motion (CPM). Accelerate the follower from a velocity of 0 to 8 in./s. Maintain a constant velocity of 8 in./s for 1.2 seconds. Decelerate the follower to zero velocity. Return the follower to the start position. Cycle time needs to be exactly 3 seconds. No information is given for the method to accelerate or decelerate, so you need to make some assumptions on the time involved for acceleration and deceleration. Develop plots of motion, velocity, acceleration, and jerk.

Reference

Rothbart, H.A. (1956) *Cams: Design, Dynamics, and Accuracy*, John Wiley & Sons, Inc., New York.

10

Vibration Theory

10.1 Introduction

Vibration exists in everyday situations such as the swaying of trees in the wind and the flow of tides. Vibration of your eardrum allows you to hear sound. You regularly experience some form of vibration in your daily lives, and you could probably give a relatively good definition of vibration. It is very likely that you have already learned some basic concepts of vibration in previous college courses, or potentially even before college. The topic was probably first formally introduced in physics, where you learned about oscillatory motion and waves. Vibration is an important subset of dynamics, and basic concepts of vibration are often covered in engineering dynamics texts. It is also very common to see examples related to vibrations in math courses, such as differential equations.

Before starting our current study of vibrations, we must first define vibration. Vibration is an oscillatory motion about an equilibrium position that repeats itself at equal time intervals. The first key term is oscillatory, which is used to describe the type of motion. Oscillatory motion is simply motion in which a body moves repeatedly back and forth (e.g., a swinging or oscillating pendulum in a clock). Problems in vibrations deal with motion of bodies in response to applied disturbances (e.g., initial displacements or applied forces) in the presence of some form of restoring force (e.g., gravity or springs). Problems can range from analysis of a basic pendulum to complex analysis of the motion of a skyscraper in response to an earthquake.

The study of vibrations requires knowledge of multiple areas of engineering and mathematics. Essential engineering subject areas are statics and dynamics, although other important engineering subject areas include mechanics of materials and fluid mechanics. Essential mathematics subject areas include calculus, differential equations, and linear algebra (for multidegree-of-freedom systems).

Vibration is very common and often undesirable. For example, vibration in machinery generates noise and can lead to component failure. Excessive vibration can often lead to catastrophic failure. A well-known example is the collapse of the Tacoma Narrows suspension bridge. The bridge was completed in the summer of 1940. Large vertical oscillations of the roadway were observed during high wind. In fact, the large motion of the bridge became a tourist attraction. The bridge collapsed on November 7, 1940 from high winds in a powerful storm.

In the study of vibrations, as with most engineering topics, the system to be investigated must be represented as a mathematical model. A discrete system is a system with the mass and

Machine Analysis with Computer Applications for Mechanical Engineers, First Edition. James Doane.
© 2016 John Wiley & Sons, Ltd. Published 2016 by John Wiley & Sons, Ltd.
Companion Website: www.wiley.com/go/doane0215

spring elements separated. In other words, a discrete model consists of a finite number of degrees of freedom. A continuous system is a system with an infinite number of degrees of freedom. A vibrating string is an example of a continuous system. Continuous systems are often represented by a simplified discrete model. This chapter will focus only on discrete systems.

This chapter will focus on single degree-of-freedom discrete systems. The main intent is to provide a general foundation of knowledge in order to better study machine dynamics. Several good vibrations textbooks are listed in the references for any student interested in a more thorough coverage of vibration theory.

10.2 System Components

It is often possible to idealize mechanical systems in terms of linear springs, masses, and dashpots. For a linear elastic spring (satisfies Hooke's law), the deformation is proportional to the applied force.

$$F = kx \tag{10.1}$$

The constant k is known as the spring constant or spring modulus and has the units N/m or lb/feet. The spring constant depends on properties of the spring. The stiffer the spring, the higher the spring constant.

It is not uncommon to have a system consisting of combinations of linear springs. The system can then be simplified by replacing the combination of springs with one spring of equivalent stiffness. An equivalent spring is a spring that replaces a combination of springs while maintaining the original system properties. The stiffness of the equivalent spring is known as the equivalent stiffness and is typically denoted as k_e.

Spring combinations can be arranged in two different ways. First, we will consider springs arranged in parallel, as shown in Figure 10.1a.

From the free body diagram shown in Figure 10.1b,

$$W = k_1 \delta_{st} + k_2 \delta_{st} \tag{10.2}$$

If we replace the two springs with one spring with an equivalent stiffness, the resulting free body diagram will be as shown in Figure 10.1c. Summation of forces yields

$$W = k_e \delta_{st} \tag{10.3}$$

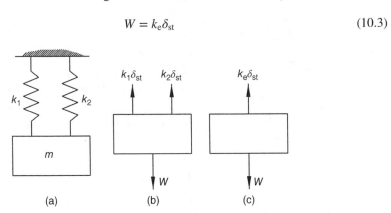

Figure 10.1   (a) System with springs in parallel. (b) Free body diagram. (c) Free body diagram for system with equivalent spring

The right-hand sides of Equations 10.2 and 10.3 must be equal.

$$k_e \delta_{st} = k_1 \delta_{st} + k_2 \delta_{st}$$
$$k_e = k_1 + k_2$$

(10.4)

The concept can be expanded to multiple springs in parallel. The equivalent stiffness of n springs in parallel is equal to the sum of all individual spring stiffness values.

$$k_e = \sum_{i=1}^{n} k_i$$

(10.5)

Next, we will examine a combination of springs arranged in series. The system is shown in Figure 10.2a in the original unstretched state. Once a weight is added, the system reaches equilibrium, as shown in Figure 10.2b.

The total static deformation δ_{st} of the mass is the sum of each individual spring deformation.

$$\delta_{st} = \delta_1 + \delta_2$$

(10.6)

From statics, it can be determined that each spring is subjected to the same force, which is equal to the weight.

$$W = k_1 \delta_1 \Rightarrow \delta_1 = \frac{W}{k_1}$$
$$W = k_2 \delta_2 \Rightarrow \delta_2 = \frac{W}{k_2}$$

(10.7)

The free body diagram in Figure 10.2c is for the equivalent spring. From summation of forces,

$$W = k_e \delta_{st} \Rightarrow \delta_{st} = \frac{W}{k_e}$$

(10.8)

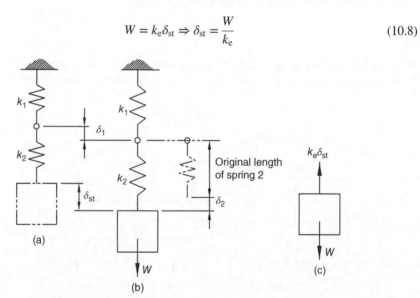

(a)   (b)   (c)

Figure 10.2 (a) System with springs in series in the unstretched position. (b) System in equilibrium position. (c) Free body diagram for system with equivalent spring

Substituting Equations 10.7 and 10.8 into Equation 10.6 gives

$$\frac{W}{k_e} = \frac{W}{k_1} + \frac{W}{k_2}$$
$$\frac{1}{k_e} = \frac{1}{k_1} + \frac{1}{k_2} \qquad (10.9)$$

Expanding the concept to a case with n springs will result in the following summation:

$$\frac{1}{k_e} = \sum_{i=1}^{n} \frac{1}{k_i} \qquad (10.10)$$

Example Problem 10.1

The system shown in Figure 10.3 consists of a mass and five springs. The spring constants are

$$k_1 = 25 \text{ lb/in.}$$
$$k_2 = 40 \text{ lb/in.}$$
$$k_4 = k_5 = 60 \text{ lb/in.}$$

Assuming the mass only moves in the vertical direction, determine the stiffness of an equivalent spring that would replace the five springs.

Solution: The system consists two sets of springs in parallel. Those two sets are then in series, as shown in Figure 10.4.

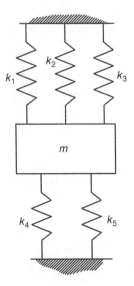

Figure 10.3

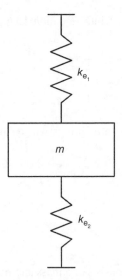

Figure 10.4

The top three springs are in parallel and can be replaced with one spring with an equivalent stiffness k_{e_1}.

$$k_{e_1} = k_1 + k_2 + k_3$$

$$k_{e_1} = 25\,\frac{\text{lb}}{\text{in.}} + 40\,\frac{\text{lb}}{\text{in.}} + 25\,\frac{\text{lb}}{\text{in.}} = 90\,\frac{\text{lb}}{\text{in.}}$$

Similarly, the lower set of springs in parallel can be replaced with one spring:

$$k_{e_2} = k_4 + k_5 = 60\,\frac{\text{lb}}{\text{in.}} + 60\,\frac{\text{lb}}{\text{in.}} = 120\,\frac{\text{lb}}{\text{in.}}$$

The system now consists of two springs (or equivalent springs) in series. The final equivalent spring stiffness, as shown in Figure 10.5, can then be determined.

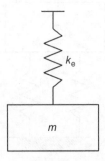

Figure 10.5

$$\frac{1}{k_e} = \frac{1}{k_{e_1}} + \frac{1}{k_{e_2}} = \frac{1}{90\ \text{lb/in.}} + \frac{1}{120\ \text{lb/in.}}$$

Answer: $k_e = 51.4\dfrac{\text{lb}}{\text{in.}}$

10.3   Frequency and Period

The period of motion is the time required to complete one oscillation (cycle) and is typically measured in seconds (or seconds per cycle). If motion is periodic and has a period T, then $x(t) = x(t + T)$ for all values of time. As will be seen later, if a motion is periodic, it can be expressed by sine or cosine functions. Because sine and cosine functions have a period of 2π, the period can also be expressed as

$$T = \frac{2\pi}{\omega} \tag{10.11}$$

where ω is the circular or angular natural frequency.

A very important term in vibrations is the natural frequency. The natural frequency is the number of cycles a single degree-of-freedom system freely vibrates every second. The natural frequency is denoted by f and has units of Hertz or cycles per second (cps).

$$1\ \text{Hz} = 1\ \text{cycle per second} = 1\ \text{s}^{-1} \tag{10.12}$$

The frequency is the reciprocal of period:

$$f = \frac{1}{T} \tag{10.13}$$

10.4   Undamped Systems

10.4.1   Equations of Motion

We will begin with the most basic problem type in vibrations, which is undamped free vibration. Consider the massless spring shown in Figure 10.6a in the unstretched position. For now we will assume that the spring follows Hooke's law (linear force deflection relationship) with spring constant k.

Adding a mass m will cause the spring to stretch a distance x_{st} until it reaches static equilibrium, which is the point when its weight W is balanced by the spring force. The free body diagram for the static equilibrium condition is shown in Figure 10.6b. From summation of forces $\sum F = 0$, we can determine that $W = kx_{st}$ for the static equilibrium condition.

Now, consider pulling the mass a distance x beyond the static equilibrium position. Under our assumption that the spring is linear, Hooke's law states that the spring will exert a restoring force on the mass. The magnitude of the restoring force will be the spring constant times the distance the spring is deflected past its equilibrium position. The direction of the restoring force will be opposite the direction of deflection. As an example, if the spring is deflected downward, the restoring force from the spring will act upward.

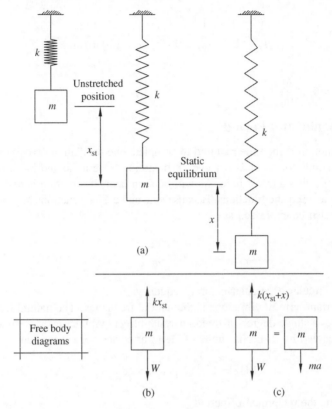

Figure 10.6   Spring–mass systems and corresponding free body diagrams

If the mass is now released from the stretched distance x, the mass will oscillate around the equilibrium position in free vibration. From the free body diagram shown in Figure 10.6c, Newton's second law gives

$$\sum F = ma$$
$$W - kx_{st} - kx = m\ddot{x} \tag{10.14}$$

However, from Figure 10.6b, the term kx_{st} is equal to W causing the differential equation of motion to reduce to the following form:

$$-kx = m\ddot{x} \tag{10.15}$$

Dividing by the mass and introducing a constant called the natural circular frequency $\omega_n = \sqrt{k/m}$ (expressed in rad/s in both SI and Customary units), we get the following equation for simple harmonic motion:

$$\ddot{x} + \omega_n^2 x = 0 \tag{10.16}$$

Equation 10.16 is a second-order homogeneous differential equation with constant coefficients. The auxiliary equation will be $s^2 + \omega_n^2 = 0$ and the two roots will be the complex numbers $s_1 = i\omega_n$ and $s_2 = -i\omega_n$. Recall from differential equations that the solution for roots of the auxiliary equation that are in the form $s_{1,2} = \alpha \pm i\beta$ is $x = A_1 e^{\alpha t} \cos \beta t + A_2 e^{\alpha t} \sin \beta t$. For our roots $\alpha = 0$ and $\beta = \omega_n$, which gives the general solution to the equation of motion as

$$x(t) = A_1 \cos \omega_n t + A_2 \sin \omega_n t \qquad (10.17)$$

Equation 10.17 represents a response known as simple harmonic motion with a period of free vibration of $T = 2\pi/\omega_n$. Because the motion is harmonic, the system is often called a harmonic oscillator.

The equation for velocity can be determined from the first derivative of Equation 10.17:

$$v(t) = \dot{x}(t) = -A_1 \omega_n \sin \omega_n t + A_2 \omega_n \cos \omega_n t \qquad (10.18)$$

The constants A_1 and A_2 depend on how the motion started (initial conditions). There are two possible initial conditions associated with Equation 10.17 and 10.18. The first would be the amount of initial displacement, which is the displacement at time equal to zero.

$$x_0 = x(0) \qquad (10.19)$$

The other possible initial condition, initial velocity, is the velocity at time equal to zero.

$$v_0 = \dot{x}(0) \qquad (10.20)$$

A system can have either or both initial conditions. Figure 10.7 shows a few possible examples of initial conditions. For all cases, we will represent the positive x direction as acting down. Figure 10.7a shows a case with $v_0 = 0$ and $x_0 > 0$. This case is released from rest (no initial velocity) from a point below the equilibrium point. Figure 10.7b is a case where the initial displacement is zero and the initial velocity acting in the negative direction. Therefore, the mass starts at the equilibrium point $x_0 = 0$ and is given an upward velocity $v_0 < 0$. Figure 10.7c is a combination of two initial conditions $v_0 > 0$ and $x_0 > 0$. For this case, the mass is stretched an initial displacement below the equilibrium point and is then given an initial velocity downward (positive direction).

Next, we examine the effects of the initial conditions. Consider first having an initial displacement of x_0 and no initial velocity. At time equal to zero, the displacement must equal the initial displacement,

$$\begin{aligned} x_0 &= A_1 \cos 0 + A_2 \sin 0 \\ x_0 &= A_1(1) + A_2(0) \Rightarrow A_1 = x_0 \end{aligned} \qquad (10.21)$$

and the velocity must be zero:

$$\begin{aligned} 0 &= -A_1 \omega_n \sin 0 + A_2 \omega_n \cos 0 \\ 0 &= -A_1 \omega_n(0) + A_2 \omega_n(1) \Rightarrow A_2 = 0 \end{aligned} \qquad (10.22)$$

giving the final displacement equation (system response) describing the motion of a system with only initial displacement.

$$x(t) = x_0 \cos \omega_n t \qquad (10.23)$$

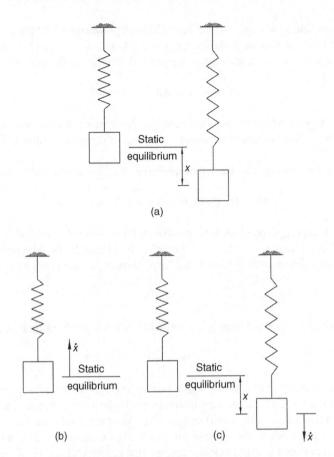

Figure 10.7　Examples of initial conditions

Next, let us examine the vibration of a system having an initial velocity of v_0 and no initial displacement. Now, the displacement must be zero at time equal to zero:

$$
\begin{aligned}
0 &= A_1 \cos 0 + A_2 \sin 0 \\
0 &= A_1(1) + A_2(0) \Rightarrow A_1 = 0
\end{aligned}
\tag{10.24}
$$

and the velocity must equal the initial velocity:

$$
\begin{aligned}
v_0 &= -A_1 \omega_n \sin 0 + A_2 \omega_n \cos 0 \\
v_0 &= -A_1 \omega_n(0) + A_2 \omega_n(1) \Rightarrow A_2 = \frac{v_0}{\omega_n}
\end{aligned}
\tag{10.25}
$$

giving the final response due to only initial velocity:

$$
x(t) = \frac{v_0}{\omega_n} \sin \omega_n t
\tag{10.26}
$$

The previous cases are specific cases for having only one initial condition. The most general case is one that has both initial velocity and initial displacement. The constants A_1 and A_2 are determined from initial conditions to be

$$A_1 = x_0$$
$$A_2 = \frac{v_0}{\omega_n} \qquad (10.27)$$

giving the final form of the general solution for free vibration.

$$x(t) = x_0 \cos \omega_n t + \frac{v_0}{\omega_n} \sin \omega_n t \qquad (10.28)$$

Example Problem 10.2

A mass weighing 3.5 pounds is hung from a spring causing it to stretch 8 inches to a position of static equilibrium. The mass is then stretched to a point 5 inches below equilibrium and given an upward velocity of 2 ft/s. Determine the response (displacement as a function of time).

Solution: The spring constant can be determined using Hooke's law:

$$k = \frac{F}{x} = \frac{W}{x_0} = \frac{3.5 \text{ lb}}{8 \text{ in.}(1 \text{ ft}/12 \text{ in.})} = 5.25 \frac{\text{lb}}{\text{ft}}$$

The natural frequency is

$$\omega_n = \sqrt{\frac{k}{m}}$$

Using $m = W/g$

$$\omega_n = \sqrt{\frac{kg}{W}} = \sqrt{\frac{5.25 \text{ lb}/\text{ft}(32.2 \text{ ft}/\text{s}^2)}{3.5 \text{ lb}}} = 6.95 \frac{\text{rad}}{\text{s}}$$

From Equation 10.28,

$$x(t) = x_0 \cos \omega_n t + \frac{v_0}{\omega_n} \sin \omega_n t$$

Initial displacement is downward giving a positive initial displacement:

$$x_0 = 5 \text{ in.} = 5/12 \text{ ft}$$

Initial velocity is upward giving a negative initial velocity:

$$v_0 = -2 \frac{\text{ft}}{\text{s}}$$

Substituting initial conditions and natural frequency,

$$x(t) = \frac{5}{12} \cos(6.95t) - \frac{2}{6.95} \sin(6.95t)$$

Answer: $x(t) = 0.42 \cos(6.95t) - 0.29 \sin(6.95t)$ (ft)

Example Problem 10.3

As shown in Figure 10.8, a large weight $W_1 = 675$ lb is suspended from a spring with a spring constant $k = 700$ lb/in. A small weight $W_2 = 250$ lb is suspended from the large weight by a cable and reaches a state of static equilibrium. Determine the response of the large weight if the cable is suddenly cut.

Solution: The small weight causes an initial displacement of

$$x_0 = \frac{F}{k} = \frac{250 \text{ lb}}{700 \text{ lb/in.}} = 0.357 \text{ in.}$$

The natural frequency is

$$\omega_n = \sqrt{\frac{k}{m}} = \sqrt{\frac{kg}{W}}$$

$$\omega_n = \sqrt{\frac{700 \text{ lb/in.} (12 \text{ in.}/\text{ft}) (32.2 \text{ ft/s}^2)}{675 \text{ lb}}}$$

$$\omega_n = 20.02 \frac{\text{rad}}{\text{s}}$$

From Equation 10.28,

$$x(t) = x_0 \cos \omega_n t + \frac{v_0}{\omega_n} \sin \omega_n t$$

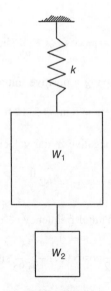

Figure 10.8

The initial velocity is equal to zero giving

$$x(t) = x_0 \cos \omega_n t$$

Answer: $x(t) = 0.357 \cos(20.02t)$ (in.)

10.4.2 Graphical Representation of Initial Conditions

The system response can be represented graphically using rotating vectors. Figure 10.9 shows the graphical representation of Equation 10.23, which is the response when the system is subjected only to initial displacement. The vertical axis represents the displacement x, and the horizontal axis is a timescale. However, it is often more convenient to have the horizontal axis represent total angular displacement ($\omega_n t$) of the rotating vector. The vector x_0 is shown in its initial position and rotates counterclockwise at a constant angular velocity of ω_n rad/s. After the vector rotates through an angular displacement θ_k, the displacement will be x_k located in the horizontal axis position equal to $\omega_n t_k$. At time zero, the displacement is at positive x_0 and slope (velocity) of zero, which are the initial conditions.

A graphical representation of Equation 10.26, response of a system to initial velocity, is shown in Figure 10.10. The initial position is zero, but the curve starts with a slope based on initial velocity.

The most general form is when motion begins at some intermediate point representing having both initial displacement and velocity. Figure 10.11 shows a graphical representation of Equation 10.28. Initial displacement is the vector x_0 and the initial velocity is the vector v_0/ω_n. The resultant of the two vectors has a magnitude A equal to the amplitude of motion. The curve begins at a location x_0 on the vertical axis and has an initial slope based on initial velocity. If the

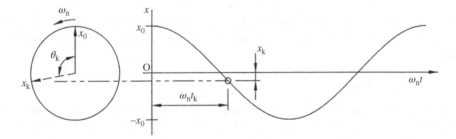

Figure 10.9   Graphical representation of equation of motion for initial displacement

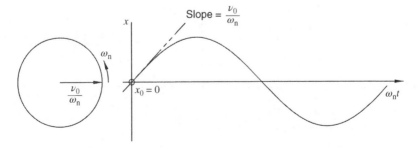

Figure 10.10   Graphical representation of equation of motion for initial velocity

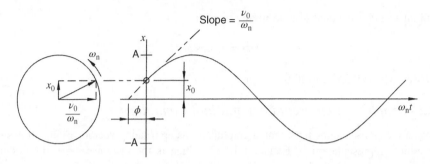

Figure 10.11 Graphical representation of equation of motion for initial displacement and initial velocity

curve were continued to the left, it would intersect the horizontal axis at an offset equal to the phase angle ϕ.

Velocity and acceleration can also be represented graphically. The velocity of the harmonic motion described by $x(t) = x_0 \cos \omega_n t$ is

$$\dot{x}(t) = -x_0 \omega_n \sin \omega_n t \tag{10.29}$$

and can be represented graphically, as shown in Figure 10.12, by a vector of length $x_0 \omega_n$ rotating in the same direction as the displacement vector. The velocity vector is rotated $90°$ ahead of the position vector.

The acceleration of the harmonic motion described by $x(t) = x_0 \cos \omega_n t$ is

$$\ddot{x}(t) = -x_0 \omega_n^2 \cos \omega_n t \tag{10.30}$$

and is represented graphically by a vector of length $x_0 \omega_n^2$ rotating in the same direction as the displacement vector (Figure 10.13). The acceleration vector is rotated $180°$ ahead of the position vector.

10.4.3 Energy Methods

Energy methods can also be used to develop equations of motion because the system can be considered to be conservative. The total energy (sum of kinetic energy and potential energy) in a conservative system remains constant.

$$T + U = \text{constant} \tag{10.31}$$

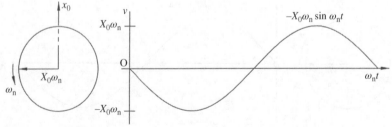

Figure 10.12 Graphical representation of velocity

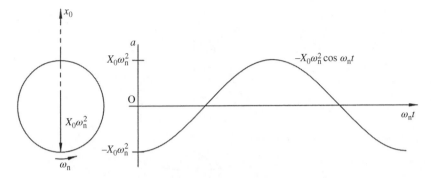

Figure 10.13   Graphical representation of acceleration

The kinetic energy is the energy stored in the mass due to its velocity:

$$T = \frac{1}{2}m\dot{x}^2 \tag{10.32}$$

and the potential energy is elastic energy (strain energy) of the deformed spring:

$$U = \frac{1}{2}kx^2 \tag{10.33}$$

Because the total energy is constant, the time rate of change of energy is equal to zero:

$$\frac{d}{dt}(T + U) = 0 \tag{10.34}$$

Equation 10.34 can be used as an alternative method to develop the equation of motion. As an illustration, consider a simple single degree-of-freedom spring mass system. Substituting Equations 10.32 and 10.33 into Equation 10.34,

$$\frac{d}{dt}\left(\frac{1}{2}m\dot{x}^2 + \frac{1}{2}kx^2\right) = 0$$
$$(m\dot{x})\ddot{x} + (kx)\dot{x} \quad\; = 0$$
$$m\ddot{x} + kx \quad\quad\quad\; = 0$$

which is the equation of motion.

Example Problem 10.4

Using energy methods, determine the equation of motion for the pendulum shown in Figure 10.14.

Solution: Referring to Figure 10.15, the change in elevation is

$$h = L - L\cos\theta$$

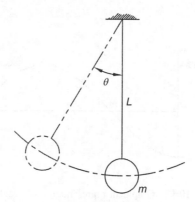

Figure 10.14

Gravitational potential energy is then

$$U = mgh = mg(L - L\cos\theta)$$
$$U = mgL(1 - \cos\theta)$$

The velocity of the mass is determined using

$$v = r\omega$$
$$v = L\dot{\theta}$$

Kinetic energy is

$$T = \frac{1}{2}mv^2 = \frac{1}{2}m\left(L\dot{\theta}\right)^2$$

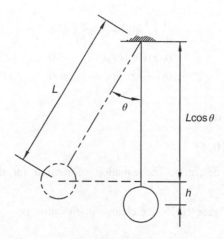

Figure 10.15

Conservation of energy

$$\frac{d}{dt}\left(\frac{1}{2}m(L\dot{\theta})^2 + mgL(1 - \cos\theta)\right) = 0$$
$$m(L^2\dot{\theta})\,\ddot{\theta} + mgL(\sin\theta)\dot{\theta} \qquad = 0$$
$$L\ddot{\theta} + g(\sin\theta) \qquad\qquad = 0$$

Answer: $\ddot{\theta} + \dfrac{g}{L}(\sin\theta) = 0$

We can linearize the equation by using the small-angle approximation

$$\sin\theta \approx \theta$$

Answer: $\ddot{\theta} + \dfrac{g}{L}\theta = 0$

It is often desired to obtain a system's natural frequency without fully developing the equation of motion. Lord Rayleigh developed a method that allows for a calculation of the fundamental frequency without development of the full equation of motion. For Rayleigh's energy method, we first state conservation of energy as

$$T_1 + U_1 = T_2 + U_2 \qquad\qquad (10.35)$$

If we specify the location of state 1 so that it represents the state of maximum potential energy, then $T_1 = 0$ and $U_1 = U_{max}$. Similarly, if state 2 represents the state of maximum kinetic energy, $T_2 = T_{max}$ and $U_2 = 0$. Equation 10.35 becomes

$$T_{max} = U_{max} \qquad\qquad (10.36)$$

Equation 10.36 can be very useful when calculating the natural frequency of a system. The process is best illustrated in an example.

Example Problem 10.5

Use Rayleigh's energy method to find the natural frequency of the system shown in Figure 10.16. Assume the pulley to be frictionless and massless.

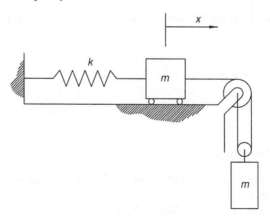

Figure 10.16

Solution: First, we need to find the maximum energy values. The maximum kinetic energy occurs at the maximum velocity. Due to the constrained motion, the hanging mass will move at half the velocity of the rolling mass.

$$T_{max} = \frac{1}{2}m\dot{x}_{max}^2 + \frac{1}{2}m\left(\frac{\dot{x}_{max}}{2}\right)^2$$

The maximum elastic potential energy will occur when the deformation reaches a maximum value. Noting that the gravitational potential energy of the hanging mass is canceled by the work done by the restoring force of the spring, the maximum potential energy is

$$U_{max} = \frac{1}{2}kx_{max}^2$$

Equating the maximum kinetic energy and the maximum potential energy,

$$\frac{1}{2}m\dot{x}_{max}^2 + \frac{1}{2}m\left(\frac{\dot{x}_{max}}{2}\right)^2 = \frac{1}{2}kx_{max}^2$$

$$m\dot{x}_{max}^2 + \frac{1}{4}m\dot{x}_{max}^2 = kx_{max}^2$$

$$\frac{5}{4}m\dot{x}_{max}^2 = kx_{max}^2$$

The motion will be simple harmonic oscillation; therefore,

$$x(t) = A\sin(\omega_n t + \phi) \Rightarrow x_{max} = A$$
$$\dot{x}(t) = A\omega_n \cos(\omega_n t + \phi) \Rightarrow \dot{x}_{max} = A\omega_n$$

Substituting maximum displacement and velocity gives

$$\frac{5}{4}mA^2\omega_n^2 = kA^2$$

The amplitude can then be canceled out indicating that the natural frequency is independent of the amplitude of motion:

$$\omega_n^2 = \left(\frac{4}{5m}\right)k$$

Answer: $\omega_n = \sqrt{\dfrac{4k}{5m}}$

10.5 Torsional Systems

Consider a torsional system, as shown in Figure 10.17. The disk has a moment of inertia I and the shaft is weightless.

Similar to the concept of spring stiffness, the shaft has a torsional stiffness, which is the torque T (pound-inches or Newton meters) required to cause an angular deflection of θ

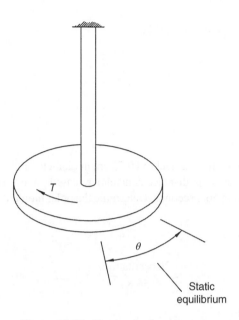

Figure 10.17 Torsional vibration system

(radians). Therefore, torsional stiffness has units of lb-in./rad (or N m/rad):

$$k_t = \frac{T}{\theta} \tag{10.37}$$

For a solid circular shaft having a shear modulus G and length l, the torsional stiffness is

$$k_t = \frac{GJ}{l} \tag{10.38}$$

where the polar moment of inertia of a solid circular shaft is

$$J = \frac{\pi d^4}{32} \tag{10.39}$$

Substituting the polar moment of inertia into Equation 10.38 gives

$$k_t = \frac{G\pi d^4}{32l} \tag{10.40}$$

Newton's second law gives the equation of motion for a torsional system.

$$I\ddot{\theta} + k_t\theta = 0 \tag{10.41}$$

Dividing by the moment of inertia gives

$$\ddot{\theta} + \frac{k_t}{I}\theta = 0 \tag{10.42}$$

Therefore, the natural frequency would be

$$\omega_n = \sqrt{\frac{k_t}{I}} \qquad\qquad (10.43)$$

Example Problem 10.6

A circular disk is suspended from a steel rod 0.75 cm in diameter and 3.5 meters in length. The disk is given an initial angular displacement and released. It is determined that the disk completes 20 oscillations in 56.8 seconds. Determine the polar moment of inertia of the circular disk. Use $G = 80\,\text{GPa}$.

Solution: The frequency is

$$f = \frac{20\ \text{oscillations}}{56.8\ \text{s}} = 0.352$$

and the circular frequency is

$$\omega_n = 2\pi f = 2\pi(0.352) = 2.212 \frac{\text{rad}}{\text{s}}$$

The torsional stiffness of the shaft is

$$k_t = \frac{G\pi d^4}{32l}$$

$$k_t = \frac{(80 \times 10^9\,\text{N/m}^2)\pi(0.0075\ \text{m})^4}{32(3.5\ \text{m})} = 7.1 \frac{\text{N\,m}}{\text{rad}}$$

The moment of inertia of the circular disk can be determined from Equation 10.43:

$$\omega_n = \sqrt{\frac{k_t}{I}} \Rightarrow I = \frac{K_t}{\omega_n^2}$$

$$I = \frac{7.1}{2.212^2}$$

Answer: $I = 1.45\ \text{kg m}^2$

10.6 Damped Systems

Free vibration is somewhat unrealistic because there are no retarding forces to bring the system to rest over time. Therefore, the next problem type to consider is that of damped free vibration. As before, the term free vibration signifies that there is no force function being applied to the system. However, this time the system will include damping forces that eventually bring it to rest. There are three sources of damping: fluid resistance damping, hysteretic damping, and coulomb damping. This section will only focus on fluid resistance damping.

10.6.1 Equations of Motion

Fluid damping can be viscous (damping force is proportional to velocity) or turbulent (damping force is proportional to velocity squared). In viscous damping, a damper, or a dashpot, is the device that resists the motion. The damping occurs as the piston moves within the fluid. Other types of damping can often be approximated using viscous damping, especially for small motions.

In the analysis of systems with viscous damping, the loss of mechanical energy is accounted for by a factor known as the viscous damping coefficient c, which is dependent on the physical geometry of the dashpot (surface area and film thickness) and fluid properties (viscosity). The viscous damping coefficient has units of N s/m or lb s/ft. For viscous damping, forces are proportional to the magnitude of velocity ($F_d = -c\dot{x}$). The value of the damping coefficient is negative because the direction of the damping force will always oppose that of velocity.

Consider the spring–mass system shown in Figure 10.18a. From the free body diagram in Figure 10.18b, Newton's second law yields the following equation with damping considered:

$$\sum F = ma \Rightarrow -c\dot{x} - kx = m\ddot{x} \tag{10.44}$$

Dividing by the mass gives the following general equation:

$$\ddot{x} + \frac{c}{m}\dot{x} + \frac{k}{m}x = 0 \tag{10.45}$$

The constant $\omega_n^2 = k/m$ has previously been defined as the natural circular frequency; however, the term c/m must be considered. It is now convenient to define a damping ratio $\varsigma = c/2m\omega_n$. The term ς (zeta) is a measure of damping severity and is nondimensional. Making both substitutions gives the equation of motion:

$$\ddot{x} + 2\varsigma\omega_n\dot{x} + \omega_n^2 x = 0 \tag{10.46}$$

Solution of the second-order homogeneous differential equation begins with the characteristic equation of $s^2 + 2\varsigma\omega_n s + \omega_n^2 = 0$, which yields two roots of $s_{1,2} = \omega_n\left(-\varsigma \pm \sqrt{\varsigma^2 - 1}\right)$. Because $0 \le \varsigma \le \infty$, the value of $\varsigma^2 - 1$ under the radical can be positive, negative, or zero. Therefore, the solution to the characteristic equation gives three possible cases depending on

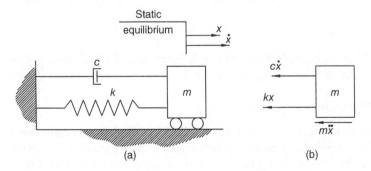

Figure 10.18 (a) Spring–mass system with viscous damping. (b) Free body diagram

the value of $\varsigma^2 - 1$. Depending on that value, the system will be overdamped, critically damped, or underdamped. Each of the three possible cases will now be examined in detail.

10.6.2 Critically Damped Systems

For a critically damped system, the value under the radical equals zero.

$$\varsigma^2 - 1 = 0$$
$$\left(\frac{c}{2m\omega_n}\right)^2 = 1 \tag{10.47}$$

The value of c that satisfies Equation 10.47 is known as the critical damping value and can be determined by solving for c:

$$c_{cr} = 2m\omega_n$$
$$c_{cr} = \frac{c}{\varsigma} \tag{10.48}$$

For a critically damped system, $\varsigma = 1$ and the roots of the characteristic equation are $s_1 = s_2 = -\omega_n$. Recall from differential equations that for a case of repeated real roots, the general solution is in the form $x = A_1 e^{st} + A_2 t e^{st}$. Substituting $s_1 = s_2 = -\omega_n$ gives

$$x(t) = e^{-\omega_n t}(A_1 + A_2 t) \tag{10.49}$$

Once again, the constants need to be determined from initial conditions. Constant A_1 is determined by setting time equal to zero in Equation 10.49 to get

$$x_0 = A_1 \tag{10.50}$$

From the time derivative of Equation 10.49 at time equal to zero,

$$\dot{x}(t) = -\omega_n e^{-\omega_n t}(A_1 + A_2 t) + A_2 e^{-\omega_n t}$$
$$v_0 = -\omega_n A_1 + A_2 = -\omega_n x_0 + A_2 \tag{10.51}$$
$$v_0 + \omega_n x_0 = A_2$$

Using initial condition, the general response of a critically damped system is

$$x(t) = e^{-\omega_n t}[x_0 + (v_0 + \omega_n x_0)t] \tag{10.52}$$

Figure 10.19 shows example plots for three different cases of initial velocity. From the plots, it can be seen that the mass will pass through the position of equilibrium at most one time.

10.6.3 Overdamped Systems

An overdamped system is one in which $\varsigma > 1$. This type of system is nonoscillatory, referred to as aperiodic, and therefore there is no period related with the motion. For cases when $\varsigma > 1$, the roots of the characteristic equation $s_{1,2} = \omega_n\left(-\varsigma \pm \sqrt{\varsigma^2 - 1}\right)$ are distinct real roots. Recall from differential equations that the solution for cases of distinct real roots is $x = A_1 e^{s_1 t} + A_2 e^{s_2 t}$.

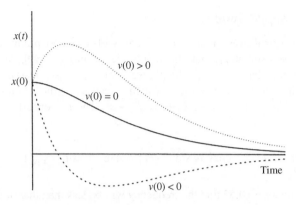

Figure 10.19   Critically damped motion plots

Substituting the roots gives

$$x(t) = A_1 e^{\omega_n\left(-\varsigma+\sqrt{\varsigma^2-1}\right)t} + A_2 e^{\omega_n\left(-\varsigma-\sqrt{\varsigma^2-1}\right)t} \tag{10.53}$$

The constants, one of which is increasing while the other is decreasing, are determined from initial conditions:

$$\begin{aligned} A_1 &= \frac{v_0 + \omega_n\left(\varsigma + \sqrt{\varsigma^2-1}\right)x_0}{2\omega_n\sqrt{\varsigma^2-1}} \\[2mm] A_2 &= \frac{-v_0 - \omega_n\left(\varsigma - \sqrt{\varsigma^2-1}\right)x_0}{2\omega_n\sqrt{\varsigma^2-1}} \end{aligned} \tag{10.54}$$

Figure 10.20 shows the plot of the response for an overdamped system. Both terms of Equation 10.54 are plotted individually (dashed lines) and the total response (solid line) is the sum of the individual terms.

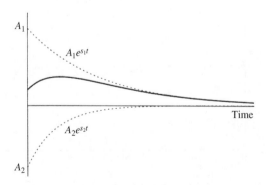

Figure 10.20   Overdamped system response

10.6.4 Underdamped Systems

Probably, the most critical case for vibrating systems will be an underdamped system ($c < c_{cr}$). The motion will be oscillatory, but the amplitude of motion will decay exponentially. Underdamped systems have $\varsigma < 1$, which give roots $s_{1,2} = -\varsigma \omega_n \pm \omega_n \sqrt{\varsigma^2 - 1}i$ that are complex. Recall that $i = \sqrt{-1}$, which causes the roots to become $s_{1,2} = -\varsigma \omega_n \pm \omega_n \sqrt{(\varsigma^2 - 1)(-1)} = -\varsigma \omega_n \pm \omega_n \sqrt{1 - \varsigma^2}$. The response for underdamped systems is therefore

$$x(t) = A_1 e^{-\varsigma \omega_n t} \cos\left(\omega_n \sqrt{1 - \varsigma^2} t\right) + A_2 e^{-\varsigma \omega_n t} \sin\left(\omega_n \sqrt{1 - \varsigma^2} t\right) \tag{10.55}$$

It can be seen in Equation 10.55 that the frequency will be less than that of free vibration. The damped natural frequency is determined by

$$\omega_d = \omega_n \sqrt{1 - \varsigma^2} = \omega_n \sqrt{1 - \left(\frac{c}{c_{cr}}\right)^2} \tag{10.56}$$

The damped period is

$$\tau_d = \frac{2\pi}{\omega_d} \tag{10.57}$$

The effects of damping on the natural frequency can be seen by rearranging Equation 10.56. Dividing by ω_n gives $\omega_d/\omega_n = \sqrt{1 - \varsigma^2}$, which is plotted in Figure 10.21. When ζ is zero, the ratio is equal to 1, indicating that the damped natural frequency is equal to the undamped natural frequency. As the damping ratio increases, the ratio decreases because the damped natural frequency is less than the undamped natural frequency. The decrease in frequency is small unless the damping ratio approaches 1, which is the state of critical damping. The

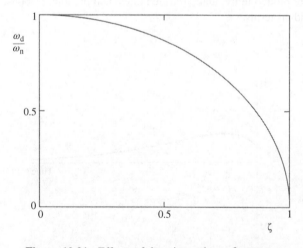

Figure 10.21 Effects of damping ratio on frequency

damping is typically small in mechanical systems resulting in a case where the damped natural frequency is nearly equal to the undamped natural frequency.

Equation 10.55 can then be expressed in the form

$$x(t) = e^{-\varsigma\omega_n t}(A_1 \cos \omega_d t + A_2 \sin \omega_d t) \qquad (10.58)$$

Notice that the terms within the parentheses are simply the response for a free vibration system. Therefore, Equation 10.58 represents free vibration that is exponentially decaying by a factor of $e^{-\varsigma\omega_n t}$. As with free vibration, the integration constants A_1 and A_2 depend on how the motion started. Introducing initial conditions gives the final response:

$$x(t) = e^{-\varsigma\omega_n t}\left(x_0 \cos \omega_d t + \frac{v_0 + x_0\varsigma\omega_n}{\omega_d} \sin \omega_d t \right) \qquad (10.59)$$

Equation 10.59 can also be written in the alternative form using phase angle:

$$x(t) = A e^{-\varsigma\omega_n t} \sin(\omega_d t + \phi_d) \qquad (10.60)$$

where

$$A = \sqrt{x_0^2 + \left(\frac{v_0 + x_0\varsigma\omega_n}{\omega_d} \right)^2}$$

$$\phi_d = \tan^{-1}\left(\frac{x_0\omega_d}{v_0 + \varsigma\omega_n x_0} \right)$$

Figure 10.22 shows a plot of underdamped motion.

Example Problem 10.7

The system shown in Figure 10.23 has a mass of 18 kg, the spring constant is 40 N/m, and the damping constant is 60 N s/m. Determine if the system is underdamped, critically damped, or overdamped. Then, determine the system response for an initial displacement of 0.4 m to the right.

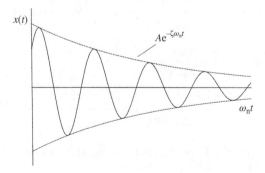

Figure 10.22 Underdamped system response

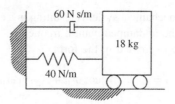

Figure 10.23

Solution: The natural frequency

$$\omega_n = \sqrt{\frac{k}{m}} = \sqrt{\frac{40 \text{ N/M}}{18 \text{ kg}}} = 1.49 \frac{\text{rad}}{\text{s}}$$

The damping ratio

$$\varsigma = \frac{c}{2m\omega_n} = \frac{60 \text{ N s/m}}{2(18 \text{ kg})(1.49 \text{ rad/s})} = 1.12$$

Because the damping ratio is greater than 1, the system is overdamped.
  Answer: $\varsigma = 1.12 > 1$ overdamped
  The response will be

$$x(t) = A_1 e^{\omega_n \left(-\varsigma + \sqrt{\varsigma^2 - 1}\right)t} + A_2 e^{\omega_n \left(-\varsigma - \sqrt{\varsigma^2 - 1}\right)t}$$

$$x(t) = A_1 e^{1.49 \left(-1.12 + \sqrt{1.12^2 - 1}\right)t} + A_2 e^{1.49 \left(-1.12 - \sqrt{1.12^2 - 1}\right)t}$$

$$x(t) = A_1 e^{-0.92t} + A_2 e^{-2.42t}$$

The constants are determined using Equation 2.54:

$$A_1 = \frac{v_0 + \omega_n \left(\varsigma + \sqrt{\varsigma^2 - 1}\right)x_0}{2\omega_n \sqrt{\varsigma^2 - 1}}$$

$$A_1 = \frac{0 + 1.49 \left(1.12 + \sqrt{1.12^2 - 1}\right)(0.4)}{2(1.49)\sqrt{1.12^2 - 1}} = 0.647$$

$$A_2 = \frac{-v_0 - \omega_n \left(\varsigma - \sqrt{\varsigma^2 - 1}\right)x_0}{2\omega_n \sqrt{\varsigma^2 - 1}}$$

$$A_2 = \frac{0 - 1.49 \left(1.12 - \sqrt{1.12^2 - 1}\right)(0.4)}{2(1.49)\sqrt{1.12^2 - 1}} = -0.247$$

The constants are substituted into the response equation to give
Answer: $x(t) = 0.647 e^{-0.92t} - 0.247 e^{-2.42t}$
Figure 10.24 shows the plot of the system response.

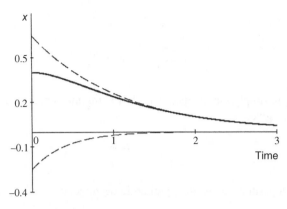

Figure 10.24

10.7 Logarithmic Decrement

The exact value of the damping constant c is often unknown. However, it can be determined experimentally by measuring successive peak amplitudes. Consider the response of a system to initial conditions, as shown in Figure 10.25. The first three peak amplitudes (x_1, x_2, x_3) could be measured experimentally.

Successive amplitudes will have a logarithmic relationship for viscously damped motion. The decay of motion is expressed by the logarithmic decrement, which is the natural logarithm of the ratio between two successive maximum amplitudes:

$$\delta = \ln \frac{x_i}{x_{i+1}} \tag{10.61}$$

Assume a maximum displacement x_i occurs at a time t_i. At that maximum displacement, $\sin(\omega_d t_i + \phi_d) = 1$ and Equation 10.60 reduces to $x_i = Ae^{-\varsigma \omega_n t_i}$. Similarly, maximum displacement x_{i+1} occurs at a time t_{i+1} giving $x_{i+1} = Ae^{-\varsigma \omega_n t_{i+1}}$. The ratio of successive peak amplitudes

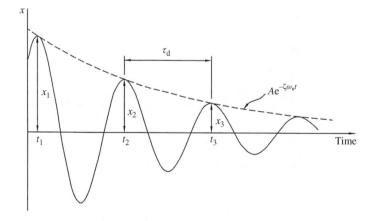

Figure 10.25   Decay of successive peak amplitudes for underdamped system

is then calculated:

$$\frac{x_i}{x_{i+1}} = \frac{Ae^{-\varsigma\omega_n t_i}}{Ae^{-\varsigma\omega_n t_{i+1}}} = e^{-\varsigma\omega_n(t_i - t_{i+1})} \tag{10.62}$$

Notice that $t_{i+1} - t_i$ is simply the damped period. Making the substitution of damped period from Equation 10.57 gives

$$\frac{x_i}{x_{i+1}} = e^{\varsigma\omega_n(2\pi/\omega_d)} \tag{10.63}$$

Taking the natural logarithm gives the logarithmic decrement:

$$\delta = \ln \frac{x_i}{x_{i+1}} = \varsigma\omega_n \frac{2\pi}{\omega_d} = \frac{2\pi\varsigma}{\sqrt{1 - \varsigma^2}} \tag{10.64}$$

If the damping is small ($x_i \approx x_{i+1}$), we can assume $\omega_d \approx \omega_n$ and Equation 10.64 gives

$$\delta \approx 2\pi\varsigma \tag{10.65}$$

To show the validity of the small damping assumption, Figure 10.26 shows logarithmic decrement as a function of damping ratio. The solid line is a plot of Equation 10.64 and the dashed line is a plot of Equation 10.65. It can be seen that for small values of ζ, the plots are nearly identical.

For some systems, x_i and x_{i+1} may be too close to accurately measure difference experimentally. For such a system, Equation 10.64 can be modified to give

$$\delta = \frac{1}{n} \ln \frac{x_i}{x_{i+n}} \tag{10.66}$$

using peak amplitudes that are n cycles apart.

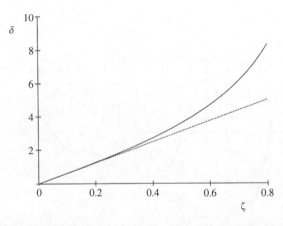

Figure 10.26   Logarithmic decrement (δ) as a function of damping ratio (ζ)

Example Problem 10.8

A damped vibrating system's amplitude decreases 10% on each cycle of motion. Find the damping coefficient c of the system if $k = 150$ lb/in. and $m = 20$ lb s^2/in.

Solution: The undamped natural frequency

$$\omega_n = \sqrt{\frac{k}{m}} = \sqrt{\frac{150}{20}} = 2.739 \frac{\text{rad}}{\text{s}}$$

Logarithmic decrement gives

$$\delta = \ln \frac{x_i}{x_{i+1}} = \ln\left(\frac{1}{0.9}\right) = 0.105$$

The damping ratio is determined from $\delta \approx 2\pi\varsigma$

$$\varsigma = \frac{\delta}{2\pi} = \frac{0.105}{2\pi} = 0.017$$

The damping coefficient can then be determined from the definition of damping ratio:

$$\varsigma = c/2m\omega_n$$
$$c = 2m\omega_n\varsigma = 2(20)(2.739)(0.017)$$

Answer: $c = 1.837 \dfrac{\text{lb s}}{\text{in.}}$

Example Problem 10.9

An underdamped vibrating system has a 5.3 kg block, a spring ($k = 230$ N/m), and a dashpot. The ratio of two consecutive amplitudes is $1.0 : 0.7$. The block is given an initial displacement of 6 cm and an initial velocity of 1.2 m/s. Determine the response and plot displacement for four cycles.

Solution: The undamped natural frequency would be

$$\omega_n = \sqrt{\frac{k}{m}} = \sqrt{\frac{230 \, \text{N/m}}{5.3 \, \text{kg}}} = 6.5876 \frac{\text{rad}}{\text{s}}$$

Assuming $\omega_d \approx \omega_n$,

$$\ln \frac{x_i}{x_{i+1}} \approx \varsigma(2\pi) \Rightarrow \varsigma \approx \frac{1}{2\pi} \ln\left(\frac{1.0}{0.7}\right) \approx 0.0568$$

From the definition of damping ratio $\varsigma = c/2m\omega_n$,

$$c \approx 2m\omega_n\varsigma \approx 2(5.3 \, \text{kg})(6.5876 \, \text{rad/s})(0.0568) \approx 3.9639$$

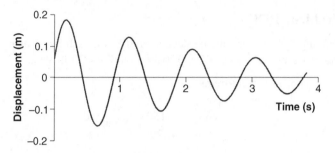

Figure 10.27

This gives the following damped natural frequency:

$$\omega_d = \omega_n \sqrt{1 - \varsigma^2} = 6.5876\sqrt{1 - 0.0568^2} = 6.5770$$

From Equation 10.59,

$$x(t) = e^{-\varsigma\omega_n t}\left(x_0 \cos \omega_d t + \frac{v_0 + x_0\varsigma\omega_n}{\omega_d} \sin \omega_d t\right)$$

$$x(t) = e^{-0.0568(6.5876)t}\left(0.06 \cos(6.577t) + \frac{1.2 + 0.06(0.0568)(6.5876)}{6.577} \sin(6.577t)\right)$$

Answer: $x(t) = e^{-0.3742t}(0.06 \cos(6.577t) + 0.1859 \sin(6.577t))$

The damped period would be

$$\tau_d = \frac{2\pi}{\omega_d} = \frac{2\pi}{6.577} = 0.9553 \text{ s}$$

The first four cycles would occur in $4(0.9553) = 3.8213$ s. The plot of the response is shown in Figure 10.27.

10.8   Forced Vibration: Harmonic Forcing Functions

In the previous sections, we discussed simple harmonic oscillation without any type of forcing function. Systems in free vibration will oscillate at the system's natural frequency, therefore the resulting motion is fairly easy to determine. Free vibrations are important and have several practical applications. However, it is common to have a system that is continually excited by a forcing function. The forcing function, which generally varies with time, can be generated in several ways. Some examples include external forces, support motion, wind, impact, and rotating system unbalance.

A system's response to any forcing function depends greatly on the type of excitation. This chapter will focus on harmonic excitation and some general basic excitations. Harmonic excitation is the most basic, and the system will eventually vibrate at the excitation frequency. We also begin with harmonic forcing functions because a complete understanding of harmonic excitation is a prerequisite to more complex loading cases. Response of undamped systems will be covered first followed by the response of damped systems. Periodic forcing functions are also included in this section, primarily because periodic functions can often be approximated

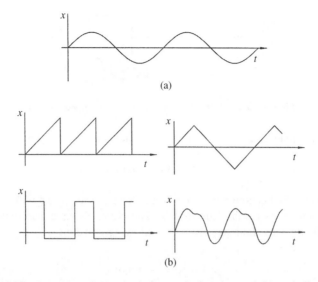

Figure 10.28   Examples of (a) a basic harmonic function and (b) periodic functions

with a series of harmonic functions using Fourier analysis. Using Fourier analysis, the system's response to periodic loading can be determined by superposition.

10.8.1 Harmonic versus Periodic Functions

This chapter focuses on harmonic forcing functions (Appendix C covers topics on periodic forcing functions), but we must understand the difference between harmonic and periodic functions. A periodic function $f(x)$ with period p is one in which

$$f(x) = f(x \pm p) = f(x \pm 2p) = \cdots = f(x \pm np) \tag{10.67}$$

where n is any integer. In other words, a periodic function repeats after an interval p. The shape of periodic functions can vary, as shown in Figure 10.28b.

Harmonic functions are a special case of periodic functions. In basic terms, a harmonic function is a function that varies as a sine or cosine function (or has the basic shape of a sine or cosine function). Harmonic functions are also called sinusoidal functions because all sine and cosine functions can be produced by shifting and stretching a basic sine function. A simple harmonic function is shown in Figure 10.28a.

10.8.2 Equations of Motion for Harmonic Excitation

The first class of forcing functions we will examine is harmonic forcing functions. Harmonic forcing functions are very common in engineering practice, and therefore one of the most important to study. Mechanical and structural systems often contain (or support) rotating machinery, which produce harmonic excitations due to rotating parts having eccentricities.

We will first examine the response of an undamped system to an externally applied harmonic force. Figure 10.29 shows an undamped system with an external force $F(t)$ that varies with time. There are numerous ways the external force could be applied to the system, and we will discuss some options in later sections of this chapter.

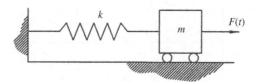

Figure 10.29   Forced vibration of an undamped system

The differential equation of motion comes from Newton's second law in the x-direction:

$$m\ddot{x} + kx = F(t) \tag{10.68}$$

The forcing function $F(t)$ can be any function, but for this section we will consider the case when that function is harmonic (sinusoidal). Therefore, using a harmonic force $F(t) = F_0 \sin(\omega t)$ and dividing by the mass gives the following equation:

$$\ddot{x} + \omega_n^2 x = \frac{F_0}{m} \sin(\omega t) \tag{10.69}$$

where F_0 is the amplitude of the force and ω is the frequency of the applied force. It is extremely important to distinguish between the two frequency terms in Equation 10.69. The natural frequency $\omega_n = \sqrt{k/m}$ is a system property and was discussed in detail in the previous sections. The term ω is a property of the applied force and has nothing to do with the system properties. Also, the force function could be $F(t) = F_0 \cos(\omega t)$. The solution would be developed exactly the same to that presented below with the substitution of $\cos(\omega t)$ for $\sin(\omega t)$ in the results.

As you know from the study of differential equations, the solution of the second-order differential equation10.69 will be the sum of two parts: the complementary solution x_c (the homogeneous solution developed in Section 10.4) and the particular solution x_p (any solution that satisfies the nonhomogeneous equation of motion). For undamped motion, the particular solution will be in the form of the force function. Because the forcing function was a sine function, the trial particular solution will be

$$x_p(t) = X \sin \omega t \tag{10.70}$$

where X is the unknown amplitude (in units of length) of the particular solution and ω is the frequency of the forcing function (not the natural frequency ω_n of the system). We will also need the second derivative of the particular solution, which is

$$\ddot{x}_p(t) = -X\omega^2 \sin \omega t \tag{10.71}$$

Substituting Equations 10.70 and 10.71 into Equation 10.69 gives

$$-X\omega^2 \sin \omega t + \omega_n^2 (X \sin \omega t) = \frac{F_0}{m} \sin(\omega t)$$

$$X(\omega_n^2 - \omega^2) = \frac{F_0}{m} \tag{10.72}$$

$$X = \frac{F_0/k}{1 - (\omega/\omega_n)^2}$$

Substituting the amplitude shown in Equation 10.72 into the particular solution shown in Equation 10.70 gives the particular solution:

$$x_p(t) = \frac{F_0/k}{1 - (\omega/\omega_n)^2} \sin \omega t \tag{10.73}$$

Notice that the particular solution has no reference to the initial conditions. Recall that the homogeneous solution gives the complimentary solution:

$$x_c(t) = A_1 \cos \omega_n t + A_2 \sin \omega_n t \tag{10.74}$$

The total solution is the sum of complimentary and particular solutions $x = x_c + x_p$ to give

$$x(t) = A_1 \cos \omega_n t + A_2 \sin \omega_n t + \frac{F_0/k}{1 - (\omega/\omega_n)^2} \sin \omega t \tag{10.75}$$

Equation 10.75 describes the complete motion of the system over time. Initially, the free and forced oscillations exist together, which can result in rather confusing motion. The first two terms of Equation 10.75 depend on initial conditions and are referred to as the transient solution, even though damping is not present for this case. If any small amount of damping was present, the first two terms would decay and only the last term would survive. Therefore, the last term (particular solution) is typically of primary importance and is referred to as the steady-state solution. Steady-state vibration is motion that remains (as long as the forcing function remains) after the transient vibration dies away. An example would be vibrations associated with continuous operation of machinery.

If the system starts at rest ($x_0 = v_0 = 0$), then the constants in Equation 10.75 become

$$A_1 = 0 \qquad A_2 = \frac{(-F_0/k)(\omega/\omega_n)}{1 - (\omega/\omega_n)^2}$$

and Equation 10.75 becomes

$$x(t) = \frac{F_0/k}{1 - (\omega/\omega_n)^2} \left(\sin \omega t - \frac{\omega}{\omega_n} \sin \omega_n t \right) \tag{10.76}$$

10.8.3 Resonance

It is important to further explore the peak amplitude of the steady-state solution, which is given in Equation 10.72. It is more convenient to express Equation 10.72 in terms of a nondimensional amplitude ratio, which is $X/(F_0/k)$. We can rearrange Equation 10.72 to get the following expression for amplitude ratio (also called magnification factor):

$$\frac{X}{F_0/k} = \frac{1}{1 - (\omega/\omega_n)^2} \tag{10.77}$$

It is of interest to examine the effects of the forcing function frequency ω on the amplitude ratio. At $\omega = 0$, the force simply becomes F_0 and the amplitude ratio becomes 1 because the amplitude is equal to the static amplitude:

$$\frac{X}{F_0/k}\bigg|_{\omega=0} = \frac{1}{1-(0/\omega_n)^2} = 1 \tag{10.78}$$

As $\omega \to \infty$, the amplitude becomes 0, as shown in Equation 10.79. The system becomes motionless because it does not have time to respond to a force function with infinite frequency.

$$\frac{X}{F_0/k}\bigg|_{\omega\to\infty} = \lim_{\omega\to\infty} \frac{1}{1-(\omega/\omega_n)^2} = 0 \tag{10.79}$$

It is especially important to examine the behavior of the system when $\omega \to \omega_n$.

$$\frac{X}{F_0/k}\bigg|_{\omega\to\omega_n} = \lim_{\omega\to\omega_n} \frac{1}{1-(\omega/\omega_n)^2} = \infty \tag{10.80}$$

Therefore, in the absence of damping, the amplitude increases without bound as the forcing frequency approaches the system's natural frequency. This condition of large amplitude motion at or near the natural frequency is known as resonance. A plot of the amplitude ratio as a function of the dimensionless frequency ratio ω/ω_n is shown in Figure 10.30. The infinite amplitude ratio (resonance) occurs when $\omega = \omega_n$, or a frequency ratio is equal to 1.

For $\omega < \omega_n$ the amplitude ratio is positive, and for $\omega > \omega_n$ the ratio is negative. Positive values indicate that the vibration is in phase with the force, and negative values indicate that the vibration is 180° out of phase with the force. It is very common to plot the absolute value (or magnitude) of the amplitude ratio, as shown in Figure 10.31.

Due to strength limitations of structural materials, failure occurs if the amplitude becomes extremely large. Therefore, it is very important in engineering design to be aware of resonance frequencies.

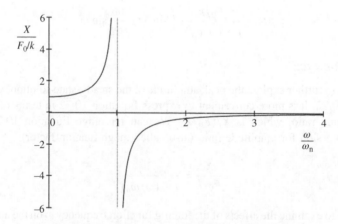

Figure 10.30 Amplitude ratio for an undamped system

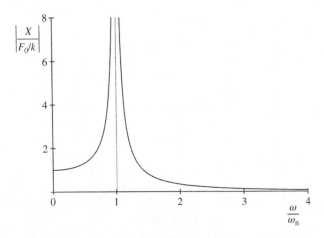

Figure 10.31 Absolute value of amplitude ratio for undamped system

Example Problem 10.10

The undamped system shown in Figure 10.32 is excited by a harmonic forcing function $F(t) = F_0 \sin \omega t$, where the amplitude F_0 is 8 lb. Determine the range of ω for which the magnitude of the steady-state response is less than 2 inches.

Solution: The natural frequency of the system is

$$\omega_n = \sqrt{\frac{k}{m}} = \sqrt{\frac{10.5\,\text{lb/in.}(12\,\text{in.}/ft)}{90\,\text{lb}/32.2\,\text{ft/s}^2}} = 6.71\,\frac{\text{rad}}{\text{s}}$$

The steady-state amplitude (in inches) is determined from

$$X = \frac{F_0/k}{1 - (\omega/\omega_n)^2} = \frac{8\,\text{lb}/(10.5\,\text{lb/in.})}{1 - (\omega/6.71)^2} = \frac{0.76\,\text{in.}}{1 - (\omega/6.71)^2}$$

Figure 10.33 shows a plot of the steady-state amplitude along with dashed lines representing the maximum allowed amplitude of 2 inches.

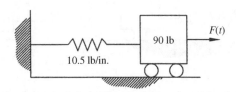

Figure 10.32

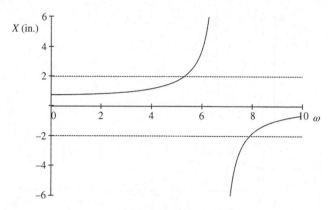

Figure 10.33

We then need to determine the value of ω for amplitudes of 2 inches and -2 inches:

$$X = 2 = \frac{0.76 \text{ in.}}{1 - (\omega/6.71)^2} \Rightarrow \omega = 5.28$$

$$X = -2 = \frac{0.76 \text{ in.}}{1 - (\omega/6.71)^2} \Rightarrow \omega = 7.89$$

Answer: $\omega < 5.28 \text{ rad/s}$ or $\omega > 7.89 \text{ rad/s}$

10.8.4 Damped Response to Harmonic Excitation

We will now consider the case when viscous damping is added into the system. Figure 10.34a shows a system with viscous damping that is excited by an external harmonic force $F(t)$.

Using Newton's second law and the free body diagram in Figure 10.34b, we can develop the differential equation of motion:

$$\sum F = ma$$
$$-c\dot{x} - kx + F(t) = m\ddot{x} \tag{10.81}$$

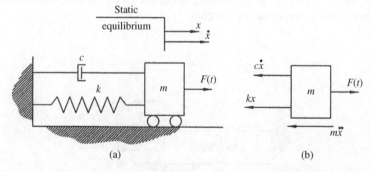

Figure 10.34 System with viscous damping. (a) Spring–mass system. (b) Corresponding free body diagram

Dividing by the mass gives the following general equation:

$$\ddot{x} + \frac{c}{m}\dot{x} + \frac{k}{m}x = \frac{F(t)}{m}$$ (10.82)

The forcing function again causes the equation of motion to be a nonhomogeneous second-order differential equation. We can now substitute the natural circular frequency and damping ratio into the equation of motion. We will again use the harmonic forcing function $F(t) = F_0 \sin(\omega t)$.

$$\ddot{x} + 2\varsigma\omega_n\dot{x} + \omega_n^2 x = \frac{F_0}{m}\sin(\omega t)$$ (10.83)

The solution to Equation 10.83 will again be the combination of the complimentary solution and any particular solution that is a solution to the complete equation of motion. For undamped motion, the particular solution only needed a single sine function, as determined in Equation 10.73. However, once damping is introduced into the system, the particular solution must be in the more general form:

$$x_p = A \cos \omega t + B \sin \omega t$$ (10.84)

and the first two time derivatives are

$$\begin{aligned} \dot{x} &= -A\omega \sin \omega t + B\omega \cos \omega t \\ \ddot{x} &= -A\omega^2 \cos \omega t - B\omega^2 \sin \omega t \end{aligned}$$ (10.85)

Substituting the particular solution and derivatives into Equation 10.83 gives

$$\begin{aligned} \left[-A\omega^2 \cos \omega t - B\omega^2 \sin \omega t\right] + 2\varsigma\omega_n\left[-A\omega \sin \omega t + B\omega \cos \omega t\right] \\ +\omega_n^2[A \cos \omega t + B \sin \omega t] = \frac{F_0}{m}\sin(\omega t) \end{aligned}$$ (10.86)

For Equation 10.86 to be true for all values of time, the cosine terms on the left-hand side of the equation must equal the cosine terms on the right-hand side of the equation and the sine terms must be equal on both sides of the equation. Therefore, Equation 10.86 will give us two equations that can be solved to get our terms A and B. Collecting the cosine terms on both sides of the equation gives

$$\left(\omega_n^2 - \omega^2\right)A + (2\varsigma\omega_n\omega)B = 0$$ (10.87)

and collecting sine terms gives

$$(-2\varsigma\omega_n\omega)A + \left(\omega_n^2 - \omega^2\right)B = \frac{F_0}{m}$$ (10.88)

This gives two equations with the unknowns A and B. In matrix form, the equations become

$$\begin{bmatrix} -2\varsigma\omega_n\omega & \omega_n^2 - \omega^2 \\ \omega_n^2 - \omega^2 & 2\varsigma\omega_n\omega \end{bmatrix}\begin{bmatrix} A \\ B \end{bmatrix} = \begin{bmatrix} F_0/m \\ 0 \end{bmatrix}$$ (10.89)

Using Cramer's rule to solve for the unknowns gives

$$A = \frac{\begin{vmatrix} F_0/m & \omega_n^2 - \omega^2 \\ 0 & 2\varsigma\omega_n\omega \end{vmatrix}}{\begin{vmatrix} -2\varsigma\omega_n\omega & \omega_n^2 - \omega^2 \\ \omega_n^2 - \omega^2 & 2\varsigma\omega_n\omega \end{vmatrix}}$$

$$A = \frac{-(2\varsigma\omega_n\omega)F_0/m}{\left(\omega_n^2 - \omega^2\right)^2 + (2\varsigma\omega_n\omega)^2}$$

(10.90)

and

$$B = \frac{\begin{vmatrix} -2\varsigma\omega_n\omega & F_0/m \\ \omega_n^2 - \omega^2 & 0 \end{vmatrix}}{\begin{vmatrix} -2\varsigma\omega_n\omega & \omega_n^2 - \omega^2 \\ \omega_n^2 - \omega^2 & 2\varsigma\omega_n\omega \end{vmatrix}}$$

$$B = \frac{\left(\omega_n^2 - \omega^2\right)F_0/m}{\left(\omega_n^2 - \omega^2\right)^2 + (2\varsigma\omega_n\omega)^2}$$

(10.91)

It is often more convenient to express the particular solution in the alternative form:

$$x_p = X \sin(\omega t - \phi)$$

(10.92)

In Equation 10.92, X is the unknown amplitude, ω is the frequency of the forcing function, and ϕ is the phase angle. The phase angle is necessary because the displacement vector will lag the force vector (the motion occurs after the force application). The first two time derivatives give

$$\dot{x}_p = \omega X \cos(\omega t - \phi)$$

(10.93)

$$\ddot{x}_p = -\omega^2 X \sin(\omega t - \phi)$$

(10.94)

The particular solution shown in Equation 10.92 is known as the steady-state solution. The steady-state amplitude is determined using values obtained from Equations 10.90 and 10.91 by $X = \sqrt{A^2 + B^2}$. Simplifying the expression gives the following steady-state amplitude equation:

$$X = \frac{F_0/m}{\sqrt{\left(\omega_n^2 - \omega^2\right)^2 + (2\varsigma\omega_n\omega)^2}}$$

(10.95)

It is again important to note that the steady-state amplitude does not depend on initial conditions. The amplitude is dependent on the magnitude and frequency of the forcing function as well as on the natural frequency of the system. You would intuitively expect the steady-state amplitude to become larger for larger forces, which can be confirmed by Equation 10.95. Also, you would expect the steady-state amplitude to decrease if the damping of the system becomes large. This intuition can also be verified in Equation 10.95.

It is once again convenient to put the steady-state amplitude into the form of a dimensionless ratio. The dimensionless ratio is called the amplitude ratio or the magnification factor. Dividing Equation 10.95 by the static displacement gives the amplitude ratio:

$$\frac{X}{F_0/k} = \frac{1}{\sqrt{\left(1-(\omega/\omega_n)^2\right)^2 + (2\varsigma(\omega/\omega_n))^2}} \tag{10.96}$$

The amplitude ratio varies with the damping ratio and the frequency ratio.

As with undamped systems, at $\omega = 0$ the amplitude becomes 1 (the system is in static equilibrium). The amplitude approaches zero as $\omega \to \infty$. Unlike undamped systems, the amplitude ratio is always finite, and at $\omega = \omega_n$, the amplitude ratio is inversely proportional to the damping ratio giving

$$\left.\frac{X}{F_0/k}\right|_{\omega \to \omega_n} = \lim_{\omega \to \omega_n} \frac{1}{\sqrt{\left(1-(\omega/\omega_n)^2\right)^2 + (2\varsigma(\omega/\omega_n))^2}} = \frac{1}{2\varsigma} \tag{10.97}$$

Figure 10.35 shows plots of amplitude ratio of damped systems with different damping ratios.

The amplitude ratio for a system with damping ratio of 0.15 is shown in Figure 10.36. Close inspection of the peak value of amplitude ratio shows that the peak value does not occur at a frequency ratio of 1, but instead occurs slightly to the left of that value.

The location of peak amplitude shifts farther to the left with increasing magnitude of damping ratio. The actual peak amplitude ratio can be determined by taking the derivative of Equation 10.96 with respect to frequency ratio. Results of the differentiation will give the location of peak amplitude ratio to be

$$\frac{\omega}{\omega_n} = \sqrt{1 - 2\varsigma^2} \tag{10.98}$$

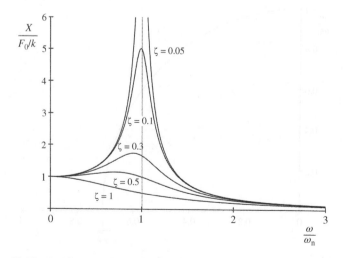

Figure 10.35   Amplitude ratio for a damped system for different levels of damping

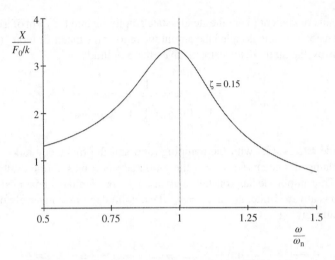

Figure 10.36 Location of maximum amplitude ratio

Equation 10.98 is plotted in Figure 10.37. It is seen in Figure 10.37, and it can be determined from Equation 10.98, that no peak amplitude ratio is present for $\varsigma > 1/\sqrt{2}$.

The phase angle (angle the response lags the forcing function) in Equation 10.92 is determined from

$$\tan \phi = \frac{-A}{B}$$

$$\phi \quad = \tan^{-1}\left(\frac{2\varsigma(\omega/\omega_n)}{1 - (\omega/\omega_n)^2} \right) \tag{10.99}$$

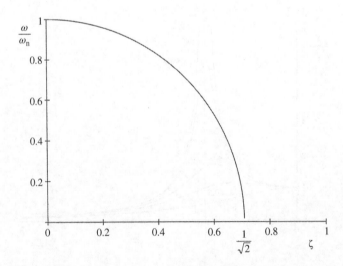

Figure 10.37 Location of peak amplitude ratio versus damping ratio

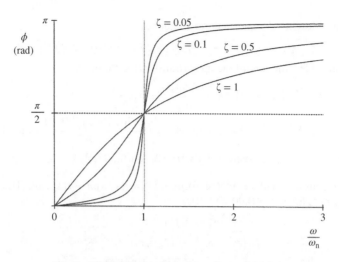

Figure 10.38 Phase angle for a damped system

The plot of the phase angle is shown in Figure 10.38. At $\omega = 0$, the phase angle becomes zero and the system moves in phase with the forcing function. It can be seen that at resonance, the phase angle is 90° for all values of ς. At $\omega \to \infty$ the phase angle becomes π, so the system moves 180° out of phase with the forcing function.

The total system response is the sum of the transient and steady-state solutions. The transient solution, or complimentary solution, is determined using Equation 10.58 if we assume an underdamped case. Combining that response with the steady-state response from Equation 10.92 gives

$$x(t) = e^{-\varsigma\omega_n t}(A_1 \cos \omega_d t + A_2 \sin \omega_d t) + X \sin(\omega t - \phi) \tag{10.100}$$

Example Problem 10.11

An underdamped vibrating system has a 4 kg block, a spring ($k = 200$ N/m), and a dashpot ($c = 9$ N s/m). The block is given an initial displacement of 200 cm and is subjected to a forcing function $F(t) = 8 \cos(10t)$ in Newtons. Determine the response and plot displacement for the first 5 seconds of motion.

Solution: The undamped natural frequency would be

$$\omega_n = \sqrt{\frac{k}{m}} = \sqrt{\frac{200\,\text{N/m}}{4\,\text{kg}}} = 7.0711\,\frac{\text{rad}}{\text{s}}$$

From the definition of damping ratio,

$$\varsigma = \frac{c}{2m\omega_n} = \frac{9}{2(4)(7.0711)} = 0.1591$$

This gives the following damped natural frequency:

$$\omega_d = \omega_n \sqrt{1 - \varsigma^2} = 7.0711 \sqrt{1 - 0.1591^2} = 6.981$$

From Equation 10.59, the complimentary (transient) solution is

$$x_c(t) = e^{-\varsigma \omega_n t} \left(x_0 \cos \omega_d t + \frac{v_0 + x_0 \varsigma \omega_n}{\omega_d} \sin \omega_d t \right)$$

$$x_c(t) = e^{-0.1591(7.0711)t} \left(0.2 \cos(6.981t) + \frac{0.2(0.1591)(7.0711)}{6.981} \sin(6.981t) \right)$$

$$x_c(t) = e^{-1.125t}(0.2 \cos(6.577t) + 0.0322 \sin(6.577t))$$

The particular solution will be in the form $x_p(t) = A \cos \omega t + B \sin \omega t$. The constants are determined from Equations 10.90 and 10.91:

$$A = \frac{\left(\omega_n^2 - \omega^2 \right) F_0 / m}{\left(\omega_n^2 - \omega^2 \right)^2 + (2\varsigma \omega_n \omega)^2}$$

$$= \frac{\left(7.07^2 - 10^2 \right) 8 / 4}{\left(7.07^2 - 10^2 \right)^2 + (2(0.16)(7.07)(10))^2}$$

$$= -0.033$$

$$B = \frac{(2\varsigma \omega_n \omega) F_0 / m}{\left(\omega_n^2 - \omega^2 \right)^2 + (2\varsigma \omega_n \omega)^2}$$

$$= \frac{(2(0.16)(7.07)10)8 / 4}{\left(7.07^2 - 10^2 \right)^2 + (2(0.16)(7.07)10)^2}$$

$$= 0.015$$

The particular solution (steady-state solution) is then

$$x_p(t) = A \cos \omega t + B \sin \omega t$$
$$x_p(t) = -0.033 \cos(10t) + 0.015 \sin(10t)$$

The plots for the homogeneous and particular solution are shown in Figure 10.39 for the first 5 seconds. The total, which is the actual response, is the sum of the homogeneous and particular

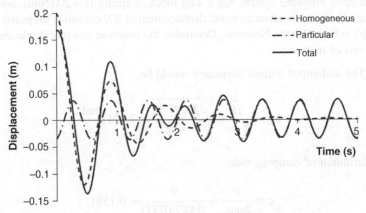

Figure 10.39

solution. You can see that after about 4 seconds, the homogeneous solution is nearly zero, and the response follows the particular (steady-state) solution.

10.8.5 Harmonic Support Motion with Viscous Damping

Another way of forcing motion is through motion of the support, which for this section will be harmonic. As a simple example, you can cause a pendulum to swing by moving the end of the string. Structures move due to ground motion caused by earthquakes or heavy traffic. Consider a case shown in Figure 10.40. The support is moving as a function $y(t)$.

From Newton's second law,

$$m\ddot{x} + c(\dot{x} - \dot{y}) + k(x - y) = 0 \tag{10.101}$$

Next, we will separate inputs (values for y) and outputs (values of x). Keeping the inputs on the left-hand side of the equation and moving the inputs to the right-hand side gives

$$m\ddot{x} + c\dot{x} + kx = c\dot{y} + ky \tag{10.102}$$

Define the support motion as the harmonic function $y(t) = y_0 \sin \omega t$.

$$m\ddot{x} + c\dot{x} + kx = c\omega y_0 \cos \omega t + ky_0 \sin \omega t \tag{10.103}$$

Dividing by the mass gives

$$\ddot{x} + 2\varsigma\omega_n\dot{x} + \omega_n^2 x = 2\varsigma\omega_n\omega y_0 \cos \omega t + \omega_n^2 y_0 \sin \omega t \tag{10.104}$$

Equation 10.104 is again a second-order differential equation. The complimentary (transient) solution, assuming an underdamped case, was previously developed and given in Equation 10.58. The steady-state solution will again begin in the form shown in Equation 10.84. The solution will again give a steady-state amplitude X. Converting that steady-state amplitude to a dimensionless ratio will give the term in Equation 10.105, known as

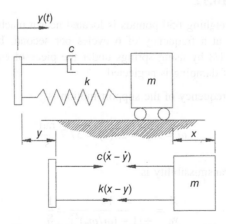

Figure 10.40 System with support motion

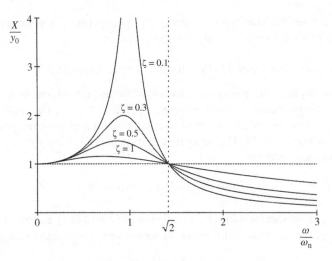

Figure 10.41   Displacement transmissibility

the transmissibility ratio:

$$\frac{X}{y_0} = \frac{\sqrt{1 + (2\varsigma(\omega/\omega_n))^2}}{\sqrt{\left(1 - (\omega/\omega_n)^2\right)^2 + (2\varsigma(\omega/\omega_n))^2}} \tag{10.105}$$

The plot of the transmissibility ratio is shown in Figure 10.41.
For an undamped system, the transmissibility ratio is

$$\frac{X}{y_0} = \frac{1}{\pm\left[1 - (\omega/\omega_n)^2\right]} \quad \text{for } \varsigma = 0 \tag{10.106}$$

Example Problem 10.12

A piece of equipment weighing 650 pounds is located near a machine producing harmonic vibration in the ground at a frequency of 6 cycles per second. It is desired to limit the transmitted vibration to 1/8 by using springs under the piece of equipment. Determine the stiffness of the springs if damping is neglected.

Solution: The circular frequency of the support motion is

$$\omega = 2\pi f = 2\pi(6) = 37.7 \frac{\text{rad}}{\text{s}}$$

For zero damping, the transmissibility is

$$\frac{X}{y_0} = \frac{1}{\pm\left[1 - (\omega/\omega_n)^2\right]} = \frac{1}{8}$$

This can be rearranged to solve for the frequency ratio:

$$-1 + \left(\frac{\omega}{\omega_n}\right)^2 = 8 \Rightarrow \frac{\omega}{\omega_n} = 3$$

The required natural frequency of the system is then determined.

$$\omega_n = \frac{\omega}{3} = \frac{37.7}{3} = 12.6 \frac{rad}{s}$$

The required spring stiffness can now be determined from the equation for natural frequency of the system:

$$\omega_n = \sqrt{\frac{kg}{W}} \Rightarrow k = \frac{\omega_n^2 W}{g}$$

$$k = \frac{\left(12.6 \frac{rad}{s}\right)^2 650 \, lb}{\left(12 \frac{in.}{ft}\right)\left(32.2 \frac{ft}{s^2}\right)}$$

Answer: $k = 265.5 \frac{lb}{in.}$

Example Problem 10.13

The damped system shown in Figure 10.42 oscillates due to the motion of a small support block, which has a vertical motion described by $y(t) = y_0 \sin \omega t$. Derive the differential equation of motion for mass m.

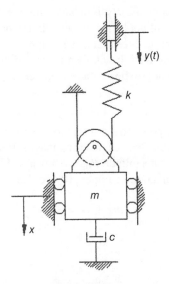

Figure 10.42

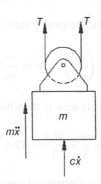

Figure 10.43

Solution: The free body diagram is shown in Figure 10.43 for an arbitrary displacement in the positive (downward) direction.

The spring will be stretched by an amount of $2x - y$, giving a tension equal to

$$T = k(2x - y)$$

Newton's second law gives

$$m\ddot{x} + c\dot{x} + 2[k(2x - y)] = 0$$
$$m\ddot{x} + c\dot{x} + 4kx \qquad = 2ky$$

Dividing by the mass and substituting $y = y_0 \sin \omega t$ gives the differential equation of motion.

$$\textit{Answer: } \ddot{x} + \frac{c}{m}\dot{x} + \frac{4k}{m}x = \frac{2ky_0}{m}\sin \omega t$$

10.9   Response of Undamped Systems to General Loading

It is fairly common in forced vibration analysis to have a forcing function $f(t)$ that is nonperiodic or discontinuous. This section will not be a complete discussion of general loading cases. The chapter will examine the solutions to three basic loading types: constant load (or step load), ramp load, and exponentially decaying load. It will then be shown that combinations of these three basic loading types can closely approximate more complex impulsive loading conditions.

10.9.1   Constant Force

We will first examine system response to a constant force or rectangular step force function that is suddenly applied to the system. Consider a spring mass system, as shown in Figure 10.44a, subjected to a constant force of magnitude F_0, as shown in Figure 10.44b.

Referring to the free body diagram in Figure 10.44c, Newton's second law gives

$$m\ddot{x} + kx = F_0 \tag{10.107}$$

Dividing by the mass gives

$$\ddot{x} + \omega_n^2 x = \frac{F_0}{m} \tag{10.108}$$

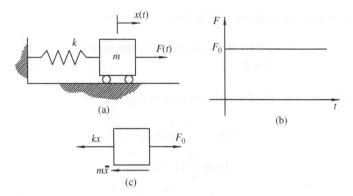

Figure 10.44   Rectangular step forcing function. (a) Undamped system. (b) Forcing function. (c) Free body diagram

Equation 10.108 is a nonhomogeneous second-order differential equation. Because the term on the right-hand side is a constant, the trial particular solution is in the form $x_p(t) = C$. The first and second derivatives of the particular solution both are equal to 0. Substituting the trial solution into Equation 10.108 gives

$$\frac{k}{m} C = \frac{F_0}{m}$$
$$C = \frac{F_0}{k}$$

Therefore, the particular solution is

$$x_p(t) = \frac{F_0}{k} \tag{10.109}$$

and the homogeneous solution is

$$x_h(t) = A \cos \omega_n t + B \sin \omega_n t \tag{10.110}$$

The total solution is the sum of the homogeneous and particular solutions:

$$x(t) = A \cos \omega_n t + B \sin \omega_n t + \frac{F_0}{k} \tag{10.111}$$

and the constants A and B depend on initial conditions.

First, consider the case when the system starts from rest. At time equal to 0, the displacement will be 0.

$$x(0) = 0 = A \cos(0) + B \sin(0) + \frac{F_0}{k}$$
$$A = \frac{-F_0}{k} \tag{10.112}$$

At time equal to zero, the velocity will be zero:

$$\dot{x}(0) = 0 = -A\omega_n \sin(0) + B\omega_n \cos(0)$$
$$B = 0$$

(10.113)

Substituting the constants into the response equation gives

$$x(t) = \frac{-F_0}{k} \cos \omega_n t + \frac{F_0}{k}$$
$$x(t) = \frac{F_0}{k}(1 - \cos \omega_n t)$$

(10.114)

Equation 10.114 is plotted in Figure 10.45.

You will notice that the constant force extends the spring a distance F_0/k and the system oscillates about that equilibrium position. The maximum displacement for the oscillating system is twice the amplitude of static displacement.

Now, we can consider a more general case to determine the constants A and B. The general case will allow for a system that is subjected to both initial displacement and initial velocity. For an initial displacement x_0 at time equal to 0,

$$x(0) = x_0 = A \cos(0) + B \sin(0) + \frac{F_0}{k}$$
$$A = x_0 - \frac{F_0}{k}$$

(10.115)

For an initial velocity v_0 at time equal to zero,

$$\dot{x}(0) = v_0 = -A\omega_n \sin(0) + B\omega_n \cos(0)$$
$$B = \frac{v_0}{\omega_n}$$

(10.116)

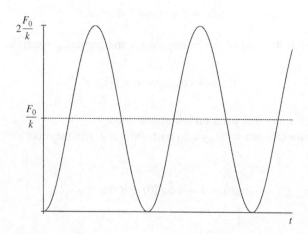

Figure 10.45 Undamped system response to a suddenly applied constant force of magnitude F_0

Substituting the constants into the response equation gives

$$x(t) = \left(x_0 - \frac{F_0}{k}\right)\cos\omega_n t + \frac{v_0}{\omega_n}\sin\omega_n t + \frac{F_0}{k} \qquad (10.117)$$

10.9.2 Ramp Load

The undamped system shown in Figure 10.46a is now subjected to a linearly increasing load, as shown in Figure 10.46b. The force increases linearly with a slope h. It will be assumed that the force does not exceed the elastic limit of the spring.

The free body diagram is shown in Figure 10.46c, and from Newton's second law

$$m\ddot{x} + kx = ht \qquad (10.118)$$

Dividing by the mass gives

$$\ddot{x} + \omega_n^2 x = \frac{h}{m}t \qquad (10.119)$$

The particular solution to Equation 10.119 will be in the form $x_p(t) = A_1 t + A_2$ and the second derivative will equal zero. Substituting the trial solution and second derivative gives

$$\frac{k}{m}(A_1 t + A_2) = \frac{h}{m}t \qquad (10.120)$$
$$(A_1 k)t + A_2 k = ht$$

From Equation 10.120, we need to equate the terms on the left and right sides of the equations, and we can determine that

$$A_2 = 0$$
$$A_1 = \frac{h}{k}$$

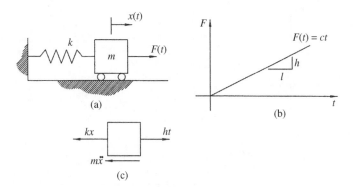

Figure 10.46 Ramp forcing function. (a) Undamped system. (b) Forcing function. (c) Free body diagram

Substituting the constants into the particular solution gives

$$x_p(t) = \frac{h}{k}t \qquad (10.121)$$

The homogeneous solution is again $x_h(t) = A \cos \omega_n t + B \sin \omega_n t$, and the total solution is the sum of the homogeneous and particular solutions:

$$x(t) = A \cos \omega_n t + B \sin \omega_n t + \frac{h}{k}t \qquad (10.122)$$

Assuming that the system starts from rest,

$$x(0) = 0 = A \cos(0) + B \sin(0) + \frac{h}{k}(0)$$

$$A = 0 \qquad (10.123)$$

$$\dot{x}(t) = 0 = -A\omega_n \sin(0) + B\omega_n \cos(0) + \frac{h}{k}$$

$$B = \frac{-h}{k\omega_n} \qquad (10.124)$$

Substituting the constants into Equation 10.122 gives

$$x(t) = \frac{-h}{k\omega_n} \sin \omega_n t + \frac{h}{k}t$$

$$x(t) = \frac{h}{k\omega_n}(\omega_n t - \sin \omega_n t) \qquad (10.125)$$

Equation 10.125 is shown in Figure 10.47. The system oscillates about a line (dashed line in the figure) with slope h/k.

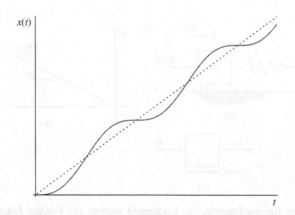

Figure 10.47 Undamped system response to a ramp force $F(t) = ht$

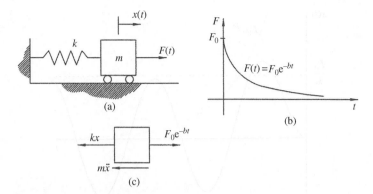

Figure 10.48 Exponentially decaying forcing function. (a) Undamped system. (b) Forcing function. (c) Free body diagram

10.9.3 Exponentially Decaying Motion

Next, consider the undamped system shown in Figure 10.48a subjected to the exponentially decaying load shown in Figure 10.48b.

From Figure 10.48c, Newton's second law gives

$$m\ddot{x} + kx = F_0 e^{-bt} \tag{10.126}$$

Dividing by the mass gives

$$\ddot{x} + \omega_n^2 x = \frac{F_0}{m} e^{-bt} \tag{10.127}$$

and the particular solution will be in the form $x_p(t) = C e^{-bt}$. Substituting the particular solution and its second derivative $\ddot{x} = C b^2 e^{-bt}$ into Equation 10.126 gives

$$C b^2 e^{-bt} + \omega_n^2 \left(C e^{-bt}\right) = \frac{F_0}{m} e^{-bt}$$
$$\left(b^2 + \frac{k}{m}\right) C e^{-bt} = \frac{F_0}{m} e^{-bt} \tag{10.128}$$

Therefore, $C = F_0/(mb^2 + k)$ and the particular solution is

$$x_p(t) = \frac{F_0}{mb^2 + k} e^{-bt} \tag{10.129}$$

Noting that the homogeneous solution is $x_h(t) = A \cos \omega_n t + B \sin \omega_n t$, the total solution is

$$x(t) = A \cos \omega_n t + B \sin \omega_n t + \frac{F_0}{mb^2 + k} e^{-bt} \tag{10.130}$$

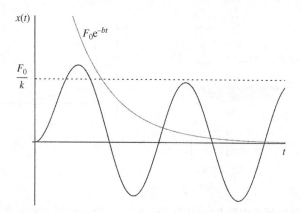

Figure 10.49 Undamped system response to an exponentially decaying force

We will again have the system start without initial conditions, which results in the following system response:

$$x(t) = \frac{F_0}{k\left(1 + (b^2/\omega_n^2)\right)} \left(\frac{b}{\omega_n} \sin \omega_n t - \cos \omega_n t + e^{-bt}\right) \qquad (10.131)$$

The maximum amplitude of motion will depend on the ratio b/ω_n. However, for initial conditions equal to zero, the maximum amplitude will never exceed $2F_0/k$. The response is shown in Figure 10.49.

10.9.4 *Combination of the Basic Forcing Functions*

The three basic loading conditions just discussed may seem limited. However, we can combine the basic functions to mimic more complex loading conditions. It is common, for example, to have a gradually applied constant force. Such a force, shown in Figure 10.50a, can be accomplished by subtracting an exponentially decaying force from a constant force. The system response to the gradually applied constant force can then be determined using superposition. Another common force function would be an impulsive force, which can

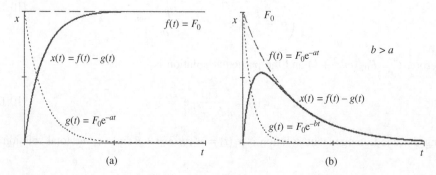

Figure 10.50 Possible combinations of basic forcing functions. (a) Constant force and exponentially decaying force. (b) Two exponentially decaying forces

sometimes be approximated by the sum of two exponentially decaying force functions, as seen in Figure 10.50b.

Example Problem 10.14

A 20 pound piston, as shown in Figure 10.51a, has a diameter of 1.5 inches and is subjected to a pressure that gradually increases from 0 to 25 psi, as shown in Figure 10.51b. The opposite side is open to the atmosphere and has a spring with a stiffness of 17.5 lb/in. Approximate the pressure curve by subtracting an exponentially decaying function from a constant function and determine the total system response.

Solution: The natural frequency of the system is

$$\omega_n = \sqrt{\frac{k}{m}} = \sqrt{\frac{17.5 \, \text{lb/in.}(12 \, \text{in.}/\text{ft})}{20 \, \text{lb}/32.2 \, \text{ft}/s^2}} = 18.39 \, \frac{\text{rad}}{\text{s}}$$

The plot of pressure shows that the pressure is approximately equal to 20 psi at a time of 0.5 seconds. Therefore, the exponentially decaying plot, as shown in Figure 10.52, will need to

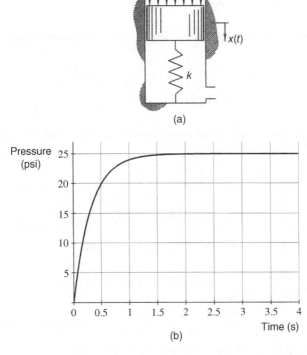

(a)

(b)

Figure 10.51

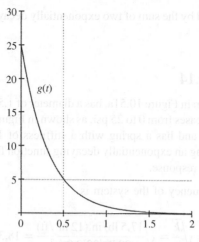

Figure 10.52

have a value of 5 at 0.5 seconds.

$$g(t) = 25e^{-bt}$$
$$5 = 25e^{-b(0.5)} \Rightarrow b = 3.22$$

The pressure plot is then approximated by

$$P(t) = 25 - 25e^{-3.22t} \quad (\text{psi})$$

The force function is determined by multiplying the pressure function by the area of the piston:

$$F(t) = P(t)\frac{\pi d^2}{4} = (25 - 25e^{-3.22t})\frac{\pi(1.5)^2}{4}$$
$$F(t) = 44.18 - 44.18e^{-3.22t} \quad (\text{lb})$$

The system response can be determined using superposition. The response to the constant force would be

$$x_1(t) = \frac{F_0}{k}(1 - \cos \omega_n t) = \frac{44.18 \text{ lb}}{17.5 \text{ lb/in.}}(1 - \cos(18.39t))$$
$$x_1(t) = 2.52(1 - \cos(18.39t)) \quad (\text{in.})$$

The response to the exponentially decaying force would be

$$x_2(t) = \frac{F_0}{k(1 + (b^2/\omega_n^2))}\left(\frac{b}{\omega_n}\sin \omega_n t - \cos \omega_n t + e^{-bt}\right)$$
$$x_2(t) = \frac{44.18 \text{ lb}}{17.5 \text{ lb/in.}(1 + (3.22^2/18.39^2))}\left(\frac{3.22}{18.39}\sin 18.39t - \cos 18.39t + e^{-3.22t}\right)$$
$$x_2(t) = 2.45(0.18 \sin 18.39t - \cos 18.39t + e^{-3.22t}) \quad (\text{in.})$$

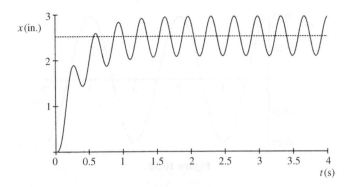

Figure 10.53

From superposition, the total system response is then

$$x(t) = x_1(t) - x_2(t)$$
$$x(t) = 2.52(1 - \cos(18.39t)) - 2.45(0.18 \sin 18.39t - \cos 18.39t + e^{-3.22t})$$

Answer: $x(t) = 2.52 - 0.44 \sin 18.39t - 0.07 \cos 18.39t - 2.45e^{-3.22t}$ (in.)
Figure 10.53 shows a plot of the system response.

10.10   Review and Summary

This chapter serves as a development of the foundational knowledge of vibration theory required for machine dynamics. Another area of importance is the concept of multidegree-of-freedom systems, which was not discussed. For more detail on the topics covered in this chapter, and for discussion of topics beyond the scope of this chapter, please see the list of textbooks provided in the section Further Reading. Though there are many great books on the topic of vibration, the list provided offers a good variety of topics.

Problems

P10.1   A 5 kg mass suspended from a spring oscillates in simple harmonic motion. The amplitude is 20 cm and the maximum acceleration is 6 m/s². What is the stiffness of the spring?

Answer: $k = 150 \dfrac{N}{m}$

P10.2   A spring mass system (no damping) is given an initial displacement and then released from rest. The system response is shown in Figure 10.54 for the first 5 seconds of motion. Determine the value of initial displacement and plot the velocity for the first 5 seconds of motion labeling the maximum value.

P10.3   A single degree-of-freedom system has harmonic motion with maximum amplitude of velocity equal to 17 in./s. Determine the displacement amplitude if the period is 0.15 seconds.

Answer: $A = 0.4$ in.

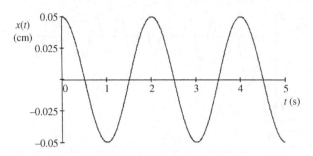

Figure 10.54

P10.4   A block with unknown mass is attached to a spring with an unknown stiffness. The system is given an initial displacement and released from rest. The system's response is shown in Figure 10.55a with time in seconds and displacement in inches. A second spring with unknown stiffness is added in series. The system is given initial displacement and released from rest. The system's response is shown in Figure 10.55b. What is the approximate period of motion (in seconds) that would occur if the two springs were rearranged in parallel?

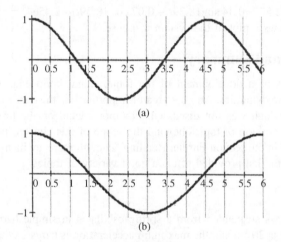

Figure 10.55

P10.5   A 4.3 kg mass is suspended from a spring causing it to stretch 3.5 cm to a position of static equilibrium. The mass is then stretched to a point 2 cm below equilibrium and released. Determine and plot the response. What is the maximum velocity?

P10.6   A 7 kg mass is suspended from two springs in series. The upper spring has a spring constant of 25 N/m and the lower spring has a spring constant of 18 N/m. From the equilibrium position, the mass is given an initial velocity downward and reaches a maximum displacement of 5 cm. Determine and plot the response.

P10.7   Assuming that slipping does not occur, determine the mass m that must be added to the system shown in Figure 10.56 to increase the period of motion 13%. If the coefficient of friction between the blocks is $\mu_s = 0.25$, validate the assumption that

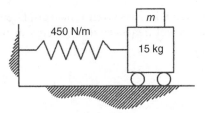

Figure 10.56

slipping does not occur if the 15 kg mass is given an initial displacement of 7 mm and released from rest.

Answer: m = 4.15 kg

P10.8   For the system shown in Figure 10.57, determine the value(s) of k that cause the system to be (a) critically damped, (b) underdamped, and (c) overdamped.

P10.9   The system from Problem P10.8 has a spring constant of 25 N/m. Determine and plot the response if the system is released from rest at a position 0.2 meters downward.

P10.10   A viscous damper is added to the system described in Problem P10.5. Given the same initial conditions, it is noticed that the amplitude decreases approximately 4% on each cycle of motion. Determine the damping constant and plot the response.

P10.11   A damped system has a 6.8 kg mass suspended from a spring and dashpot. Analysis of the system shows that the amplitude of motion decays from 0.42 mm to 0.26 mm in one cycle with a time of 1.4 seconds between peaks. Determine the spring stiffness and the damping constant for the system.

$$Answer: \begin{array}{l} k = 138\dfrac{N}{m} \\[2mm] c = 4.7\dfrac{N\,s}{m} \end{array}$$

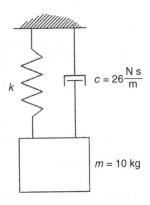

Figure 10.57

P10.12 The mass shown in Figure 10.58 is given a small displacement and then released from rest. Determine the natural frequency using energy methods.

Answer: $\omega_n = \sqrt{\dfrac{5k}{4m}}$

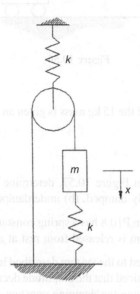

Figure 10.58

P10.13 A square swimming platform shown in Figure 10.59 is supported by four cylindrical drums and floats in water. Determine the period of small vertical oscillation of the platform about the equilibrium position shown. Assume the bottom of the platform always remains above water.

Answer: $\rho_w = 64 \text{ lb/ft}^3$

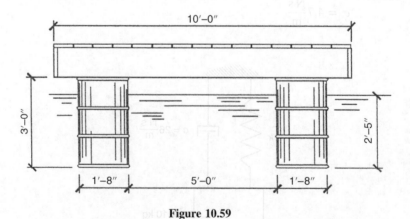

Figure 10.59

P10.14 A resonance test is performed on a damped system under an excitation force that is harmonic. During the test, it is noted that the amplitude at resonance is exactly two times the amplitude at an excitation frequency 20% greater than resonance. What is the damping ratio of the system?

Answer: $\varsigma = 0.1375$

P10.15 Which of the systems shown in Figure 10.60 is undergoing resonance?

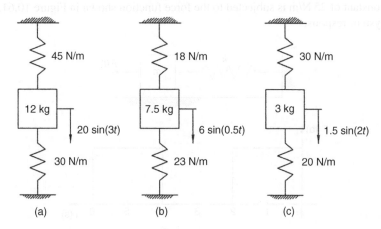

(a)   (b)   (c)

Figure 10.60

P10.16 A 4.3 kg mass is suspended by a spring with a constant of 500 N/m. The mass is excited by a harmonic forcing function $F(t) = F_0 \sin(\omega t)$. Determine the amplitude of motion if $F_0 = 12$ N and $\omega = 5.5$ rad/s.

P10.17 Repeat Problem P10.16 for $F_0 = 18$ N and $\omega = 12$ rad/s.

P10.18 Determine the amplitude of motion for the system shown in Figure 10.60a if the bottom 30 N/m spring is replaced by a spring with a constant of 50 N/m.

P10.19 A pendulum consists of a 2.6 kg mass suspended from a 500 mm string. Determine the amplitude of motion for the mass if the supporting end of the string has harmonic motion defined by $x(t) = x_0 \sin \omega t$, where $x_0 = 75$ mm and $\omega = 2.8$ rad/s.

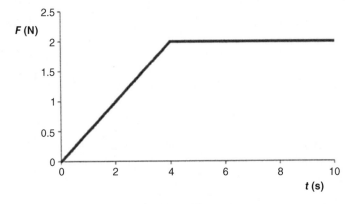

Figure 10.61

P10.20 A 74 N weight is suspended by a spring and dashpot. The spring stiffness is 1350 N/m and the viscous damping constant is 86 N s/m. A harmonic forcing function has a peak amplitude of 7 N. Determine the resonant frequency and the amplitude of motion at resonance.

P10.21 An undamped spring–mass system consists of a 1.5 kg mass and a spring with a constant of 25 N/m is subjected to the force function shown in Figure 10.61. Plot the system response.

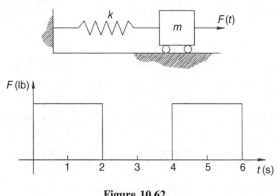

Figure 10.62

P10.22 As shown in Figure 10.62, a 40 pound block is attached to a spring with $k = 5$ lb/in. and damping is negligible. The force applied to the mass varies periodically as shown, with a maximum force magnitude of 12 pounds.. Determine the response of the block using Fourier series (see Appendix C).

Further Reading

Hartog, J.P.D. (1985) *Mechanical Vibrations*, Dover Publications, New York.
Meirovitch, L. (1986) *Elements of Vibration Analysis*, McGraw-Hill, New York.
Palm, W.J. (2007) *Mechanical Vibration*, John Wiley & Sons, Inc., New York.
Steidel, R.F. (1989) *An Introduction to Mechanical Vibrations*, John Wiley & Sons, Inc., New York.

11

Dynamic Force Analysis

11.1 Introduction

As defined in Chapter 1, a machine is a device designed to provide force or do work. To fully design machine elements, it is essential to determine the forces acting on those elements. In a linkage mechanism, all links, pins, and bearings must be designed to withstand forces within the mechanism. Cam mechanisms must be designed to transmit forces and not exceed allowable contact stress. This chapter will discuss methods for finding the forces within a linkage mechanism and will briefly introduce the concept of gear force analysis. Once forces are determined, a complete stress analysis can be used to design the connecting pins and bearings. Force analysis is also required in order to size the motor necessary to drive the linkage mechanism. This chapter will only discuss the forces and torques involved. The actual stress analysis and motor design, although extremely important, is beyond the scope of this text. Machine design textbooks cover the important aspects of design for allowable stresses.

The processes used to determine forces are primarily based on Newtonian mechanics. Dynamic analysis of machines, however, differs from most problems faced in an engineering dynamics course. Typical dynamics problems have known forces applied to a rigid body and the resulting motion must be determined. For typical machine analysis problems, accelerations are known and the resulting forces are to be determined.

Methods of dynamic force analysis for planar linkage mechanisms require Newton's second law to be written for each moving link in the system. It is typically appropriate to express Newton's law as three equations (Equation 11.1):

$$\sum F_x = ma_x \quad \sum F_y = ma_y \quad \sum M = I_G \alpha \tag{11.1}$$

Once the equations are written for all moving links, the set of equations must all be solved simultaneously. It will be shown that dynamic force analysis for a simple four-bar mechanism will result in nine equations with nine unknowns: eight pin reaction force components and the required input torque.

There are different methods for dynamic force analysis. Regardless of the method used, complete force analysis of a linkage mechanism is a complicated and tedious process. The process requires assumptions (e.g., rigid members and negligible friction) and often several

Machine Analysis with Computer Applications for Mechanical Engineers, First Edition. James Doane.
© 2016 John Wiley & Sons, Ltd. Published 2016 by John Wiley & Sons, Ltd.
Companion Website: www.wiley.com/go/doane0215

iterations. Although calculations can be done by hand, the process tends to be very time consuming (especially if forces are required for multiple positions of the mechanism). Computer software, especially those with built-in matrix operations, can greatly aid in the calculations and expedite the process. With computers, the forces can be calculated at several positions of the mechanism. Computer applications will be discussed at the end of the chapter.

11.2 Superposition Method of Force Analysis

11.2.1 Introduction

The first method presented will be the superposition method. This method is fairly well suited for hand calculations, but for practical purposes it would be limited to a few positions of the mechanism. The superposition method is a good introductory method and is a good prerequisite to matrix methods to follow. The method of superposition can only be used for linear systems, which will have a linear relationship between applied and output forces. Though the motion is often nonlinear, many mechanisms can be treated as linear systems in force analysis.

The basic principle of the superposition method is that the net effect of multiple forces acting on a linear system is equal to the vector summation of the effects of each force acting individually on the system. In other words, the superposition method solves a complicated problem by breaking it down into several more basic problems. Figure 11.1 will illustrate the concept of the superposition method using a simple static force analysis, although the concepts will soon be expanded to mechanisms in motion. Consider the mechanism shown in Figure 11.1a with two applied forces. The torque on link 2 is required to maintain static equilibrium of the system.

The superposition method states that the original loading case can be separated into simplified cases, as shown in Figure 11.1b. The summation of the results of these simplified cases will give the same results as the original case. For example, the total required torque T_2 shown in Figure 11.1a will equal the sum of the torques T_{2_1} and T_{2_2} for the two simplified cases shown in Figure 11.1b.

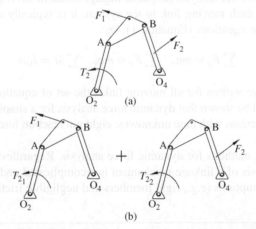

Figure 11.1 Concept of superposition. (a) Original loading scenario. (b) Summation of two simplified loading scenarios

The superposition method is very useful in graphical force analysis. Each member in the resulting simplified loading cases can often be analyzed as two- or three-force members (see Supplementary Concept 11.1 for more details), which easily allows for drawing force polygons. The force polygons could be drawn and measured by hand with reasonable accuracy. However, CAD software can be used to generate the scaled force polygons, and force magnitudes and directions can be determined from the software. The general process is best illustrated with the use of an example.

Supplementary Concept 11.1

Two-Force Members

A two-force member, such as the one shown in Figure 11.2a, is in static equilibrium if

- the two forces are collinear, and
- the two forces are equal in magnitude and opposite in direction.

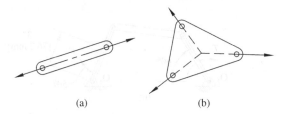

(a)   (b)

Figure 11.2

Three-Force Members

A three-force member, such as the one shown in Figure 11.2b, is in static equilibrium if

- the three forces intersect at a common point, and
- the three forces create a closed force polygon.

Example Problem 11.1

For the mechanism shown in Figure 11.3, determine the required input torque T_2 for static equilibrium. Link 2 is 4.5 inches (114.3 mm), link 3 is 6 inches (152.4 mm), and link 4 is 5 inches (127 mm). Force $F = 25$ N is applied to link 3 and force $P = 18$ N is applied to link 4 in the directions shown.

Solution: The method of superposition will be used, and the problem is reduced into two simplified loading cases, as shown in Figure 11.4a and b.

We will begin with the simplified loading case shown in Figure 11.4a. The free body diagrams for that loading case are shown in Figure 11.5. Link 4 is a two-force member and link 3 is a three-force member.

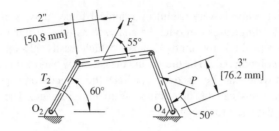

Figure 11.3

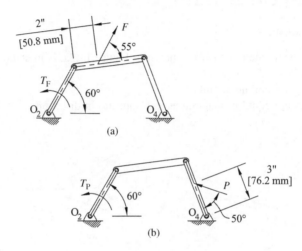

Figure 11.4

Scaling the force polygon shown in Figure 11.5b gives the magnitudes

$$F_A = 21.0395 \text{ N} \quad F_B = 7.0273 \text{ N}$$

and the directions are shown. The force at A is then shown in Figure 11.5c acting on link 2. The force at O_2 will be of the same magnitude, but opposite in direction. These two forces will create a moment couple that must be opposed by the input torque:

$$T_F = 0.0241 \text{ m} (21.0395 \text{ N}) = 0.507 \text{ N m}$$

Free body diagrams for the second loading case are shown in Figure 11.6. Link 4 is a three-force member and link 3 is a two-force member. Scaling the force polygon in Figure 11.6b gives the magnitudes:

$$F_{O_4} = 11.0260 \text{ N} \quad F_B = 8.5170 \text{ N}$$

The free body diagram of link 3 shown in Figure 11.6c shows that the force at A is equal and opposite to that at B. Figure 11.6d shows that the input torque must overcome the moment couple. The negative sign is because the actual torque will be clockwise:

$$T_P = 0.0909 \text{ m} (8.5170 \text{ N}) = -0.774 \text{ N m}$$

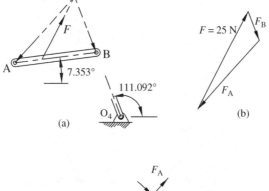

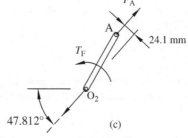

(a)

(b)

(c)

Figure 11.5

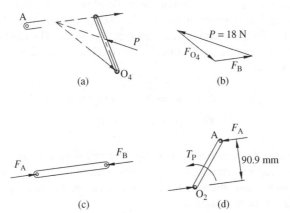

(a)

(b)

(c)

(d)

Figure 11.6

The final torque is the sum of both scenarios:

$$T_2 = T_F + T_P = 0.507 \, \text{N m} - 0.774 \, \text{N m}$$

Answer: $T_2 = 0.267 \, \text{N m}$ (cw)

11.2.2 Equivalent Offset Inertia Force

The process of superposition works well for static loading cases, as illustrated in the previous example. With some simplifications of the inertial loadings, it can work equally as well for

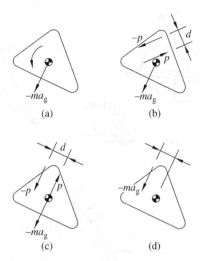

Figure 11.7 Equivalent offset inertia force. (a) Free body diagram of rigid body. (b) Inertial moment replaced with couple. (c) Preferred location of couple. (d) Equivalent offset inertia force

dynamic loading. When using the superposition method for dynamic force analysis, it is convenient to use equivalent offset inertia force that will replace the translational and rotational inertial forces with one equivalent inertial force. To illustrate the concept, consider the free body diagram of a rigid body shown in Figure 11.7a. There are no external forces acting on the body, so only the inertial force and inertial moment are shown. The negative signs are due to d'Alembert's principle, which is an alternative form of Newton's equation to reduce a dynamic problem into a static problem.

$$F = ma_g \Rightarrow F + \left(-ma_g\right) = 0$$
$$T = I_g\alpha \Rightarrow T + \left(-I_g\alpha\right) = 0$$

(11.2)

The inertial moment $-I_g\alpha$ can be replaced by an equivalent moment couple, as shown in Figure 11.7b. The couple is created by two forces p and $-p$ located a distance d apart. The moment couple must cause the same rotational direction as $-I_g\alpha$ and the mathematical requirement for the moment couple is

$$pd = I_g\alpha$$

(11.3)

An infinite number of combinations of the force p and distance d exist that will satisfy Equation 11.3. Also, the moment couple can be placed at any location.

It is convenient to pick the arrangement of the moment couple shown in Figure 11.7c where one of the force vectors of the couple located at the center of mass is in the opposite direction of $-ma_g$. If that particular force p is picked to equal the magnitude ma_g, then the two vectors cancel out and according to Equation 11.3 the distance d must be

$$d = \frac{I_g\alpha}{ma_g}$$

(11.4)

Such substitutions leave the equivalent offset inertia force shown in Figure 11.7d. The direction of the acceleration vector will be opposite the direction of the true acceleration of the center of mass. The offset distance will be that determined from Equation 11.4. The inertial force must be offset on the side that creates the same rotation direction as the original $-I_g\alpha$ term.

Example Problem 11.2

The rigid link shown in Figure 11.8 has a mass of $m = 0.8$ kg and a moment of inertia $I_g = 80$ kg mm^2. The translational and rotational accelerations are $a_g = 4.2$ mm/s^2 and $\alpha = 0.6$ rad/s^2 in the directions shown. Replace the inertial forces with an equivalent offset inertia force.

Solution: The inertial forces will be in the opposite direction of the actual accelerations shown. The translational and rotational inertia values are as follows:

$$ma_g = 0.8 \text{ kg} \times 4.2 \frac{\text{mm}}{\text{s}^2} = 3.36 \frac{\text{kg mm}}{\text{s}^2}$$

$$I_g\alpha = 80 \text{ kg mm}^2 \times 0.6 \frac{\text{rad}}{\text{s}^2} = 48 \frac{\text{kg mm}^2}{\text{s}^2}$$

Figure 11.9 shows the equivalent inertia force, which is equal in magnitude to the translational inertia force. The equivalent force is offset a distance determined using Equation 11.4.

$$d = \frac{I_g\alpha}{ma_g} = \frac{48 \text{ kg mm}^2/\text{s}^2}{3.36 \text{ kg mm}/\text{s}^2} = 14.29 \text{ mm}$$

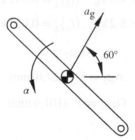

Figure 11.8

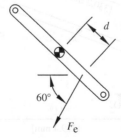

Figure 11.9

The equivalent inertia force will be

$$F_e = ma_g = 3.36 \times 10^{-3} \text{ N}$$

11.2.3 Superposition Method for Dynamic Force Analysis of a Four-Bar Mechanism

Now that the general concepts of superposition and equivalent offset inertia forces have been developed, we can illustrate the full procedure of dynamic force analysis. The inertial forces for each link can be reduced to an equivalent offset inertia force. This gives a set of equivalent static loading cases that can be analyzed individually to determine joint forces and input torque. Also, you would need to add additional static loading cases to account for any externally applied loads. Use caution not to round the results too greatly for each individual loading case. If rounding errors are high in the individual cases, a summation of the results will only cause an accumulation of those rounding errors. The general procedure of the superposition method for a complete linkage mechanism is best described with the use of an example problem.

Example Problem 11.3

The four-bar mechanism shown in Figure 11.10 has the dimensions shown. The mass properties of each link are given below. The crank rotates counterclockwise at a constant rate of 50 rad/s. Determine the required input torque T_2 for the instant shown where $\theta_2 = 140°$.

$$m_2 = 2.4 \text{ kg} \quad (I_g)_2 = 0.001 \text{ kg m}^2$$

$$m_3 = 4.6 \text{ kg} \quad (I_g)_3 = 0.031 \text{ kg m}^2$$

$$m_4 = 6.2 \text{ kg} \quad (I_g)_4 = 0.023 \text{ kg m}^2$$

$$O_2g_2 = 0$$

$$Ag_3 = 5'' \ (127.0 \text{ mm})$$

$$O_4g_4 = 4'' \ (101.6 \text{ mm})$$

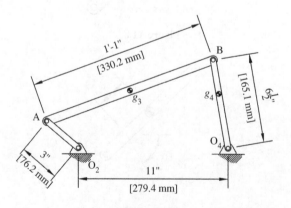

Figure 11.10

Solution: The angular positions of links 3 and 4 can be determined from position analysis discussed in Chapter 3 to be

$$\theta_3 = 20.153° \quad \theta_4 = 99.690°$$

A complete kinematic analysis of the mechanism is required to develop the necessary acceleration terms. All velocity and acceleration terms can be found using methods discussed in Chapter 4 or computer methods discussed in Chapter 6.

$$\vec{v}_A = -2.449\hat{i} - 2.919\hat{j} \quad \text{(m/s)}$$

$$\vec{a}_A = 145.931\hat{i} - 122.451\hat{j} \quad \text{(m/s)}$$

$$\omega_3 = 7.59\frac{\text{rad}}{\text{s}} \quad \text{(ccw)} \quad \omega_4 = 20.35\frac{\text{rad}}{\text{s}} \quad \text{(ccw)}$$

$$\alpha_3 = 247.36\frac{\text{rad}}{\text{s}^2} \quad \text{(ccw)} \quad \alpha_4 = 543.29\frac{\text{rad}}{\text{s}^2} \quad \text{(cw)}$$

The acceleration of the link center of gravity locations can now be determined. The center of gravity of link 2 is located at the fixed pivot point, so its acceleration is zero. The acceleration of the center of gravity of link 3 is

$$\vec{a}_{g_3} = \vec{a}_A + \vec{a}_{g_3/A}$$

$$\vec{a}_{g_3} = \vec{a}_A + \vec{\omega}_3 \times \left(\vec{\omega}_3 \times \vec{r}_{g_3/A}\right) + \vec{\alpha}_3 \times \vec{r}_{g_3/A}$$

The position vector to locate the center of gravity relative to point A is

$$\vec{r}_{g_3/A} = 0.119\hat{i} + 0.044\hat{j} \quad \text{(m)}$$

Substitution into the acceleration equation gives

$$\vec{a}_{g_3} = \begin{matrix} 145.931\hat{i} - 122.451\hat{j} + 7.59\hat{k} \times \left(7.59\hat{k} \times \left(0.119\hat{i} + 0.044\hat{j}\right)\right) \\ + 247.36\hat{k} \times \left(0.119\hat{i} + 0.044\hat{j}\right) \end{matrix}$$

$$\vec{a}_{g_3} = 128.19\hat{i} - 95.524\hat{j} \quad \text{(m/s}^2\text{)}$$

The acceleration of the center of gravity of link 4 is

$$\vec{a}_{g_4} = \vec{\omega}_4 \times \left(\vec{\omega}_4 \times \vec{r}_{g_4/O_4}\right) + \vec{\alpha}_4 \times \vec{r}_{g_4/O_4}$$

The position vector is

$$\vec{r}_{g_4/O_4} = -0.017\hat{i} + 0.100\hat{j} \quad \text{(m)}$$

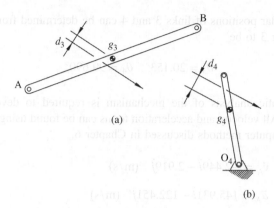

(a)

(b)

Figure 11.11

The acceleration is

$$\vec{a}_{g_4} = \begin{aligned} &20.35\hat{k} \times (20.35\hat{k} \times (-0.017\hat{i} + 0.100\hat{j})) \\ &+ 543.29 \times (-0.017\hat{i} + 0.100\hat{j}) \end{aligned}$$

$$\vec{a}_{g_4} = 61.49\hat{i} - 32.20\hat{j} \quad (\text{m/s}^2)$$

The equivalent offset inertia forces can now be determined for each link. Figure 11.11a shows the equivalent inertial force for link 3. Equation 11.4 gives

$$d_3 = \frac{(I_g)_3 \alpha_3}{m_3 a_{g_3}}$$

Using the acceleration magnitude of

$$a_{g_3} = \sqrt{128.19^2 + (-95.524)^2} = 159.86 \frac{\text{m}}{\text{s}^2}$$

gives the following:

$$d_3 = \frac{0.031 \text{ kg m}^2 \left(247.36 \frac{\text{rad}}{\text{s}^2}\right)}{4.6 \text{ kg} \left(159.86 \frac{\text{m}}{\text{s}^2}\right)}$$

$$d_3 = 0.01 \text{ m}$$

Figure 11.11b shows the equivalent inertial force for link 4. From Equation 11.4,

$$d_4 = \frac{(I_g)_4 \alpha_4}{m_3 a_{g_4}}$$

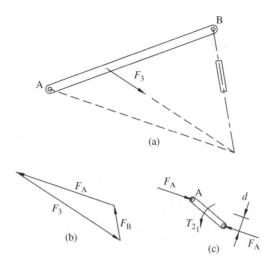

F_3

(a)

F_A

F_3

F_B

(b)

F_A

A

T_{2_1}

d

F_A

(c)

Figure 11.12

$$d_4 = \frac{0.023 \text{ kg m}^2(543.29 \text{ rad/s})}{6.2 \text{ kg}(69.42 \text{ m/s}^2)}$$

$$d_4 = 0.029 \text{ m}$$

The method of superposition can now be used for the force analysis of the linkage system. Figure 11.12 shows the force analysis for the inertial force on link 3. Link 3 is a three-force member, as shown in Figure 11.12a. Because link 4 is a two-force member, the direction of the force at B is known. Once the required direction of the force at A is determined, a force polygon can be developed, as shown in Figure 11.12b. The inertial force acting on link 3 is

$$F_3 = m_3 a_{g_3} = 4.6 \text{ kg}\left(159.86 \frac{\text{m}}{\text{s}^2}\right) = 735 \text{ N}$$

Scaling the force polygon gives

$$F_A = 613 \text{ N}$$

Figure 11.12c shows the force applied at point A. The reaction force at O_2 must be equal and opposite, which creates a moment couple that must be counteracted by an applied torque. The distance between the two forces is measured to be 0.0276 meters, causing the moment couple to be

$$T_{2_1} = d \cdot F_A = 0.0276 \text{ m} (613 \text{ N}) = 16.9 \text{ N m}$$

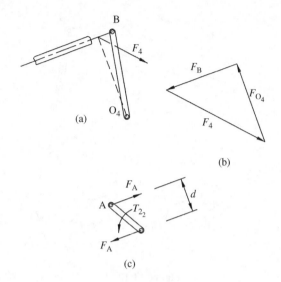

Figure 11.13

The process is repeated using the inertial force on link 4, as shown in Figure 11.13. Link 4 is now a three-force member, which allows us to determine the magnitude of the force at point B. The inertial force is

$$F_4 = m_4 a_{g_4} = 6.2\,\text{kg}\left(69.42\,\frac{\text{m}}{\text{s}^2}\right) = 430\,\text{N}$$

Scaling the force polygon shown in Figure 11.13b gives

$$F_B = 291\,\text{N}$$

Because link 3 is now a two-force member, the magnitude of the force at point A will equal the magnitude of the force at point B.

Figure 11.13c shows the force applied to link 2. The distance between the force vectors is measured to be 0.066 meters giving a torque of

$$T_{2_2} = d \cdot F_A = 0.066\,\text{m}\,(291\,\text{N}) = 19.2\,\text{N m}$$

The total torque is determined by superposition:

$$T_2 = T_{2_1} + T_{2_2} = 16.9\,\text{N m} + 19.2\,\text{N m}$$

Answer: $T_2 = 36.1\,\text{N m}$ (ccw)

11.3 Matrix Method Force Analysis

11.3.1 General Concepts

The superposition method is graphical and visual, but requires a lot of time. Graphical methods also tend to have a disadvantage of accuracy, and they are also often limited to a few positions of the mechanism. Because it is frequently required to have force analysis for a complete cycle of motion, graphical methods are not generally practical for a detailed analysis. Repeated analysis of complete cycles is also required in cases where optimization is important (which is common). For these reasons, it is desired to have a more analytical approach, preferably one that can easily be programmed for computer solution.

The matrix method will solve all forces for a given position with a set of equations. A general planar four-bar mechanism will result in nine unknowns: two unknown forces at each of the four revolute joints and one unknown input that is typically an input torque on the driving link (note that only one input force can be described for a single degree-of-freedom system). Each of the three movable links will give three equations of motion: two summation of forces equations and one summation of moments equation. Therefore, the result will be a set of nine simultaneous equations. The matrix method solves the nine equations simultaneously with the use of matrix operations (see Chapter 6 for information on solving simultaneous equations with matrix operations) and, though the equations can be solved by hand, it is best suited for computer solution.

11.3.2 Four-Bar Linkage

Figure 11.14a shows a general four-bar mechanism. In general, there will be one required input that could be applied in many different ways, but for this example there will be an input torque

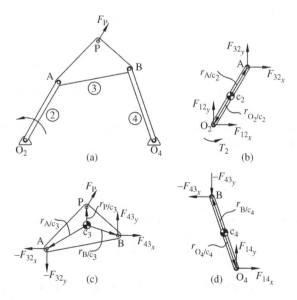

Figure 11.14 Dynamic force analysis. (a) Four-bar mechanism. (b) Free body diagram for link 2. (c) Free body diagram for link 3. (d) Free body diagram for link 4

applied to link 2. For this particular scenario, there is an external force (F_P) applied to the coupler at point P. To begin the dynamic force analysis, each link will now be removed and analyzed individually for forces. We will begin with link 2, which is shown in Figure 11.14b, removed from the total system. The link will have a total of four unknown forces (F_{12_x}, F_{12_y}, F_{32_x}, F_{32_y}) and one unknown input torque (T_2).

The center of gravity for link 2, labeled as c_2, must be located. Referring to Figure 11.14b, Newton's law can then be used to get three equations. The first two equations come from summation of forces:

$$\sum F_x = ma_x \Rightarrow F_{12_x} + F_{32_x} = m_2(a_x)_2 \tag{11.5}$$

$$\sum F_y = ma_y \Rightarrow F_{12_y} + F_{32_y} = m_2(a_y)_2 \tag{11.6}$$

The acceleration terms in Equations 11.5 and 11.6 are acceleration of the center of mass of link 2. The third equation is developed using summation of moments about the center of gravity:

$$\sum M = I_G \alpha$$

$$\vec{r}_{O_2/c_2} \times \vec{F}_{12} + \vec{r}_{A/c_2} \times \vec{F}_{32} + \vec{T}_2 = (I_G)_2 \vec{\alpha}_2 \tag{11.7}$$

Equation 11.7 is kept in vector form and is the cross product of position vectors and forces. Completing the cross products gives

$$\left[(r_{O_2/c_2})_x \hat{i} + (r_{O_2/c_2})_y \hat{j} \right] \times \left[F_{12_x} \hat{i} + F_{12_y} \hat{j} \right] + \\ \left[(r_{A/c_2})_x \hat{i} + (r_{A/c_2})_y \hat{j} \right] \times \left[F_{32_x} \hat{i} + F_{32_y} \hat{j} \right] + T_2 \hat{k} = (I_G)_2 \alpha_2 \hat{k} \tag{11.8}$$

$$(r_{O_2/c_2})_x F_{12_y} \hat{k} - (r_{O_2/c_2})_y F_{12_x} \hat{k} + (r_{A/c_2})_x F_{32_y} \hat{k} - (r_{A/c_2})_y F_{32_x} \hat{k} + T_2 \hat{k} = (I_G)_2 \alpha_2 \hat{k} \tag{11.9}$$

Equation 11.9 can now be written in scalar form because all of the vectors are along the same axis, which is the $\hat{k}$ unit vector.

$$-(r_{O_2/c_2})_y F_{12_x} + (r_{O_2/c_2})_x F_{12_y} - (r_{A/c_2})_y F_{32_x} + (r_{A/c_2})_x F_{32_y} + T_2 = (I_G)_2 \alpha_2 \tag{11.10}$$

Equations 11.5, 11.6, and 11.10 give the final three equations of motion for link 2.

The process is now repeated for link 3. Referring to Figure 11.14c, link 3 will have an additional two unknown forces (F_{43_x} and F_{43_y}). Summation of forces will give

$$\sum F_x = ma_x \Rightarrow F_{43_x} - F_{32_x} + F_{P_x} = m_3(a_x)_3 \tag{11.11}$$

$$\sum F_y = ma_y \Rightarrow F_{43_y} - F_{32_y} + F_{P_x} = m_3(a_y)_3 \tag{11.12}$$

Notice that forces F_{32_x} and F_{32_y} are the same as those for link 2, but in the opposite direction. Summation of moments of link 3 will be similar to that of link 2. There is no input torque, but for this particular example there is an additional force at point P.

$$\sum M = I_G \alpha$$

$$\vec{r}_{B/c_3} \times \vec{F}_{43} + \vec{r}_{A/c_3} \times -\vec{F}_{32} + \vec{r}_{P/c_3} \times \vec{F}_P = (I_G)_3 \vec{\alpha}_3 \qquad (11.13)$$

$$\left[(r_{B/c_3})_x \hat{i} + (r_{B/c_3})_y \hat{j} \right] \times \left[F_{43_x} \hat{i} + F_{43_y} \hat{j} \right] + \left[(r_{A/c_3})_x \hat{i} + (r_{A/c_3})_y \hat{j} \right]$$
$$\times \left[-F_{32_x} \hat{i} - F_{32_y} \hat{j} \right] \left[(r_{P/c_3})_x \hat{i} + (r_{P/c_3})_y \hat{j} \right] \times \left[F_{P_x} \hat{i} + F_{P_y} \hat{j} \right] = (I_G)_3 \alpha_3 \hat{k} \qquad (11.14)$$

$$(r_{B/c_3})_x F_{43_y} \hat{k} - (r_{B/c_3})_y F_{43_x} \hat{k} - (r_{A/c_3})_x F_{32_y} \hat{k} + (r_{A/c_3})_y F_{32_x} \hat{k} + (r_{P/c_3})_x F_{P_y} - (r_{P/c_3})_y F_{P_x} \hat{k}$$
$$= (I_G)_3 \alpha_3 \hat{k} \qquad (11.15)$$

Expressing Equation 11.15 in scalar form gives the final moment equation for link 3:

$$(r_{A/c_3})_y F_{32_x} - (r_{A/c_3})_x F_{32_y} + (r_{B/c_3})_x F_{43_y} - (r_{B/c_3})_y F_{43_x} - (r_{P/c_3})_y F_{P_x} + (r_{P/c_3})_x F_{P_y}$$
$$= (I_G)_3 \alpha_3 \qquad (11.16)$$

Finally, equations are developed for link 4. Referring to Figure 11.14d, link 4 introduces the last two unknown forces (F_{14_x} and F_{14_y}). Summation of forces will give

$$\sum F_x = ma_x \Rightarrow F_{14_x} - F_{43_x} = m_4(a_x)_4 \qquad (11.17)$$

$$\sum F_y = ma_y \Rightarrow F_{14_y} - F_{43_y} = m_4(a_y)_4 \qquad (11.18)$$

and summation of moments will give

$$\sum M = I_G \alpha$$

$$\vec{r}_{O_4/c_4} \times \vec{F}_{14} + \vec{r}_{B/c_4} \times -\vec{F}_{43} = (I_G)_4 \vec{\alpha}_4 \qquad (11.19)$$

$$\left[(r_{O_4/c_4})_x \hat{i} + (r_{O_4/c_4})_y \hat{j} \right] \times \left[F_{14_x} \hat{i} + F_{14_y} \hat{j} \right] + \left[(r_{B/c_4})_x \hat{i} + (r_{B/c_4})_y \hat{j} \right] \times \left[-F_{43_x} \hat{i} - F_{43_y} \hat{j} \right]$$
$$= (I_G)_4 \alpha_4 \hat{k} \qquad (11.20)$$

$$(r_{O_4/c_4})_x F_{14_y} \hat{k} - (r_{O_4/c_4})_y F_{14_x} \hat{k} - (r_{B/c_4})_x F_{43_y} \hat{k} + (r_{B/c_4})_y F_{43_x} \hat{k} = (I_G)_4 \alpha_4 \hat{k} \qquad (11.21)$$

Expressing Equation 11.21 in scalar form gives

$$-(r_{O_4/c_4})_y F_{14_x} + (r_{O_4/c_4})_x F_{14_y} + (r_{B/c_4})_y F_{43_x} - (r_{B/c_4})_x F_{43_y} = (I_G)_4 \alpha_4 \qquad (11.22)$$

Now that equations have been developed for each link individually, we can now combine all equations in one complete set. Putting Equations 11.5, 11.6, 11.10–11.12, 11.16–11.18 and 11.22 together in matrix form gives

$$
\begin{bmatrix}
1 & 0 & 1 & 0 & 0 & 0 & 0 & 0 & 0 \\
0 & 1 & 0 & 1 & 0 & 0 & 0 & 0 & 0 \\
(-r_{O_2/c_2})_y & (r_{O_2/c_2})_x & (-r_{A/c_2})_y & (r_{A/c_2})_x & 0 & 0 & 0 & 0 & 1 \\
0 & 0 & -1 & 0 & 1 & 0 & 0 & 0 & 0 \\
0 & 0 & 0 & -1 & 0 & 1 & 0 & 0 & 0 \\
0 & 0 & (r_{A/c_3})_y & (-r_{A/c_3})_x & (-r_{B/c_3})_y & (r_{B/c_3})_x & 0 & 0 & 0 \\
0 & 0 & 0 & 0 & -1 & 0 & 1 & 0 & 0 \\
0 & 0 & 0 & 0 & 0 & -1 & 0 & 1 & 0 \\
0 & 0 & 0 & 0 & (r_{B/c4})_y & (-r_{B/c4})_x & (-r_{O_4/c4})_y & (r_{O_4/c4})_x & 0
\end{bmatrix}
\begin{bmatrix}
F_{12_x} \\
F_{12_y} \\
F_{32_x} \\
F_{32_y} \\
F_{43_x} \\
F_{43_y} \\
F_{14_x} \\
F_{14_y} \\
T_2
\end{bmatrix}
$$

$$
=
\begin{bmatrix}
m_2(a_x)_2 \\
m_2(a_y)_2 \\
(I_G)_2 \alpha_2 \\
m_3(a_x)_3 - (F_P)_x \\
m_3(a_y)_3 - (F_P)_y \\
(I_G)_3 \alpha_3 - (r_{P/c_3})_x F_{P_y} + (r_{P/c_3})_y F_{P_x} \\
m_4(a_x)_4 \\
m_4(a_y)_4 \\
(I_G)_4 \alpha_4
\end{bmatrix}
\tag{11.23}
$$

Equation 11.23 is the matrix form of nine equations with nine unknowns. Though the solution would be tedious by hand, computer software can be used to quickly solve for the nine unknowns. For help using software for the solution, refer to Section 6.2.4 that discusses methods for using MATLAB® to solve sets of simultaneous equations. Equation 11.23 has the general form:

$$
[A]_{9x9}[x]_{1x9} = [B]_{1x9}
\tag{11.24}
$$

Once matrices $[A]$ and $[B]$ are developed (noting that all terms in those matrices are known values), the unknown values can be determined in MATLAB® using $[x] = [A]\backslash[B]$.

Several points should be discussed about the matrix method when it comes to a complete design process for the linkage mechanism. It will generally be desired to do a force analysis for a complete cycle of the linkage motion. Therefore, a complete acceleration analysis is required before starting the force analysis. The system of nine equations will need to be developed for multiple positions within the cycle. The components of all position vectors will change with each new position, as well as the acceleration terms. It should also be noted that in a true design process, the link geometries may not be completely known. The required mass, moment of inertia, and center of gravity for each link cannot be determined until the forces are known. Values of link mass, moment of inertia, and center of gravity must be estimated before starting the process of force analysis. Once the forces have been determined, the designer must verify that the estimated linkage properties were sufficient for the calculated forces. If they are not, then the process must be repeated. Therefore, a full design process will typically require multiple iterations of force analysis. The necessity of iterative solutions for a complete design practically forces the use of computer software. Section 11.8 will discuss the use of computers for force analysis.

Example Problem 11.4

Determine the joint forces and driving torque for the linkage mechanism in Example Problem 11.3 using the matrix method.

Solution: Recall from Example Problem 11.3 the mass properties of each link.

$$m_2 = 2.4\,\text{kg} \quad (I_g)_2 = 0.001\,\text{kg m}^2$$
$$m_3 = 4.6\,\text{kg} \quad (I_g)_3 = 0.031\,\text{kg m}^2$$
$$m_4 = 6.2\,\text{kg} \quad (I_g)_4 = 0.023\,\text{kg m}^2$$

We will begin by developing the three equations for link 2. From Equations 11.5 and 11.6,

$$F_{12_x} + F_{32_x} = m_2(a_x)_2$$
$$F_{12_y} + F_{32_y} = m_2(a_y)_2$$

From Example Problem 11.3, we know the center of gravity of link 2 is fixed, therefore will have zero acceleration. The summation of force equations becomes

$$F_{12_x} + F_{32_x} = 0$$
$$F_{12_y} + F_{32_y} = 0$$

Equation 11.10 gives the summation of moments for link 2:

$$-(r_{O_2/c_2})_y F_{12_x} + (r_{O_2/c_2})_x F_{12_y} - (r_{A/c_2})_y F_{32_x} + (r_{A/c_2})_x F_{32_y} + T_2 = (I_G)_2 \alpha_2$$

From Example Problem 11.3, we can determine the positional coordinate for the center of mass:

$$(r_{O_2/c_2})_y = (r_{O_2/c_2})_x = 0$$

$$(r_{A/c_2})_x = 0.0762 \cos 140 = -0.058$$

$$(r_{A/c_2})_y = 0.0762 \sin 140 = 0.049$$

Recalling that the angular acceleration of link is zero, the final summation of moments equation becomes

$$-0.049 F_{32_x} - 0.058 F_{32_y} + T_2 = 0$$

The values obtained can now be placed into the matrix equation (11.23). The process can be repeated for links 3 and 4. The only modifications to Equation 11.23 would be any reference to the force applied at point P, because our current example does not contain that force. Matrix A is

$$A = \begin{bmatrix}
1 & 0 & 1 & 0 & 0 & 0 & 0 & 0 & 0 \\
0 & 1 & 0 & 1 & 0 & 0 & 0 & 0 & 0 \\
\left(-r_{O_2/c_2}\right)_y & \left(r_{O_2/c_2}\right)_x & \left(-r_{A/c_2}\right)_y & \left(r_{A/c_2}\right)_x & 0 & 0 & 0 & 0 & 1 \\
0 & 0 & -1 & 0 & 1 & 0 & 0 & 0 & 0 \\
0 & 0 & 0 & -1 & 0 & 1 & 0 & 0 & 0 \\
0 & 0 & \left(r_{A/c_3}\right)_y & \left(-r_{A/c_3}\right)_x & \left(-r_{B/c_3}\right)_y & \left(r_{B/c_3}\right)_x & 0 & 0 & 0 \\
0 & 0 & 0 & 0 & -1 & 0 & 1 & 0 & 0 \\
0 & 0 & 0 & 0 & 0 & -1 & 0 & 1 & 0 \\
0 & 0 & 0 & 0 & \left(r_{B/c_4}\right)_y & \left(-r_{B/c_4}\right)_x & \left(-r_{O_4/c_4}\right)_y & \left(r_{O_4/c_4}\right)_x & 0
\end{bmatrix}$$

The terms for the center of mass locations for link 2 are developed earlier. The remaining terms are as follows:

$$\left(r_{A/c_3}\right)_x = -Ag_3 \cos \theta_3 = 0.127 \cos 20.153 = -0.119$$

$$\left(r_{A/c_3}\right)_y = -Ag_3 \sin \theta_3 = 0.127 \sin 20.153 = -0.044$$

$$\left(r_{B/c_3}\right)_x = \left(r_3 - Ag_3\right) \cos \theta_3 = (0.3302 - 0.127) \cos 20.153 = 0.191$$

$$\left(r_{B/c_3}\right)_y = \left(r_3 - Ag_3\right) \sin \theta_3 = (0.3302 - 0.127) \sin 20.153 = 0.070$$

$$\left(r_{B/c_4}\right)_x = \left(r_4 - O_4g_4\right) \cos \theta_4 = (0.1651 - 0.1016) \cos 99.69 = -0.011$$

$$\left(r_{B/c_4}\right)_y = \left(r_4 - O_4g_4\right) \sin \theta_4 = (0.1651 - 0.1016) \sin 99.69 = 0.063$$

$$\left(r_{O_4/c_4}\right)_x = -O_4g_4 \cos \theta_4 = -0.1016 \cos 99.69 = 0.017$$

$$\left(r_{O_4/c_4}\right)_y = -O_4g_4 \sin \theta_4 = -0.1016 \sin 99.69 = -0.100$$

Substituting the position terms gives the final A matrix:

$$A = \begin{bmatrix}
1 & 0 & 1 & 0 & 0 & 0 & 0 & 0 & 0 \\
0 & 1 & 0 & 1 & 0 & 0 & 0 & 0 & 0 \\
0 & 0 & -0.049 & -0.058 & 0 & 0 & 0 & 0 & 1 \\
0 & 0 & -1 & 0 & 1 & 0 & 0 & 0 & 0 \\
0 & 0 & 0 & -1 & 0 & 1 & 0 & 0 & 0 \\
0 & 0 & -0.044 & 0.119 & -0.070 & 0.191 & 0 & 0 & 0 \\
0 & 0 & 0 & 0 & -1 & 0 & 1 & 0 & 0 \\
0 & 0 & 0 & 0 & 0 & -1 & 0 & 1 & 0 \\
0 & 0 & 0 & 0 & 0.063 & 0.011 & 0.100 & 0.017 & 0
\end{bmatrix}$$

Matrix B is determined next. Refer to Example Problem 11.3 for values of center of mass and angular accelerations.

$$B = \begin{bmatrix} m_2(a_x)_2 \\ m_2(a_y)_2 \\ (I_G)_2 \alpha_2 \\ m_3(a_x)_3 \\ m_3(a_y)_3 \\ (I_G)_3 \alpha_3 \\ m_4(a_x)_4 \\ m_4(a_y)_4 \\ (I_G)_4 \alpha_4 \end{bmatrix} = \begin{bmatrix} 2.4(0) \\ 2.4(0) \\ 0.001(0) \\ 4.6(128.19) \\ 4.6(-95.524) \\ 0.031(247.36) \\ 6.2(61.49) \\ 6.2(-32.20) \\ 0.023(-543.29) \end{bmatrix} = \begin{bmatrix} 0 \\ 0 \\ 0 \\ 589.67 \\ -439.41 \\ 7.67 \\ 381.24 \\ -199.64 \\ -12.49 \end{bmatrix}$$

The joint forces and input torque are determined by multiplying the inverse of matrix A and matrix B:

$$[C] = [A]^{-1}[B]$$

$$Answer: \quad \begin{bmatrix} F_{12_x} \\ F_{12_y} \\ F_{32_x} \\ F_{32_y} \\ F_{43_x} \\ F_{43_y} \\ F_{14_x} \\ F_{14_y} \\ T_2 \end{bmatrix} = \begin{bmatrix} 826.65 \\ -124.27 \\ -826.65 \\ 124.87 \\ -236.75 \\ -314.34 \\ 144.52 \\ -514.00 \\ -33.20 \end{bmatrix}$$

11.4   Sliding Joint Forces

The examples so far have focused on planar linkage mechanisms with revolute joints. This section will discuss the basic principles for force analysis of planar mechanisms that contain sliding joints. Kinematic analysis of mechanisms with sliding joints was discussed in Section 4.8. Sliding joints will have a contact force and a friction force acting in the opposite direction of the sliding motion.

Consider the general linkage mechanism shown in Figure 11.15a that contains a sliding joint. Link 2 rotates at an angular velocity ω_2 and accelerates at an angular acceleration value of α_2. Methods discussed in Chapter 4 can be used to determine the angular velocity of link 4 (link O_4B). The normal force, which is the force exerted on link 4 from the slider link 3, is shown in Figure 11.15b.

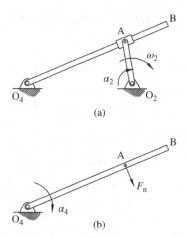

(a)

(b)

Figure 11.15   (a) Planar mechanism with revolute and sliding joints. (b) Normal force

The normal force can be determined using summation of moment equation about the fixed pivot point.

$$\sum M_{O_4} = I_4 \alpha_4 \tag{11.25}$$

Example Problem 11.5

The crank O_2A for the mechanism shown in Figure 11.16 rotates with an angular velocity of 80 rpm counterclockwise. Link 4 has a mass of 4.7 kg and the collar mass is negligible. The crank is 0.3 meters and the distance O_2O_4 is 0.4 meters. Position analysis has been completed to give the angles shown. The total length of link 4 is 1 meter and the center of gravity is located at the midpoint. Determine the force the collar exerts on link 4.

Solution: The rotational speed of the input crank must first be converted:

$$\omega_2 = 80 \frac{\text{rev}}{\text{min}} \left(\frac{2\pi \text{ rad}}{\text{rev}} \right) \left(\frac{1 \text{ min}}{60 \text{ s}} \right) = 8.38 \frac{\text{rad}}{\text{s}}$$

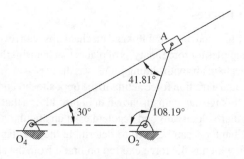

Figure 11.16

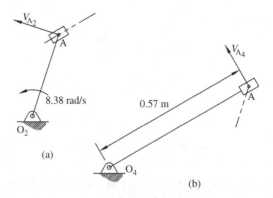

(a)

(b)

Figure 11.17

The velocity of point A on link 2 is determined from pure rotation of the crank, as shown in Figure 11.17a.

$$v_{A_2} = r_2\omega_2 = 0.3\,\text{m}\left(8.38\frac{\text{rad}}{\text{s}}\right) = 2.51\frac{\text{m}}{\text{s}}$$

Point A on link 4 is in pure rotation, as shown in Figure 11.17b. The distance from O_4 to point A can be found using law of sines:

$$\frac{O_4A}{\sin 108.19} = \frac{0.3}{\sin 30} \Rightarrow O_4A = 0.57\,\text{m}$$

The velocity equation is

$$v_{A_4} = r_4\omega_4 = 0.57 m\omega_4 \tag{1}$$

The relative velocity equation is

$$v_{A_2} = v_{A_4} + v_{A_2/A_4}$$

The velocity term v_{A_2/A_4} represents point A_2 moving with point A_4 fixed and will be oriented along the line of link 4. The velocity vector polygon, as shown in Figure 11.18, can then be used to determine the velocity magnitudes.

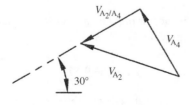

Figure 11.18

Scaling the velocity polygon gives

$$v_{A_2/A_4} = 1.67 \frac{m}{s}$$

$$v_{A_4} = 1.87 \frac{m}{s}$$

Using Equation 1,

$$\omega_4 = \frac{1.87 \text{ m/s}}{0.57 \text{ m}} = 3.28 \frac{rad}{s}$$

Next we perform the acceleration analysis, again using a vector polygon. The normal acceleration (tangential acceleration will be zero due to $\alpha_2 = 0$) of point A on link 2 will be

$$a_{A_2} = r_2\omega_2^2 = 0.3 \text{ m} \left(8.38 \frac{rad}{s} \right)^2 = 21.07 \frac{m}{s^2}$$

The magnitude of the normal component of the acceleration of point A on link 4 can also be determined:

$$\left(a_{A_4} \right)_n = r_4\omega_4^2 = 0.57 \text{ m} \left(3.28 \frac{rad}{s} \right)^2 = 6.13 \frac{m}{s^2}$$

The relative acceleration equation is determined based on the absolute acceleration given in Equation 4.33:

$$\vec{a}_A = \vec{a}_B + \dot{\vec{\omega}} \times \vec{\rho} + \vec{\omega} \times (\vec{\omega} \times \vec{\rho}) + 2\vec{\omega} \times \dot{\vec{\rho}}_{rel} + \ddot{\vec{\rho}}_{rel}$$

For our particular case, we need

$$\vec{a}_{A_2} = \vec{a}_{A_4} + \vec{a}_{A_2/A_4}$$

$$\vec{a}_{A_2} = \left(\vec{a}_{A_4} \right)_n + \left(\vec{a}_{A_4} \right)_t + \vec{a}_{A_2/A_4} + 2\vec{\omega}_4 \times \vec{v}_{A_2/A_4}$$

The acceleration vector polygon is now started, as shown in Figure 11.19, using the calculated acceleration values.

The magnitude of the Coriolis acceleration can also be determined:

$$\left| 2\vec{\omega}_4 \times \vec{v}_{A_2/A_4} \right| = 2 \left(3.28 \frac{rad}{s} \right) \left(1.67 \frac{m}{s} \right) = 10.96 \frac{m}{s^2}$$

Adding the Coriolis term to the vector polygon allows for measurement of the two unknown acceleration terms. Scaling Figure 11.19 gives

$$a_{A_2/A_4} = 9.57 \frac{m}{s^2}$$

$$\left(a_{A_4} \right)_t = 3.09 \frac{m}{s^2}$$

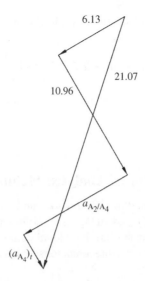

Figure 11.19

The rotational acceleration of link 4 is

$$\alpha_4 = \frac{(a_{A_4})_t}{r_4} = \frac{3.09 \text{ m/s}^2}{0.57 \text{ m}} = 5.42 \frac{\text{rad}}{\text{s}^2} \quad \text{(cw)}$$

Now that the kinematic analysis is complete, we can begin the kinetic analysis. The forces are shown in Figure 11.20.

Summation of moments about the fixed pivot point gives

$$\sum M_{O_4} = I_4 \alpha_4$$

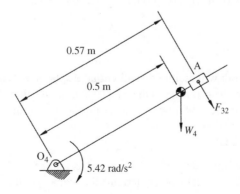

Figure 11.20

Knowing that the mass moment of inertia of a thin rod about its end is $1/3\,\text{mL}^2$,

$$\left[4.7\,\text{kg}\left(9.81\,\frac{\text{m}}{\text{s}^2}\right)\cos 30\right](0.5\,\text{m}) + F_{32}(0.57\,\text{m}) = \frac{4.7\,\text{kg}(1\,\text{m})^2}{3}\left(5.42\,\frac{\text{rad}}{\text{s}^2}\right)$$

Answer: $F_{32} = -20.1\,\text{N}$

11.5   Energy Methods of Force Analysis: Method of Virtual Work

Energy methods, such as the method of virtual work, can be very useful for force analysis for certain types of mechanisms. A major advantage is the ability to analyze a complete mechanism rather than to separate the mechanism into individual components. Keeping the mechanism fully assembled for analysis will give a direct relationship between the input forces and output forces.

This section will examine the method of virtual work. The method of virtual work has numerous applications outside machine analysis. The method is commonly used, for example, in structural analysis to determine deflection of trusses and beams and to solve statically indeterminate structures. The method of virtual work uses work U that is the dot product of force F and displacement s:

$$U = \vec{F} \cdot \vec{s} \tag{11.26}$$

Work associated with rotary motion is the product of torque and angular displacement. Work is a scalar (magnitude and sign but no direction) and has units of energy. Some forces will not work on the system. Such forces include connection forces, reaction forces, and forces acting in a direction perpendicular to motion. The method of virtual work is based on the principle that for a rigid body in equilibrium with applied external forces, the total work done by the forces will be zero for small displacement.

Because the system will be in static equilibrium, no motion will occur. The displacements will be imaginary infinitesimal displacements known as virtual displacements. Virtual work is the work done by forces or moments through the virtual displacement:

$$\delta U = \vec{F} \cdot \delta \vec{s} \tag{11.27}$$

The method of virtual work will require that all virtual displacements are represented in terms of one variable; therefore, the geometry of the mechanism should allow for simple relationships between variables. The process will be clarified with the use of an example.

Example Problem 11.6

The slider–crank mechanism shown in Figure 11.21 has a force of $F = 5.6\,\text{kN}$ applied to the slider. Using the method of virtual work, determine the moment required on link 2 to maintain static equilibrium.

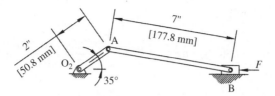

Figure 11.21

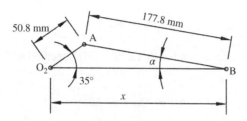

Figure 11.22

Solution: The geometry of the mechanism is shown in Figure 11.22. A relationship must be developed between variables. Law of sines gives

$$\frac{\sin \theta_2}{177.8} = \frac{\sin \alpha}{50.8}$$

$$\sin \alpha = \frac{50.8}{177.8} \sin \theta_2 = 0.286 \sin \theta_2 \tag{1}$$

Substituting the values gives

$$\sin \alpha = 0.286 \sin 35$$

$$\alpha = 9.44°$$

The slider position can be expressed as

$$x = 50.8 \cos \theta_2 + 177.8 \cos \alpha$$

Defining the virtual displacement,

$$\delta x = -50.8(\delta \theta_2) \sin \theta_2 - 177.8(\delta \alpha) \sin \alpha \tag{2}$$

Differentiating Equation 1 gives

$$\delta \alpha \cos \alpha = 0.286(\delta \theta_2) \cos \theta_2$$

$$\delta \alpha = \frac{0.286(\delta \theta_2) \cos \theta_2}{\cos \alpha}$$

Substitution into Equation 2 gives

$$\delta x = -50.8(\delta\theta_2)\sin\theta_2 - 177.8\sin\alpha\left(\frac{0.286(\delta\theta_2)\cos\theta_2}{\cos\alpha}\right)$$

Substituting known values

$$\delta x = -50.8(\delta\theta_2)\sin 35 - 177.8\sin 9.44\left(\frac{0.286(\delta\theta_2)\cos 35}{\cos 9.44}\right)$$

$$\delta x = 36(\delta\theta_2) \tag{3}$$

Equation 3 gives the relationship between variables. The method of virtual work can now be used:

$$\delta U = F\delta x + M_2\delta\theta_2 = 0$$

$$F(36\delta\theta_2) + M_2\delta\theta_2 = 0$$

$$M_2 = -F(36) = -5.6(36)$$

Answer: $M_2 = 201.6\dfrac{\text{N}}{\text{m}}$ (ccw)

11.6 Force Analysis for Slider–Crank Mechanisms Using Lumped Mass

11.6.1 Lumped Mass Assumption

To illustrate the concept of lumped mass approximation, we will examine the slider–crank mechanism shown in Figure 11.23a. At this point, we will simplify the process by having the centroid of the crank located at the fixed pivot point O_2. The motion of the coupler, as discussed in Chapter 3, is complex. However, the two endpoints have very basic motion. To simplify dynamic analysis, the coupler can be modeled as a simple rod in which the mass is lumped as two masses, which are located at both ends of the connecting rod as shown in Figure 11.23b.

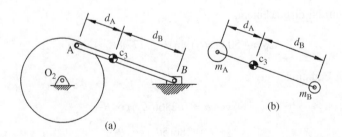

Figure 11.23 (a) Slider–crank mechanism. (b) Lumped mass model of connecting rod

The lumped model must satisfy certain conditions so that the model is dynamically equivalent to the true coupler. The sum of the two lumped masses must equal the total mass of the coupler:

$$m_A + m_B = m_3 \tag{11.28}$$

The next condition is that the location of the center of mass of the lumped masses must be the same location of the coupler's original center of mass:

$$m_A(d_A) = m_B(d_B) \tag{11.29}$$

Finally, the moment of inertia with respect to the center of mass must be the same. The same moment of inertia will ensure that the lumped mass model will cause the same angular acceleration as the original coupler:

$$m_A(d_A)^2 + m_B(d_B)^2 = (I_G)_3 \tag{11.30}$$

The distances d_A and d_B are constant and so is the moment of inertia $(I_G)_3$. Therefore, the only two parameters that can be adjusted for all three criteria are the two masses m_A and m_B. The three conditions cannot all be met by only adjusting the two variables. Therefore, we can satisfy only two criteria. Not satisfying the third condition will cause some amount of dynamic error in the lumped mass model, but our goal is to minimize the dynamic difference. It is common to meet the conditions of total mass equivalence and center of mass. Solving Equations 11.28 and 11.29 together gives

$$m_A = \left(\frac{d_B}{d_A + d_B}\right) m_3$$
$$\tag{11.31}$$
$$m_B = \left(\frac{d_A}{d_A + d_B}\right) m_3$$

The equations can be simplified by noting that $d_A + d_B$ is simply the length of the coupler L_3.

$$m_A = \frac{m_3 d_B}{L_3}$$
$$\tag{11.32}$$
$$m_B = \frac{m_3 d_A}{L_3}$$

The total accuracy of the masses determined in Equation 11.32 will depend on how closely the condition of Equation 11.30 is met.

11.6.2 Acceleration of the Slider

The acceleration of the slider can be determined by taking the second time derivative of its position. From Chapter 3, the position of the slider for an in-line slider–crank mechanism (offset equal to zero) is

$$x_B = O_2A \cos \theta_2 + AB\sqrt{1 - \left(\frac{O_2A}{AB}\right)^2 \sin^2 \theta_2} \tag{11.33}$$

The exact second derivative of Equation 11.33 is rather difficult to find analytically. Though an exact derivative could be determined using software, it is common to simplify the expression using the following binomial series:

$$\sqrt{1-s} = 1 - \frac{1}{2}s - \frac{1}{8}s^2 - \frac{1}{16}s^3 - \cdots \tag{11.34}$$

using

$$s = \left(\frac{O_2A}{AB} \sin \theta_2 \right)^2 \tag{11.35}$$

Higher order terms can be eliminated; so, using only the first two terms of the series, the approximate location of the slider would be

$$x_B \approx O_2A \cos \theta_2 + AB \left[1 - \frac{1}{2}\left(\frac{O_2A}{AB} \right)^2 \sin^2 \theta_2 \right] \tag{11.36}$$

It is also convenient to use the trigonometric identity:

$$\sin^2 \theta_2 = \frac{1}{2} - \frac{1}{2} \cos (2\theta_2) \tag{11.37}$$

Substituting Equation 11.37 into Equation 11.36 gives

$$x_B \approx O_2A \cos \theta_2 + AB \left[1 - \frac{1}{2}\left(\frac{O_2A}{AB} \right)^2 \left(\frac{1}{2} - \frac{1}{2} \cos (2\theta_2) \right) \right]$$

$$x_B \approx O_2A \cos \theta_2 + AB \left[1 - \frac{1}{4}\left(\frac{O_2A}{AB} \right)^2 + \frac{1}{4}\left(\frac{O_2A}{AB} \right)^2 \cos (2\theta_2) \right]$$

$$x_B \approx O_2A \cos \theta_2 + AB - \frac{(O_2A)^2}{4(AB)} + \frac{(O_2A)^2}{4(AB)} \cos (2\theta_2) \tag{11.38}$$

To get the expression in terms of time, we substitute $\theta_2 = \omega_2 t$ to get

$$x_B(t) \approx O_2A \cos (\omega_2 t) + AB - \frac{(O_2A)^2}{4(AB)} + \frac{(O_2A)^2}{4(AB)} \cos (2\omega_2 t) \tag{11.39}$$

The acceleration of the slider is determined by taking the second time derivative

$$a_B(t) \approx -(O_2A)\omega_2^2 \left[\cos (\omega_2 t) + \frac{O_2A}{AB} \cos (2\omega_2 t) \right] \tag{11.40}$$

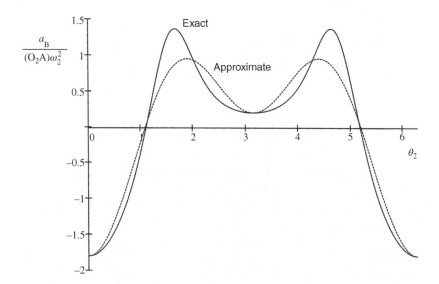

Figure 11.24 Comparison of exact and approximate acceleration for $\lambda = 0.8$

We can now examine the accuracy of Equation 11.40. Dividing Equation 11.40 by $(O_2A)\omega_2^2$ will give a dimensionless ratio. We can also define a ratio of crank length to connecting rod length as $\lambda = O_2A/AB$. Making both changes gives

$$\frac{a_B}{(O_2A)\omega_2^2} = -[\cos(\theta_2) + \lambda\cos(2\theta_2)] \tag{11.41}$$

Equation 11.41 can now be plotted for different values of λ. Figure 11.24 compares the approximate acceleration with exact acceleration for $\lambda = 0.8$. The exact acceleration, which is developed from exact differentiation of Equation 11.33, has higher peak values compared to the approximation.

An error expression can be developed to be the exact function minus the approximate function. The error expressions can then be plotted for various values of λ. Figure 11.25 shows the error for three values of λ. It can be seen that the error decreases as λ becomes smaller, which indicates that the connecting rod length increases relative to the crank length.

Example Problem 11.7

A slider–crank mechanism has a crank and a coupler that are 2 and 5.5 inches long, respectively. The mechanism has no offset and rotates at 80 rpm. Determine the approximate acceleration of the slider as a function of time and plot the results.

Solution: The rotation speed of the crank must be converted to radians per second.

$$\omega_2 = 80\,\frac{\text{rev}}{\text{min}}\left(\frac{2\pi\,\text{rad}}{\text{rev}}\right)\left(\frac{1\,\text{min}}{60\,\text{s}}\right) = 8.378\,\frac{\text{rad}}{\text{s}}$$

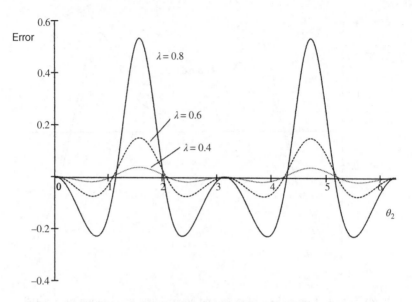

Figure 11.25   Error function (exact–approximate) for different values of λ

The acceleration equation is determined using Equation 11.40.

$$a_B(t) \approx -(O_2A)\omega_2^2\left[\cos\left(\omega_2 t\right) + \frac{O_2A}{AB}\cos\left(2\omega_2 t\right)\right]$$

$$a_B(t) \approx -(2\text{ in.})\left(8.378\frac{\text{rad}}{\text{s}}\right)^2\left[\cos\left(8.378t\right) + \frac{2\text{ in.}}{5.5\text{ in.}}\cos\left(2(8.378)t\right)\right]$$

Answer: $a_B(t) \approx -140.38[\cos\left(8.378\,t\right) + 0.36\cos\left(16.756\,t\right)]\text{ in.}/\text{s}^2$
The acceleration function is plotted in Figure 11.26 for one complete cycle.

11.7 Gear Forces

11.7.1 Introduction

Forces and gear mechanisms are critical for the complete design of the gears as well as the shafts and the bearing. Forces acting on individual gear teeth are required to examine bending stress on the gear tooth. That bending stress, although not discussed here, is typically estimated using the Lewis bending equation and is discussed in detail in machine design textbooks. The forces acting on individual teeth are also critical for analysis of surface durability, or wear, of the contact surfaces. This section will briefly introduce the concepts of gear forces. Full discussion of the topic is beyond the scope of this text. Machine design textbooks should be referred to for more information.

11.7.2 Spur Gears

Owing to their simplicity, we will begin with a simple gear train consisting to two spur gears. Figure 11.27a shows the gear and the pinion each separated into two individual free body

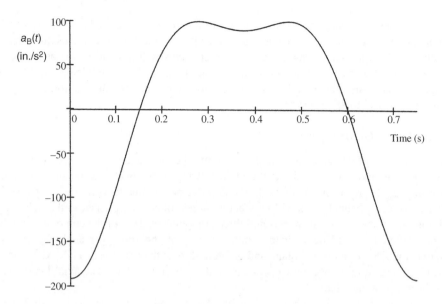

Figure 11.26

diagrams. At the contact point between the gear and the pinion, a force exists and acts in the direction of the pressure line (at an angle equal to the pressure angle). For the gear and the pinion, the force will be equal and opposite as shown. Looking at either the gear or the pinion individually, the summation of forces must be zero. Therefore, the force acting at the tooth must be counteracted by a reaction force at the bearing as shown.

Figure 11.27b shows the force acting on the gear separated into components along the centerline and perpendicular to the centerline. The force along the centerline, known as the

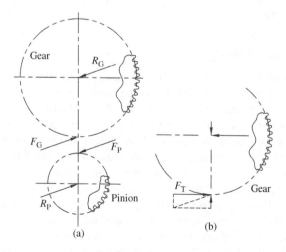

Figure 11.27 (a) Forces acting on gear and pinion. (b) Force components on gear

radial load, does not create a moment on the gear but must be considered for shaft forces. The force acting perpendicular to the centerline (labeled F_T in the figure) is the useful force and transmits the load to create a moment. Both components will have reactions at the bearing causing the shaft to be in two axes bending (in addition to any torque loads on the shaft). Though shaft analysis is beyond the scope of this text, it can be seen that the force analysis of the gears is critical to shaft design.

11.7.3 Other Gear Types

Force analysis for gears other than spur gears follows the same general concepts. However, the process becomes more complex due to the additional angles involved. The transmitted force for spur gears, as illustrated in Figure 11.27b, only develops components in the x–y plane. The force between two helical gears will now act at an additional angle equal to the helix angle. Therefore, the force will be separated into three components. Two of the components will be the transmitted load and radial load, similar to that of the spur gear analysis. The third component would be out of the plane and is referred to as the axial load. This axial load will again be countered by a reaction causing an axial force on the shaft. The axial load requires thrust bearings on the shaft.

Examination of other gear types, such as bevel gears and worm gears, will again have the additional component. As stated previously, a complete discussion of gear force analysis is beyond the scope of this text. Machine design textbooks may be referred to for additional information.

11.8 Computer Methods

As discussed in Section 11.3, the design process requires several iterations of the force analysis process. Therefore, it is best to automate that process as much as possible. The matrix method, discussed in Section 11.3 for four-bar mechanisms, is very well suited for MATLAB® or MathCAD programming. Methods discussed in Chapter 6 to develop a full kinematic analysis can be continued to perform a kinetic analysis. Programming can be done to include optimization to allow for multiple rounds of analysis to determine the linkage geometries to minimize mechanism weight.

11.9 Review and Summary

Force analysis is essential for designing machine elements. Members within the machine must be strong enough to withstand the stresses. If the members are too small, they can fail during operation. However, making the members too large will result in a lot of excess weight and cost.

This chapter focused on methods of determining joint forces and a required input force for linkage mechanisms. Graphical methods, such as superposition, allow for easy and visual solutions but are very time consuming for repeated calculations. Analytical methods, such as matrix methods, are very well suited for computer applications.

The force analysis problems discussed in this chapter are still based on some assumptions. For force analysis the links are assumed to be rigid bodies, although in reality no material will be truly rigid. Any deformations caused by the forces will be assumed small and negligible. Another assumption is that friction is negligible. As an example, the linkage mechanism is assumed to have frictionless revolute joints. The matrix method discussed in this chapter can be expanded to include friction at the joints if desired.

Problems

P11.1   Use superposition method to determine the input torque for the mechanism described in row 1 of Table 11.1. Assume the center of mass is located at the midpoint of each link.

Table 11.1   Data for problems

Row	Linkage lengths (m)	Linkage angles (°)	Linkage rotation velocity (rad/s)	Linkage rotation acceleration (rad/s^2)	Linkage mass (kg)	Moment of inertia about center of gravity (kg m^2)
1	$r_1 = 1.4$	$\theta_2 = 65$	$\omega_2 = 26.18$	$\alpha_2 = 0$	$m_2 = 4$	$I_2 = 0.04$
	$r_2 = 0.35$	$\theta_3 = 20.57$	$\omega_3 = -4.44$	$\alpha_3 = 122.46$	$m_3 = 11$	$I_3 = 1.32$
	$r_3 = 1.2$	$\theta_4 = 99.87$	$\omega_4 = 8.70$	$\alpha_4 = 250.29$	$m_4 = 7$	$I_4 = 0.33$
	$r_4 = 0.75$					
2	$r_1 = 3.0$	$\theta_2 = 45$	$\omega_2 = -62.83$	$\alpha_2 = 0$	$m_2 = 9$	$I_2 = 0.37$
	$r_2 = 0.7$	$\theta_3 = 31.65$	$\omega_3 = 43.01$	$\alpha_3 = -1397$	$m_3 = 14$	$I_3 = 1.41$
	$r_3 = 1.1$	$\theta_4 = 145.65$	$\omega_4 = -5.85$	$\alpha_4 = 2737$	$m_4 = 23$	$I_4 = 6.92$
	$r_4 = 1.9$					
3	$r_1 = 6.5$	$\theta_2 = 76$	$\omega_2 = 146.6$	$\alpha_2 = 0$	$m_2 = 15$	$I_2 = 0.61$
	$r_2 = 2.1$	$\theta_3 = 6.97$	$\omega_3 = -52.38$	$\alpha_3 = -1354$	$m_3 = 24$	$I_3 = 2.42$
	$r_3 = 4.5$	$\theta_4 = 120.56$	$\omega_4 = 104.57$	$\alpha_4 = 1514$	$m_4 = 40$	$I_4 = 12.03$
	$r_4 = 3.0$					
4	$r_1 = 9$	$\theta_2 = 18$	$\omega_2 = 68.07$	$\alpha_2 = 0$	$m_2 = 5$	$I_2 = 1.07$
	$r_2 = 1.6$	$\theta_3 = 17.45$	$\omega_3 = -17.33$	$\alpha_3 = -362$	$m_3 = 19$	$I_3 = 62.84$
	$r_3 = 6.3$	$\theta_4 = 121.63$	$\omega_4 = 0.38$	$\alpha_4 = 3427$	$m_4 = 9$	$I_4 = 5.88$
	$r_4 = 2.8$					

P11.2   Use superposition method to determine the input torque for the mechanism described in row 2 of Table 11.1. Assume the center of mass is located at the midpoint of each link.

P11.3   Use superposition method to determine the input torque for the mechanism described in row 3 of Table 11.1. Assume the center of mass is located at the midpoint of each link.

P11.4   Use superposition method to determine the input torque for the mechanism described in row 4 of Table 11.1. Assume the center of mass is located at the midpoint of each link.

P11.5   Use superposition method to determine the input torque for the mechanism described in row 4 of Table 11.1. The center of mass of link 2 is located at the fixed pivot point, the center of mass of link 3 is located at the midpoint, and the center of mass of link 4 is located at 60% of the link length from the fixed pivot point.

P11.6   Use the matrix method to determine the joint reaction forces and input torque for the mechanism described in row 1 of Table 11.1. Assume the center of mass is located at the midpoint of each link.

P11.7 Use the matrix method to determine the joint reaction forces and input torque for the mechanism described in row 2 of Table 11.1. Assume the center of mass is located at the midpoint of each link.

P11.8 Use the matrix method to determine the joint reaction forces and input torque for the mechanism described in row 3 of Table 11.1. Assume the center of mass is located at the midpoint of each link.

P11.9 Use the matrix method to determine the joint reaction forces and input torque for the mechanism described in row 4 of Table 11.1. Assume the center of mass is located at the midpoint of each link.

P11.10 Develop the equations in matrix form to determine the joint reaction forces and input torque for the statically loaded slider–crank system shown in Figure 11.21.

P11.11 The statically loaded slider–crank system shown in Figure 11.21 has the center of gravity locations for each link located at the midpoints. Link 2 has a mass of 2.1 kg, link 3 has a mass of 5.3 kg, and slider mass is 3 kg. The coefficient of static friction between the slider and the contact surface is 0.3. Develop the equations in matrix form and solve for the joint reaction forces and input torque for the static loading shown.

P11.12 Solve Example Problem 11.3 using the matrix method.

P11.13 Determine forces at the joints and the driving torque for the linkage mechanism shown in Figure 11.28a. The center of gravity for links 2 and 4 will be located at the midpoint between pin joints. The center of gravity of link 3 is in the position shown in Figure 11.28b. The kinematic data for the system is given in Table 11.2. Link 2 is 3 inches long, link 4 is 4 inches long, and the coupler is dimensioned as shown in Figure 11.28b.

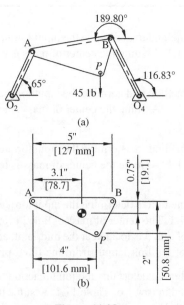

Figure 11.28

Table 11.2 Data for Problem P11.13

Link	Weight (lb)	Moment of inertia (lb-in.-s^2)	ω (rad/s)	α (rad/s^2)	a_G (in./s^2)
2	1.2	0.25	10	0	$-63.4\hat{i} - 135.9\hat{j}$
3	5.3	1.20	−4.93	−3.39	$-204.9\hat{i} - 277.5\hat{j}$
4	1.6	0.35	6.44	89.29	$-121.9\hat{i} - 154.7\hat{j}$

P11.14 A four-bar mechanism has a 14 inch ground link, a 4 inch driving link, a 12 inch coupler, and a 6.5 inch output link. The mechanism is in an open position with $\theta_2 = 70°$. Links 2, 3, and 4 have masses of 0.05, 0.3, and 0.1 lb s^2/in. respectively. The moments of inertia for links 3 and 4 are 0.43 lb in. s^2 and link 2 is 0.2 lb in. s^2. All center of gravity locations are at the linkage midpoints. Do a complete kinematic and kinetic analysis for the system if the crank rotates at a constant rate of 75 rpm clockwise. The output link has a torque of 20 in. lb counterclockwise.

P11.15 Solve Problem P11.14 if the crank is accelerating at a rate of 15 rad/s^2 and link 3 has a torque of 12 in. lb clockwise.

P11.16 The connecting rod for a slider–crank mechanism is to be modeled as lumped masses located at points A and B. The connecting rod is 1.8 meters long, has a mass of 22.6 kg, and the center of mass is located 0.6 meters from end A. Determine the mass values at each location to develop a dynamically equivalent system.

Further Reading

Budynas, R.G. and Nisbett, J.K. (2011) *Shigley's Mechanical Engineering Design*, McGraw-Hill, New York.
Greenwood, D.T. (1988) *Principles of Dynamics*, Prentice-Hall, New Jersey.

12

Balancing of Machinery

12.1 Introduction

Machine design topics discussed prior to this chapter have focused primarily on how to design machines to perform a specific task. Another important aspect to consider in the design of machinery is that the machine is properly balanced so that it operates smoothly. Forces within machine elements are ultimately transmitted to the support frame. Often the forces cause support vibration because the machine forces fluctuate with time. Vibration of the support frame can cause many problems such as excessive noise, component fatigue failure, or structural failure of the support frame. The vibrations can be drastically reduced by properly designing the machinery components to reduce or eliminate unbalanced inertia forces. This process of reducing unbalanced inertia forces is known as balancing. Rotating links, for example, can be balanced by designing the link to have the proper geometry. Adding or removing mass in specific locations will cause a geometry that is naturally balanced.

Some systems are unbalanced by design. In material handling applications, as an example, unbalance is used in vibratory feeders. More often engineers design systems to be balanced, but small undesired errors in manufacturing and installation cause unbalance. The unbalance often causes unwanted vibration in the system that can lead to excess noise or component failure. Balancing methods are then used to remove or greatly reduce the unbalance to produce smoother operation.

This chapter will first examine methods for balancing rotating systems. The two methods of balancing are static and dynamic balancing. The method used depends on the application. Static balancing is for motion that is essentially in a single plane. In other words, static balancing is appropriate for systems with axial dimensions small compared to radial dimensions. Some examples include gears, thin flywheels, and fans. Dynamic balancing, although more complicated, is the method required for complete balance. It is also required for rotating masses in more than one plane.

This chapter will also discuss concepts for linkage system balancing, although the process is far more complex. Most of the topics herein on linkage balancing will focus on single-cylinder slider–crank mechanisms. Topics of multicylinder engines will be discussed in Chapter 13. Balancing of other types of four-bar linkage mechanisms will also be introduced in this chapter.

Machine Analysis with Computer Applications for Mechanical Engineers, First Edition. James Doane.
© 2016 John Wiley & Sons, Ltd. Published 2016 by John Wiley & Sons, Ltd.
Companion Website: www.wiley.com/go/doane0215

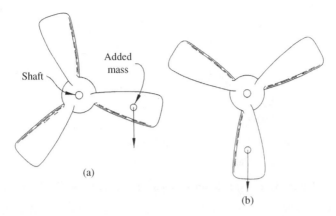

Figure 12.1   (a) Propeller with static unbalance. (b) Static equilibrium position

12.2 Static Balancing

12.2.1 Basic Concepts

The first method discussed will be static balancing of masses in pure rotation. Although the name suggests otherwise, the process is not for a system at rest. To better understand the concept of static unbalance, consider a propeller such as the one shown in Figure 12.1a. In theory, a three-blade system, such as the one shown, is inherently balanced due to geometry. In reality, even brand new blades may not all be perfectly even due to manufacturing tolerance or misalignment. As an example, assume that one blade has a small additional mass causing an imbalance. The propeller is now placed on a shaft so that it is free to rotate. If released from rest, the additional mass would cause the propeller to rotate and oscillate around the equilibrium position until it settles in the position shown in Figure 12.1b. The heavy blade will ultimately point in the downward direction. Mass could then be removed from the heavy blade and the process could be repeated until the blade is balanced. A statically balanced blade will not rotate once released from rest in any angular orientation.

Now we will illustrate the procedure of static balancing by examining a simple rotating system with a static unbalance. To initially define the inertial forces involved, let us first study a single-point mass rotating about a fixed point at a constant speed, as shown in Figure 12.2a. The mass m is attached to a massless rod and located at a radial distance r from the fixed axis of rotation.

Because the rotational speed is constant, the tangential acceleration will be zero and the only acceleration term is the normal acceleration $a_n = r\omega^2$ directed toward the center of rotation. We can apply the inertial force to the mass based on the method of d'Alembert, which is simply rearranging Newton's law by bringing inertial forces to the left-hand side of the equation:

$$\sum \vec{F} - m\vec{a} = 0 \tag{12.1}$$

For purposes of balancing, the only force of interest is the inertial force that will have a magnitude of

$$-ma_n = -mr\omega^2 \tag{12.2}$$

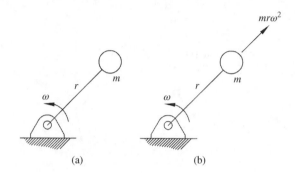

Figure 12.2   (a) Point mass in pure rotation. (b) Inertial force

The inertial force is shown in Figure 12.2b and because of the negative sign, it will always be directed away from the center of rotation.

12.2.2 Graphical Method for Rotor Balancing

To illustrate the concept of correcting a system of masses with static unbalance, let us now consider a more complex system consisting of three equal masses (all in the same transverse plane) at different radial locations rotating at a constant speed, as shown in Figure 12.3a. Each mass will have an inertial force as shown. A force polygon, shown in Figure 12.3b, can be drawn with the inertial forces.

The three inertial forces do not create a closed force polygon indicating an unbalanced system. The force required to close the force polygon (shown as the dashed line vector) would be an inertial force caused by an additional balancing mass m_b located at a radius r_b, as shown in Figure 12.3c.

It should be noticed that the rotation speed ω is a term that is present in all inertial forces. Therefore, the rotational speed will not change the overall shape of the force polygon but will only scale the vector polygon. Therefore, the required balancing mass does not depend on the rotational speed. A second important note is that the final inertia force for the balancing mass is determined as a product of the balancing mass and its radius. Any combination of mass and radius that satisfies the vector magnitude $m_b r_b$ can be used. In other words, there is not only one unique solution. Therefore, the designer has some freedom to choose the best design.

The graphical method can be accomplished by sketching scaled force polygons by hand, but the accuracy will be greatly improved by using CAD software. The following example illustrates the process.

Example Problem 12.1

The system of three masses shown in Figure 12.4 rotates at a constant speed. It is desired to balance the system by adding a mass at a radial value of $1.5r$. Use graphical methods to determine the required mass and relative angular position of the balancing mass.

Solution: Figure 12.5 shows the force polygon for the three inertial forces of the existing masses and the required inertial force of the balancing mass to close the polygon.

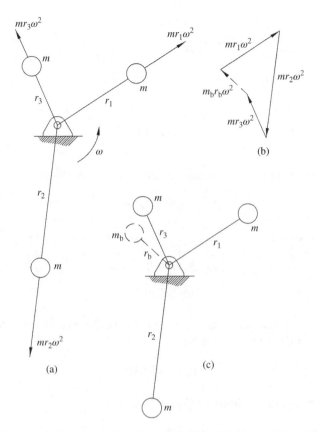

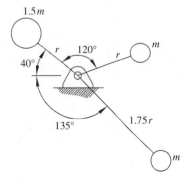

Figure 12.3 (a) System of point masses in pure rotation. (b) Force polygon of inertial forces. (c) System with balancing mass

Figure 12.4

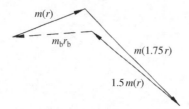

Figure 12.5

Measuring the length of the balancing inertial force vector shows

$$m_b r_b = 1.03(mr)$$

For a radial value of $1.5r$,

$$m_b(1.5r) = 1.03(mr)$$

$$m_b = 0.687m$$

The direction of the balancing mass vector can also be determined from the scaled vector polygon. The direction is measured to be

$$\theta_b = 183.827°$$

The final balanced system is shown in Figure 12.6.

12.2.3 Analytical Method for Rotor Balancing

Now that the concept of static balancing has been illustrated graphically, we can develop equations for static balancing. It has been illustrated that for a system with n rotating masses and a balancing mass, the sum of the inertial forces must equal zero. Placing the system in an arbitrary reference position, such as that shown in Figure 12.3a, we can use position vectors to

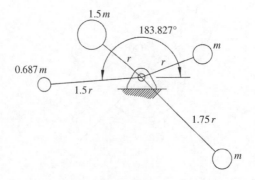

Figure 12.6

locate the masses and write the following expression. External forces are excluded from the equation because balancing is involved only with summation of inertial forces.

$$-m_b \vec{r}_b \omega^2 + \sum_{i=1}^{n} -m_i \vec{r}_i \omega^2 = 0 \tag{12.3}$$

As discussed before with the graphical method, the rotation speed can be canceled out of the equation. Dividing by ω^2 and rearranging to solve for the product of balancing mass and its radial location gives

$$m_b \vec{r}_b = \sum_{i=1}^{n} -m_i \vec{r}_i \tag{12.4}$$

Equation 12.4 is a vector equation and can now be expressed using x and y components:

$$\begin{aligned} m_b (r_b)_x &= -\sum_{i=1}^{n} m_i (r_i)_x \\ m_b (r_b)_y &= -\sum_{i=1}^{n} m_i (r_i)_y \end{aligned} \tag{12.5}$$

However, the use of polar coordinates is more convenient. The balancing mass product can be expressed as

$$m_b r_b = m_b \sqrt{(r_b)_x^2 + (r_b)_y^2} = \sqrt{\left(m_b r_{b_x}\right)^2 + \left(m_b r_{b_y}\right)^2} \tag{12.6}$$

and will be located at an angle

$$\theta_b = \tan^{-1} \frac{m_b (r_b)_y}{m_b (r_b)_x} \tag{12.7}$$

Substituting Equation 12.5 into Equations 12.6 and 12.7 gives

$$m_b r_b = \sqrt{\left(\sum_{i=1}^{n} m_i (r_i)_x\right)^2 + \left(\sum_{i=1}^{n} m_i (r_i)_y\right)^2} \tag{12.8}$$

$$\theta_b = \tan^{-1} \frac{-\sum_{i=1}^{n} m_i (r_i)_y}{-\sum_{i=1}^{n} m_i (r_i)_x} \tag{12.9}$$

A balancing mass can now be designed. Any combination of balancing mass and radial position can be selected as long as the product satisfies Equation 12.8. It must, however, be placed in the angular position to satisfy Equation 12.9.

A couple of important concepts must be discussed relating to Equation 12.9. The first is that the angle θ_b is relative to the angular positions of all masses used to develop the summations.

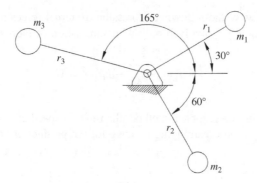

Figure 12.7

The second is that the negative signs must be maintained in order to calculate the angle in the proper quadrant. The process will be illustrated in a couple of examples.

Example Problem 12.2

The system of three masses shown in Figure 12.7 rotates at a constant speed and it is unbalanced.

$$m_1 = m_2 = 1 \quad m_3 = 1.25$$
$$r_1 = r_2 = 2.25 \quad r_3 = 2.75$$

It is desired to balance the system by removing mass from m_3. What percentage of mass must be removed from m_3 to balance the system?

Solution: Due to symmetry, it can be determined that mass m_3 is in the correct angular location to balance the remaining two masses. The x and y coordinates of each mass are required.

$$(r_1)_x = 2.25 \cos (30) = 1.949 \qquad (r_1)_y = 2.25 \sin (30) = 1.125$$
$$(r_2)_x = 2.25 \cos (-60) = 1.125 \qquad (r_2)_y = 2.25 \sin (-60) = -1.949$$
$$(r_3)_x = 2.75 \cos (165) = -2.656 \quad (r_3)_y = 2.75 \sin (165) = 0.712$$

Suppose the system consists of the masses m_1 and m_2 with m_3 being the balancing mass.

$$m_3 r_3 = m_b r_b = \sqrt{\left(\sum_{i=1}^{n} m_i (r_i)_x\right)^2 + \left(\sum_{i=1}^{n} m_i (r_i)_y\right)^2}$$

$$m_3 r_3 = \sqrt{\left(m_1 (r_1)_x + m_2 (r_2)_x\right)^2 + \left(m_1 (r_1)_y + m_2 (r_2)_y\right)^2}$$

$$m_3 r_3 = \sqrt{(1(1.949) + 1(1.125))^2 + (1(1.125) + 1(-1.949))^2} = 3.183$$

Therefore, the required mass m_3 for the system to be balanced would be

$$m_3 = \frac{3.183}{r_3} = \frac{3.183}{2.75} = 1.157$$

The percent of reduction is

$$\%\text{reduction} = \left(\frac{1.25 - 1.157}{1.25}\right) 100$$

Answer: 7.44%

Example Problem 12.3

The system of three masses shown in Figure 12.8 rotates at a constant speed of 15 rpm clockwise.

$$m_1 = 1.2\,\text{kg} \quad r_1 = 50.8\,\text{mm}$$
$$m_2 = 2.4\,\text{kg} \quad r_2 = 31.8\,\text{mm}$$
$$m_3 = 2.4\,\text{kg} \quad r_3 = 25.4\,\text{mm}$$

It is desired to balance the system using a balancing mass of 2 kg. Determine the required radial position and relative angular position of the balancing mass.

Solution: The angular position of the balancing mass is found using Equation 12.9:

$$\theta_b = \tan^{-1} \frac{-\sum_{i=1}^{n} m_i(r_i)_y}{-\sum_{i=1}^{n} m_i(r_i)_x}$$

$$\theta_b = \tan^{-1} \frac{-[1.2(50.8 \sin 37) + 2.4(31.8 \sin 120) - 2.4(25.4 \sin 80)]}{-[1.2(50.8 \cos 37) + 2.4(31.8 \cos 120) + 2.4(25.4 \cos 80)]}$$

$$\theta_b = \tan^{-1} \frac{-42.748}{-21.110} = 63.72$$

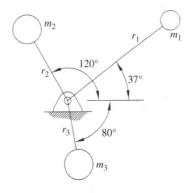

Figure 12.8

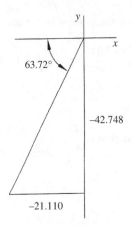

Figure 12.9

Figure 12.9 shows the location based on the x and y locations. Next, we can determine the required mass and radial location using Equation 12.8:.

$$m_b r_b = \sqrt{\left(\sum_{i=1}^{n} m_i(r_i)_x\right)^2 + \left(\sum_{i=1}^{n} m_i(r_i)_y\right)^2}$$

$$m_b r_b = \sqrt{(21.110)^2 + (42.748)^2} = 47.676$$

For a balancing mass of 2 kg,

$$2r_b = 47.676$$
$$r_b = 23.84 \text{ mm}$$

The final arrangement is shown in Figure 12.10.

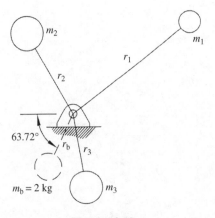

Figure 12.10

Table 12.1 Data for Example Problem 12.3

i	Mass (kg)	Radius (mm)	θ (°)	$\cos \theta$	$\sin \theta$	$m_i r_i \cos \theta$	$m_i r_i \sin \theta$
1	1.2	50.8	37	0.799	0.602	48.685	36.687
2	2.4	31.8	120	-0.500	0.866	-38.160	66.095
3	2.4	25.4	280	0.174	-0.985	10.586	-60.034
					Σ	21.110	42.748

As the systems become more complex, it can be useful to develop a tabular method to help organize the solution. For example, Table 12.1 shows the data for Example Problem 12.3 in a concise form to help organize the pertinent information. The summations in the last two columns give $\sum_{i=1}^{n} m_i(r_i)_x$ and $\sum_{i=1}^{n} m_i(r_i)_y$, which are used in the calculations as illustrated in Example Problem 12.3.

12.3 Dynamic Balancing

Static balancing is applicable to systems of rotating masses where all the masses are in a single axial plane, which results in the inertia forces being parallel vectors. Some rotating systems with masses in different planes can be approximated as single plane with minimal error if the system is short in the axial direction (the masses are approximately in a single plane). If the system becomes too long in the axial direction, dynamic balancing must be used. For example, a camshaft will have unbalanced masses in various locations along the axial direction and would require dynamic balancing. In order for a system to be dynamically balanced, the summation of forces must be zero, as with static balancing, and the summation of moments must be zero.

It is quite possible that a system is statically balanced but still be dynamically unbalanced. For example, the system shown in the x–y plane in Figure 12.11 has two equal masses located opposite one another. The inertial forces would cancel out creating a statically balanced system. However, the two masses are not in equal planes and create an unbalanced moment couple.

If possible, the system should be statically balanced before starting the process of dynamic balancing. Consider a shaft with multiple turbine blades, which need to be dynamically balanced. It is best to balance each individual set of blades before assembling the complete shaft. After the static balancing is complete, the total assembly can be dynamically balanced.

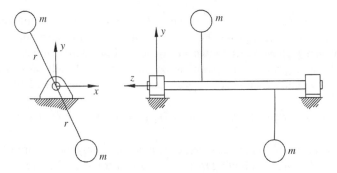

Figure 12.11 Dynamic unbalance

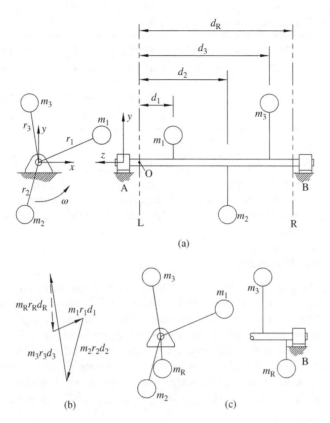

Figure 12.12 (a) Initial system. (b) Moment vector polygon. (c) Location of balancing mass m_R

To illustrate the procedure of dynamic balancing, consider the system of three masses shown in Figure 12.12a. The masses are rotating on a shaft supported by bearings at locations A and B. Unlike static balancing, the three masses are located in different planes in the z-axis direction (axial direction). To dynamically balance the system, two balancing masses will be required. Two masses are required to create a moment couple to counteract the unbalanced moment in the system. These balancing masses will be added in two planes labeled as L and R in the figure (note that the balancing planes can be located in different locations at the discretion of the designer).

We will begin with the summation of moments. It is convenient to take the summation about a point on one of the balancing planes. For this illustration, we will take moments about point O located on balancing plane L, as shown in Figure 12.12a. Inertial forces from each mass, including the balancing mass to be placed in balancing plane R, will cause a moment about point O. The total moment summation equation is given in Equation 12.10:

$$\left(m_R \vec{r}_R \omega^2\right)d_R + \left(m_1 \vec{r}_1 \omega^2\right)d_1 + \left(m_2 \vec{r}_2 \omega^2\right)d_2 + \left(m_3 \vec{r}_3 \omega^2\right)d_3 = 0 \qquad (12.10)$$

The vector equation can be expressed graphically as a vector polygon. Figure 12.12b shows the moment vector polygon after dividing by the common factor of ω^2. The vector $\left(m_R \vec{r}_R\right)d_R$ will be the vector required to close the polygon.

Dividing Equation 12.10 by ω^2 and separating into x and y components will give

$$m_R(r_R)_x = \frac{-m_1(r_1)_x d_1 - m_2(r_2)_x d_2 - m_3(r_3)_x d_3}{d_R} \qquad (12.11)$$

$$m_R(r_R)_y = \frac{-m_1(r_1)_y d_1 - m_2(r_2)_y d_2 - m_3(r_3)_y d_3}{d_R} \qquad (12.12)$$

The final mass–radius product is determined from

$$m_R r_R = \sqrt{\left[m_R(r_R)_x\right]^2 + \left[m_R(r_R)_y\right]^2} \qquad (12.13)$$

and the radial location is given by

$$\theta_R = \tan^{-1}\left(\frac{m_R(r_R)_y}{m_R(r_R)_x}\right) \qquad (12.14)$$

Once the balancing mass for balancing plane R has been determined, the balancing mass for plane L can be determined from summation of forces of the system shown in Figure 12.12c. The general process is identical to the static balancing method discussed in Section 12.2. The summation of forces must be zero for the initial three masses as well as both balancing masses. Summation of all inertial forces will give

$$-m_L\vec{r}_L\omega^2 - m_R\vec{r}_R\omega^2 - m_1\vec{r}_1\omega^2 - m_2\vec{r}_2\omega^2 - m_3\vec{r}_3\omega^2 = 0 \qquad (12.15)$$

Dividing by ω^2 and separating into x and y components,

$$m_L(r_L)_x = -m_R(r_R)_x - m_1(r_1)_x - m_2(r_2)_x - m_3(r_3)_x \qquad (12.16)$$

$$m_L(r_L)_y = -m_R(r_R)_y - m_1(r_1)_y - m_2(r_2)_y - m_3(r_3)_y \qquad (12.17)$$

The final mass–radius product is determined from

$$m_L r_L = \sqrt{\left[m_L(r_L)_x\right]^2 + \left[m_L(r_L)_y\right]^2} \qquad (12.18)$$

and the radial location is given by

$$\theta_L = \tan^{-1}\left(\frac{m_L(r_L)_y}{m_L(r_L)_x}\right) \qquad (12.19)$$

Figure 12.13 shows the system with both balancing masses and the location of balancing mass m_L in the z-axis direction.

The complete process of dynamic balancing will give products for $m_L\vec{r}_L$ and $m_R\vec{r}_R$. Any values for the two masses and position vectors can be used that satisfy the equations above. The fact that there is not only one possible solution gives the designer some freedom to optimize the

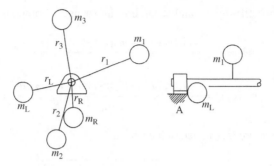

Figure 12.13 Location of balancing mass m_L

system. Also note that the locations of the two balancing planes are not uniquely defined and can be positioned in any convenient locations.

Example Problem 12.4

The system shown in Figure 12.14 consists of three masses ($m_1 = m_2 = 1.6$ kg *and* $m_3 = 3$ kg) located in different locations along the shaft. Mass 1 is located at a radial distance of 2 inches (50.8 mm), mass 2 at a radial distance of 3 inches (76.2 mm), and mass 3 at a radial distance of 1.5 inches (38.1 mm). Balancing masses, each with a mass of 2.5 kg, are to be located in the planes L and R shown. Determine the radial and angular locations of each mass to dynamically balance the system.

Solution: First the x and y components of each mass location are required:

$$(r_1)_x = 50.8 \cos 80 = 8.82 \text{ mm}$$
$$(r_1)_y = 50.8 \sin 80 = 50.03 \text{ mm}$$
$$(r_2)_x = 76.2 \cos 20 = 71.60 \text{ mm}$$
$$(r_2)_y = -76.2 \sin 20 = -26.06 \text{ mm}$$
$$(r_3)_x = -38.1 \cos 70 = -13.03 \text{ mm}$$
$$(r_3)_y = 38.1 \sin 70 = 35.80 \text{ mm}$$

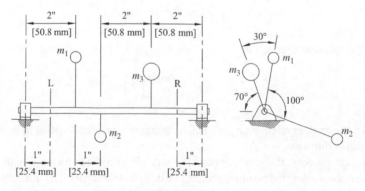

Figure 12.14

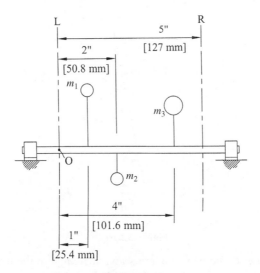

Figure 12.15

We will sum moments about point O in plane L, as shown in Figure 12.15.
Equation 12.11 gives

$$m_R(r_R)_x = \frac{-m_1(r_1)_x d_1 - m_2(r_2)_x d_2 - m_3(r_3)_x d_3}{d_R}$$

$$m_R(r_R)_x = \frac{-1.6(8.82)(25.4) - 1.6(71.60)(50.8) - 3(-13.03)(101.6)}{127} = -17.37$$

Equation 12.12 gives

$$m_R(r_R)_y = \frac{-m_1(r_1)_y d_1 - m_2(r_2)_y d_2 - m_3(r_3)_y d_3}{d_R}$$

$$m_R(r_R)_y = \frac{-1.6(50.03)(25.4) - 1.6(-26.06)(50.8) - 3(35.80)(101.6)}{127} = -85.25$$

Equation 12.13 gives the final mass–product value for the right plane mass:

$$m_R r_R = \sqrt{-17.37^2 + -85.25^2} = 87 \text{ kg mm}$$

For a balancing mass of 2.5 kg, the radial location will be

$$r_R = \frac{87 \text{ kg mm}}{2.5 \text{ kg}} = 34.8 \text{ mm}$$

The angular location of the right plane balancing mass is determined from Equation 12.14:

$$\theta_R = \tan^{-1}\left(\frac{-85.25}{-17.37}\right) = 78.48°$$

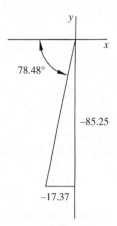

Figure 12.16

As shown in Figure 12.16, the negative terms give the angle in the third quadrant.
Therefore, the final angle from the x-axis will have 180° added to the obtained value.

Answer: $r_R = 34.8$ mm
$\theta_R = 258.48°$

Figure 12.17 shows the x–y view with the right balancing mass added.
Next we solve for the balancing mass in the left plane. Equations 12.16 and 12.17 give

$$m_L(r_L)_x = -m_R(r_R)_x - m_1(r_1)_x - m_2(r_2)_x - m_3(r_3)_x$$

$$m_L(r_L)_x = 17.37 - 1.6(8.82) - 1.6(71.60) - 3(-13.03) = -72.21 \text{ kg mm}$$

$$m_L(r_L)_y = -m_R(r_R)_y - m_1(r_1)_y - m_2(r_2)_y - m_3(r_3)_y$$

$$m_L(r_L)_y = 85.25 - 1.6(50.03) - 1.6(-26.06) - 3(35.80) = -60.50 \text{ kg mm}$$

From Equation 12.18,

$$m_L r_L = \sqrt{-72.21^2 + -60.50^2} = 94.2 \text{ kg mm}$$

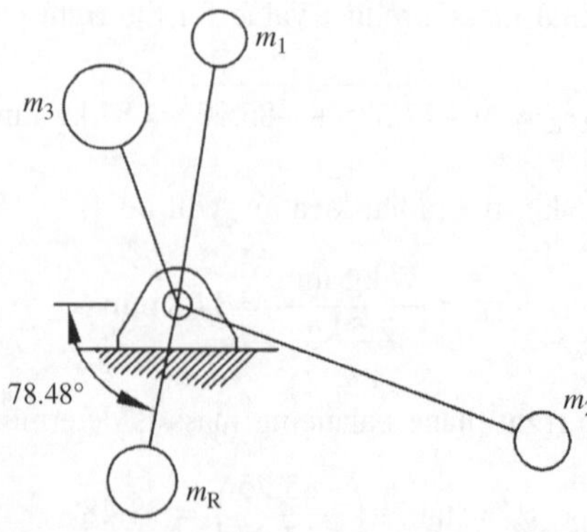

Figure 12.17

For a balancing mass of 2.5 kg,

$$r_L = \frac{94.2 \text{ kg mm}}{2.5 \text{ kg}} = 37.68 \text{ mm}$$

The angular location is determined from Equation 12.19:

$$\theta_L = \tan^{-1}\left(\frac{-60.50}{-72.21}\right) = 39.96°$$

Both terms are negative indicating that the angle is again in the third quadrant. The final angle will again have 180° added to get the reference from the positive x-axis.

Answer:
$$\begin{aligned} r_L &= 37.7 \text{ mm} \\ \theta_R &= 219.96° \end{aligned}$$

As mentioned, the axial locations for each balancing plane are chosen at the discretion of the designer. The actual plane locations may be limited due to actual space requirements. However, there is a general advantage of placing the balancing planes near the bearings (as done in the previous example). Adding the balancing masses near the bearings will cause only a small increase in the bending moment in the shaft. Locating the balancing planes toward the center of the shaft will cause a large increase in the bending moment. Also, having the two balancing planes located as far apart from each other as possible will reduce the amount of mass added.

12.4 Vibration from Rotating Unbalance

Unbalance of rotating parts, resulting in an unbalance of centrifugal forces, is a common source of forced vibrations. Sometimes the unbalance is intentional to create a vibration. However, in most applications the vibration is unwanted. This section will be an introduction to the concepts of forced vibration due to a rotating unbalance. Vibration textbooks, such as those listed in Further Reading section of Chapter 10, should be referred to for a more complete coverage of the topic.

Rotating unbalance can be modeled with mass m that has an eccentricity e. The mass rotates at a constant angular velocity ω. The full system, as shown in Figure 12.18a, has a mass M and is restrained to move in the vertical direction. Therefore, the nonrotating mass is equal to $M - m$.

From Figure 12.18b, the total vertical displacement of the eccentric mass is $x + e \sin \omega t$. The acceleration of the eccentric mass is determined from the second time derivative:

$$a_m = \frac{d^2}{dt^2}(x + e \sin \omega t) = \ddot{x} - e\omega^2 \sin \omega t \tag{12.20}$$

From the free body diagram shown in Figure 12.18c, Newton's second law in the x-direction gives the following (lateral motion is ignored):

$$-kx - c\dot{x} = (M - m)\ddot{x} + m(\ddot{x} - e\omega^2 \sin \omega t)$$

$$M\ddot{x} + c\dot{x} + kx = me\omega^2 \sin \omega t \tag{12.21}$$

Equation 12.21 is of the same form of equations that we have developed in Chapter 10 for damped systems with harmonic forcing functions. For the case of a rotating unbalance, the

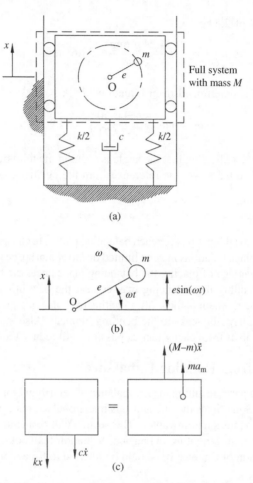

Figure 12.18   Vibration caused by rotating unbalance. (a) System model. (b) Motion of rotating mass. (c) Free body diagram

force function is $F(t) = me\omega^2 \sin \omega t$. Therefore, the solution will be identical to that developed for $F(t) = F_0 \sin \omega t$ with a force magnitude $F_0 = me\omega^2$. The magnitude of the steady-state solution is

$$X = \frac{me\omega^2}{k\sqrt{\left[1-(\omega/\omega_n)^2\right]^2 + (2\varsigma(\omega/\omega_n))^2}} \tag{12.22}$$

We can reduce Equation 12.22 to a nondimensional form called the magnification ratio:

$$\frac{MX}{me} = \frac{(\omega/\omega_n)^2}{\sqrt{\left[1-(\omega/\omega_n)^2\right]^2 + (2\varsigma(\omega/\omega_n))^2}} \tag{12.23}$$

The magnification ratio is shown in Figure 12.19 for different values of damping.

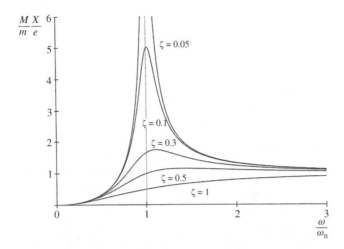

Figure 12.19 Magnification ratio

12.5 Balancing Slider–Crank Linkage Mechanisms

12.5.1 Introduction

Previous sections discussed balancing systems of masses in pure rotation. We will now move into systems, such as linkage mechanisms, that contain motion other than pure rotation. Balancing complete linkage mechanisms is far more complex than balancing simple rotating masses. In most practical examples, complete balance of a linkage mechanism may not be possible. In a general four-bar mechanism, the crank and rocker are in pure rotation and can be balanced using the methods from Section 12.2. The main complication of balancing the entire mechanism is due to the complex motion of the coupler.

We will first consider the case of balancing slider–crank mechanisms. Because slider–crank mechanisms are very common in machinery, a considerable amount of work has been done to determine methods for balancing the mechanism. Some common applications of machines that use single slider–crank mechanisms (single-cylinder mechanisms) are stamping machines or simple air compressors. Many mechanisms use multiple slider–crank mechanisms, such as multicylinder engines, in arrangements to assist in balancing. Multicylinder arrangements will be discussed in Chapter 13, but for now we will focus on mechanisms with a single slider–crank mechanism.

12.5.2 Inertial Forces

To explain the procedure, consider the slider–crank mechanism shown in Figure 12.20a. The crank rotates at a constant speed ω_2, and the slider is in line with the fixed pivot of the crank (no offset). The crank is in pure rotation and the slider will move with reciprocating linear motion. In many applications, the crank is the input into the system and the slider is the output. In other applications, such as internal combustion engines, the slider is the input and the crank is the output (turning the crank shaft in the case of internal combustion engines). The coupler, or connecting rod, will move in complex general plane motion. It is the complex motion of the connecting rod that complicates the balancing process.

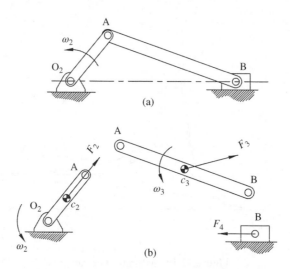

Figure 12.20 (a) Slider–crank mechanism. (b) Inertial forces of each moving link

To consider the balancing of the mechanism, we must first determine the inertial forces of each link within the mechanism. It is these inertial forces that cause the system to be unbalanced. Figure 12.20b shows the inertial forces of each link (the links have been separated to reduce confusion, but joint forces are not shown for clarity). The inertial force due to the rotating crank will maintain a constant magnitude (for constant rotation speed) but will change direction with the rotation. The inertial forces due to the coupler will change magnitude and direction in proportion to the acceleration changes. Likewise, the inertia force of the slider will stay horizontal, but will change magnitude and direction proportional to the slider acceleration. All of these inertial forces will be in directions opposite to actual accelerations.

Balancing of the inertial forces shown in Figure 12.20b will be very complex. However, the coupler can be modeled as a dynamically equivalent lumped mass system to simplify the analysis. Concepts of lumped mass assumption and force equations for slider–crank mechanisms were discussed in Chapter 11. Figure 12.21a shows the same slider–crank mechanism with the connecting rod modeled using the lumped mass assumption.

The slider–crank mechanism shown in Figure 12.21a can now be split into two distinct portions. Figure 12.21b shows the crank with the lumped mass m_A, both being in pure rotation. Because these masses are in pure rotation, we can balance their effects using static balancing. A common approach to improve balancing of a slider–crank mechanism is to add a counterweight on the rotating crank. This counterweight will balance the mass of the crank and the coupler lumped mass located at point A.

The slider mass and the lumped mass m_B are shown in Figure 12.21c with a net inertial force that is still not balanced. That reciprocating inertial force will be proportional to the acceleration of the slider. Recall from Chapter 11 that we can approximate the acceleration of the slider as

$$a_B(t) \approx -(O_2A)\omega_2^2 \left[\cos(\omega_2 t) + \frac{O_2A}{AB} \cos(2\omega_2 t) \right] \qquad (12.24)$$

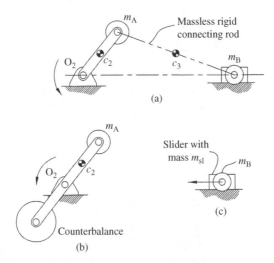

Figure 12.21 (a) Slider–crank mechanism with lumped mass. (b) Balance of crank using counterweight. (c) Unbalanced sliding mass

The inertial force from the slider and lumped mass at B will be in the opposite direction and will be equal to

$$F_s \approx m_{rec}(O_2A)\omega_2^2 \cos{(\omega_2 t)} + \frac{\lambda m_{rec}}{4}(O_2A)(2\omega_2)^2 \cos{(2\omega_2 t)} \qquad (12.25)$$

where $\lambda = O_2A/AB$ (this ratio is typically between 1/3 and 1/5 for efficient and smooth performance) and $m_{rec} = m_{sl} + m_B$. The first term in Equation 12.25 represents the fundamental frequency and is called the primary part of the shaking force, which will be the largest component. The second term in Equation 12.25 is the second harmonic and is called the secondary part of the shaking force. Higher order harmonics would exist for a more accurate representation of the acceleration, but balancing higher order terms becomes increasingly difficult. Also, the magnitude of higher order terms becomes negligible.

12.5.3 Balancing Primary Forces

Now that we can balance the inertial components of the purely rotating components, we can focus our attention on balancing the shaking forces. Equation 12.25 represents an approximate shaking force due to the unbalanced inertial forces of the sliding mass (combination of slider mass and lumped mass at point B). The larger component of the shaking force is the primary part. Our first effort to reduce the shaking force would be to attempt to balance the primary part of the shaking force.

Consider the slider–crank mechanism shown in Figure 12.22a. A second counterweight has been added to the system at a distance O_2A from the fixed rotation point of the crank (the previous counterbalance from Figure 12.21 is not shown). The counterweight has the same mass as the total reciprocating mass, which is the mass of the slider plus the lumped mass at point B. The counterweight will be in pure rotation and will generate an inertial force as shown.

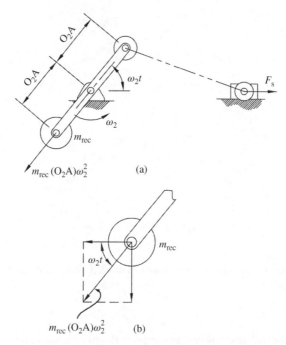

Figure 12.22 (a) Slider–crank with additional counterweight m_{rec} added to crank. (b) Horizontal and vertical components of inertial force caused by the counterweight

Figure 12.22b shows an enlarged view of the counterweight. The inertial force $m_{rec}(O_2A)\omega_2^2$ has been represented by horizontal and vertical components. It is noticed that the horizontal component $m_{rec}(O_2A)\omega_2^2 \cos(\omega_2 t)$ will balance the primary shaking force generated by the sliding mass. However, we have added a new vertical shaking force $m_{rec}(O_2A)\omega_2^2 \sin(\omega_2 t)$ to the system. A more common approach is to partially balance the horizontal shaking force while developing a smaller vertical component. The actual mass and location of the counterweight can be optimized to develop the smallest possible shaking force magnitude.

12.5.4 Illustrative Example of Slider–Crank Balancing

Now that the general process has been explained, we can illustrate the process with an actual example. Our goal is to balance the slider–crank mechanism shown in Figure 12.23a. The crank rotates at a constant speed of 750 rpm. The link dimensions are shown. The center of mass of the crank is located at 1 inch from point O_2, and the center of mass of the connecting rod is located at 3 inches from point A. The crank weight is 8 pounds, the connecting rod weight is 18 pounds, and the slider weight is 12 pounds.

We want to first model the connecting rod using lumped masses at points A and B, which is illustrated in Figure 12.23b. From Section 11.6.1,

$$W_{A_1} = \frac{18\,\text{lb}(7\,\text{in.})}{10\,\text{in.}} = 12.6\,\text{lb}$$

$$W_{B_1} = \frac{18\,\text{lb}(3\,\text{in.})}{10\,\text{in.}} = 5.4\,\text{lb}$$

(12.26)

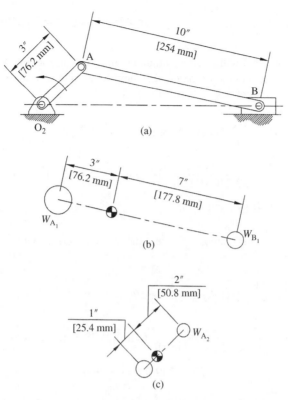

Figure 12.23 (a) Slider–crank mechanism for illustrative example. (b) Lumped mass approximation of connecting rod. (c) Lumped mass approximation of crank

The total reciprocating mass is

$$W_{\text{rec}} = W_{\text{slider}} + W_{B_1} = 12\,\text{lb} + 5.4\,\text{lb} = 17.4\,\text{lb} \qquad (12.27)$$

As shown in Figure 12.23c, the crank can also be modeled as lumped masses at points O_2 and A:

$$W_{A_2} = \frac{8\,\text{lb}(1\,\text{in.})}{3\,\text{in.}} = 2.667\,\text{lb} \qquad (12.28)$$

The total rotating weight is given by

$$W_{\text{rot}} = W_{A_1} + W_{A_2} = 12.6\,\text{lb} + 2.667\,\text{lb} = 15.267\,\text{lb} \qquad (12.29)$$

We can now determine the inertial forces for the given rotation speed of $\omega_2 = 78.54\,\text{rad/s}$. The inertial force for the rotating weight is

$$F_{\text{rot}} = \frac{W_{\text{rot}}}{g}(O_2A)\omega_2^2$$

$$F_{rot} = \frac{15.267 \text{ lb}}{32.2 \text{ ft/s}^2 (12 \text{ in.}/\text{ft})}(3 \text{ in.})(78.54 \text{ rad/s})^2 = 731.153 \text{ lb} \tag{12.30}$$

The rotational inertial force will be constant, due to the constant rotational speed, and directed away from the center of rotation of the crank. The primary and secondary shaking forces can now be determined as functions of the crank angle:

$$F_P = \frac{W_{rec}}{g}(O_2A)\omega_2^2 \cos \theta_2$$

$$F_P = \frac{17.4 \text{ lb}}{32.2 \text{ ft/s}^2 (12 \text{ in.}/\text{ft})}(3 \text{ in.})(78.54 \text{ rad/s})^2 \cos \theta_2$$

$$F_P = 833.323 \cos \theta_2 \tag{12.31}$$

For the secondary shaking force, we need the ratio of crank length to coupler length $\lambda = 3/10 = 0.3$:

$$F_S = \frac{\lambda}{4}\frac{W_{rec}}{g}(O_2A)(2\omega_2)^2 \cos 2\theta_2$$

$$F_S = \frac{0.3}{4}\frac{17.4 \text{ lb}}{32.2 \text{ ft/s}^2 (12 \text{ in.}/\text{ft})}(3 \text{ in.})(2 \cdot 78.54 \text{ rad/s})^2 \cos 2\theta_2$$

$$F_S = 249.997 \cos (2\theta_2) \tag{12.32}$$

The total shaking force is the sum of the primary and secondary forces.

The bearing force at O_2 will be a combination of the rotating inertial force and the shaking force. A very convenient way to illustrate the shaking force on the bearing is a polar diagram, as shown in Figure 12.24 (axes tick marks represent 200 pound increments). The dashed line circle represents the rotational force and has a radius of 731.153 units. Points on the plot, shown as the solid line, are determined using a vector sum of the rotational force and the primary and secondary shaking forces. Point A, for example, is the total inertia force on the crankshaft for a rotation of θ_2. The total force vector would be a vector from point O to point A.

Since the inertial forces have been determined for the unbalanced system, we can now work toward balancing the mechanism. A counterbalance will be added on the crank similar to that

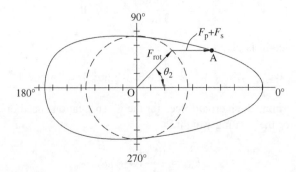

Figure 12.24 Polar diagram of unbalanced inertia force on crankshaft

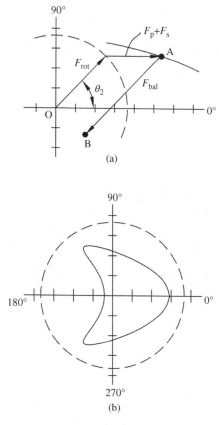

Figure 12.25 (a) Vector summation with counterbalance force. (b) Polar diagram for partially balanced system

shown in Figure 12.21b. The counterbalance will create an inertial force in the opposite direction of the lumped mass at point A. The inertial force of the counterbalance will completely balance the rotating mass at point A and will partially balance the shaking forces due to the reciprocating mass. The vector for the counterbalance force F_{bal} will be added to the previous vectors to move point A to some point B, as shown in Figure 12.25a. The optimal value of F_{bal} can be determined, but for this example we will assume it will equal the inertia force from the rotating mass at A plus approximately 60% of the maximum primary shaking force.

$$F_{bal} = F_{rot} + 0.6F_P = 731.153 \text{ lb} + 0.6(833.323 \text{ lb}) = 1231 \text{ lb} \qquad (12.33)$$

Figure 12.25b shows the new polar diagram with the counterbalance weight.

12.5.5 Lanchester Balancer

The previous sections focused on slider–crank balancing by adding counterbalance to the rotating crank. The primary shaking force could be balanced, but as a result it created a vertical shaking force. Also, the secondary shaking forces were not balanced using counterbalance.

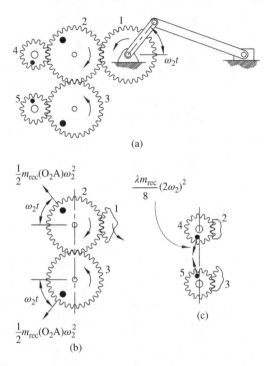

(a)

(b)

(c)

Figure 12.26 (a) Lanchester balancer arrangement. (b) Large gear set for primary force balancing. (c) Small gear set for secondary force balancing

A more complete balancing of slider–crank mechanisms can be achieved with Lanchester balancers. The balancer uses counterrotating meshed gears with eccentric masses to balance the primary and secondary shaking forces for an in-line slider–crank (zero offset). Arakelian and Makhsudyan (2010) demonstrate a procedure for Lanchester balancing for a more general slider–crank arrangement.

Consider the slider–crank mechanism shown in Figure 12.26a that consists of a typical in-line slider–crank mechanism and five gears. Gear 1 is attached to the crank and rotates at the same speed as the crank. Gear 1 drives gear 2, and gears 1–3 are all identical. Therefore, gears 2 and 3 will rotate at the same speed as the crank. Gears 4 and 5 are half the size of the other gears. Therefore, their rotation speed will be two times the speed of the crank.

The balancing of the system occurs due to the eccentric masses located on the gears. The larger gears will balance the primary shaking forces, while the smaller gears will balance the secondary shaking force. Because all eccentric masses are located in opposed pairs, no additional shaking forces are created in the y-axis direction.

Let us begin by examining the larger gear set shown in Figure 12.26b. Recall that gear 2 and 3 both will be rotating at the same speed as the crank. Each gear has an eccentrically located mass. Each eccentric mass and radial location is picked such that the inertial force from each mass is equal to $1/2m_{rec}(O_2A)\omega_2^2$. The resultant inertial force will have a horizontal component of $m_{rec}(O_2A)\omega_2^2 \cos(\omega_2 t)$, which is equal to the primary shaking force. Considering that the horizontal component is in the opposite direction of the inertial force of the sliding mass, the primary shaking force will be balanced. The vertical resultant will be zero, so no vertical shaking force will be created. Gears 2 and 3 must lie on the same vertical line so that the vertical components of the inertial forces do not create a moment couple.

Now that the primary shaking forces are balanced, we will examine the smaller gear set shown in Figure 12.26c. These gears will be rotating at twice the speed of the crank and will be used to balance the secondary shaking forces. The eccentric masses and radial locations are picked such that the inertial force from each mass will be $(\lambda m_{rec}/8)(2\omega_2)^2$. Therefore, the horizontal resultant is equal to $(\lambda m_{rec}/4)(2\omega_2)^2 \cos(2\omega_2 t)$ and balances the secondary shaking forces.

12.6 Balancing Linkage Mechanisms

12.6.1 Introduction

The previous section focused on balancing slider–crank mechanisms, which is a specific case of a four-bar linkage mechanism. This section will consider other examples of four-bar mechanisms. It is very difficult to completely balance a four-bar mechanism. Though many methods for linkage balancing have been developed, it is common for a method to focus on balancing a portion of the unbalance at the expense of other portions.

12.6.2 Global Center

Consider the general four-bar linkage mechanism shown in Figure 12.27. The individual link center of mass locations is shown. The entire mechanism, in the arbitrary position shown, will have a total global center of mass as shown. As the mechanism moves, that global center of mass will, however, generally move indicating an unbalanced system. Complete static balancing of the linkage mechanism would require that the global center remain stationary during the motion of the mechanism.

Each link center of mass can be considered as a point mass. The global center is the centroid of the three point masses. The x and y coordinates of the global center of mass will be

$$\overline{x} = \frac{x_2 m_2 + x_3 m_3 + x_4 m_4}{m_2 + m_3 + m_4} \tag{12.34}$$

$$\overline{y} = \frac{y_2 m_2 + y_3 m_3 + y_4 m_4}{m_2 + m_3 + m_4} \tag{12.35}$$

Figure 12.27   Global center of mass

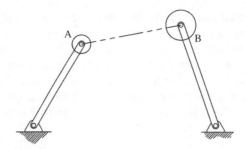

Figure 12.28   Four-bar linkage mechanism with the coupler modeled using lumped mass

The centroids are essentially weighted averages. The location of $\bar{x}$ is the weighted average where each individual x coordinate x_k is given a weight of $m_k / \sum m$. The position vector locating the global center of mass can be directly determined using the position vectors of each individual center of mass:

$$\vec{r}_g = \frac{\vec{r}_2 m_2 + \vec{r}_3 m_3 + \vec{r}_4 m_4}{m_2 + m_3 + m_4} \tag{12.36}$$

The global center location could then be calculated for multiple points within a cycle to represent the motion of the global center during operation.

Counterweights can be used to modify the global center and attempt to make the location stationary. Berkof and Lowen (1969) demonstrate a method of linearly independent vectors to develop a complete balance of certain planar linkage mechanisms.

12.6.3 Shaking Forces

Though total balancing is complicated, shaking forces for a general four-bar linkage mechanism can be relatively easily balanced. We can again model the coupler using lumped mass assumptions, as shown in Figure 12.28. Placing the lumped masses at points A and B will now generate a model of two simple rotating mass problems. Balancing masses can be added to completely balance the shaking forces. The added masses will now modify the inertial forces that change the joint forces.

12.7 Flywheels

12.7.1 Introduction

Sections 12.5 and 12.6 discussed ideas of balancing linkage mechanisms in an attempt to improve the performance of the mechanism. Balancing made the overall operation of the mechanism smoother by reducing or eliminating shaking forces. However, speed fluctuation of the driving crank will also cause unsmooth performance of the mechanism. It is common in rotating equipment to have fluctuation in the shaft rotation speed $\dot{\theta}_2$. Typically, it is desired to minimize the amount of speed fluctuation and keep the fluctuation within some allowable limit.

A common mechanical device used to reduce the speed fluctuation is called a flywheel. A flywheel is a heavy disk used to store rotational energy, and they basically operate as an

Figure 12.29 Flywheel application

energy accumulator or reservoir. The high moment of inertia of the flywheel, such as the one shown in Figure 12.29, resists changes in rotational speed. Due to their large size and typical high rotation speed, flywheels can be very dangerous. Stresses can become very large and the flywheel could rupture. Pieces from a rupture would have high energy and high velocities. Therefore, it is common to fully contain the region of the flywheel in an enclosure for safety.

A flywheel is highly efficient for short-term energy storage. The flywheel will store rotational kinetic energy that can be used at a later time. Flywheels will vary in geometry from solid disks to a spoke and rim design. Materials used for flywheels will also vary. Cast iron can be used for low speeds, but alloy steels are better for very high-speed operation.

12.7.2 Speed Fluctuation

To better understand the functionality of the flywheel, we need to examine the differences between input power and required power of a system for one complete cycle, which is the cause of speed fluctuation of the drive crank. Consider a very common case where a motor is the input power for a linkage mechanism. The motor will produce a constant driving torque. The torque T_2, which is determined using force analysis methods discussed in Chapter 11, required to drive the linkage mechanism will fluctuate with time.

The input torque provided by the motor and the required torque to drive the mechanism can be plotted together to give a result such as that shown in Figure 12.30a. It is shown that the input power will be greater than the required power during part of the cycle and the input power will be less than the required power during the remainder of the cycle. When the input torque is greater than that required, the system will increase in speed, as shown in Figure 12.30b. The system will decrease in speed when the required torque is greater than the input torque. Therefore, the fluctuation in torque will result in speed fluctuations.

The case discussed in this section was for a constant input power from a motor and fluctuating required power of the linkage mechanism. The reverse scenario can also exist where the input power fluctuates and the required power remains constant. The concepts discussed will also apply to that type of scenario.

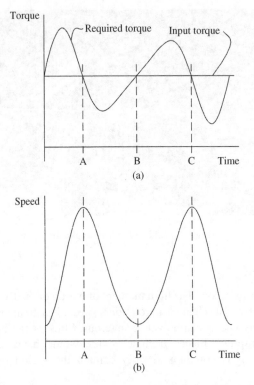

Figure 12.30 (a) General example of constant input torque and fluctuating required torque for a complete cycle of a linkage mechanism. (b) Resulting speed fluctuation

12.7.3 Flywheel Energy

Now that the concept of speed fluctuation has been introduced, we can examine the function of the flywheel to reduce speed fluctuation. We will be interested in changes in kinetic energy within the system. The kinetic energy of a rigid body in plane motion is given by

$$KE = \frac{1}{2}mv_g^2 + \frac{1}{2}I_g\omega^2 \qquad (12.37)$$

The first term is the kinetic energy due to the motion of the mass center and the second term is kinetic energy due to rotation about the mass center. The center of mass for a flywheel will remain stationary causing the first term to be equal to zero. Therefore, the kinetic energy of a flywheel will depend on its moment of inertia about its mass center I_g and its rotation speed ω. As the designer, we have control over the moment of inertia, and we will soon see how to determine the required moment of inertia. The rotational speed, however, is determined more by the operating conditions of the machine.

Setting the first term of Equation 12.37 to zero, we know that the kinetic energy of disk rotating about a fixed center is determined by

$$KE = \frac{1}{2}I_g\omega^2 \qquad (12.38)$$

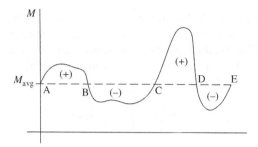

Figure 12.31   General turning moment diagram for one complete cycle

The change in the flywheel's kinetic energy can be expressed by the change in rotational speed of the disk:

$$\Delta KE = \frac{1}{2} I_g \left(\omega_{max}^2 - \omega_{min}^2 \right)$$ (12.39)

12.7.4 Fluctuation of Energy

Next we will examine the energy variation within the actual system. Referring back to Figure 12.30, we determined that the system will experience an increase in speed when the input power exceeds the required power. This increase in speed will result in an increase in the system's kinetic energy. Similarly, the speed and kinetic energy will decrease when the input power is less than the required power. The actual magnitude of the velocity change will depend on the inertia of the system.

To examine the energy fluctuations within a linkage system, consider the general turning moment diagram shown in Figure 12.31. This particular turning moment diagram may be from a general four-bar linkage mechanism, but it is important to note that the general shape of a turning moment diagram can vary drastically depending on the actual machine and application.

The energy for one complete cycle is determined from

$$E = \oint M d\theta$$ (12.40)

Assuming that a complete cycle occurs in 2π radians, the energy produced will be

$$E = \int_0^{2\pi} M d\theta$$ (12.41)

The average moment, as shown with the dashed line, will be equal to

$$M_{avg} = \frac{E}{2\pi} = \frac{1}{2\pi} \int_0^{2\pi} M d\theta$$ (12.42)

For steady-state operation (system speed is not changing from one cycle to the next), the resisting moment M_R will be equal to the average moment. Since the speed will not change from one cycle to the next, the energy at point E must equal the energy at point A.

At some points within the cycle, the driving torque is greater than the resisting torque, and at other points the driving torque is less than the resisting torque. Energy at the key locations when resisting torque is equal to the driving torque can be determined. Define the energy at point A as E_A. The remaining key point energies can be determined using integration:

$$E_B = E_A + \int_{\theta_A}^{\theta_B} (M - M_R)d\theta \tag{12.43}$$

$$E_C = E_B + \int_{\theta_B}^{\theta_C} (M - M_R)d\theta \tag{12.44}$$

$$E_D = E_C + \int_{\theta_C}^{\theta_D} (M - M_R)d\theta \tag{12.45}$$

$$E_E = E_D + \int_{\theta_D}^{\theta_E} (M - M_R)d\theta = E_A \tag{12.46}$$

The actual number of these key locations will vary depending on the particular application. The integrals will be positive when the driving torque is greater than the resisting torque, causing an increase in energy. The integrals will be negative when the driving torque is less than the resisting torque, causing a decrease in energy.

The energy will be maximum E_{max} at one of the key points, and it will be minimum E_{min} at one of the key points. The maximum fluctuation of energy will be the difference between the maximum and minimum energies.

$$(\Delta E)_{max} = E_{max} - E_{min} \tag{12.47}$$

The maximum and minimum energies will be due to kinetic energy of the flywheel as well as the kinetic energy of the linkages within the system. However, the energy in the flywheel is generally very large compared to the energy stored in the linkages. Therefore, Equation 12.47 can be written as

$$(\Delta E)_{max} = \frac{1}{2}I_g\left(\omega_{max}^2 - \omega_{min}^2\right) \tag{12.48}$$

12.7.5 Flywheel Design

We will now move to the process of determining the required moment of inertia for a desired amount of speed fluctuation. It is convenient to modify Equation 12.48:

$$(\Delta E)_{max} = \frac{1}{2}I_g(\omega_{max} - \omega_{min})(\omega_{max} + \omega_{min}) \tag{12.49}$$

Defining the average running speed as

$$\omega_{avg} = \frac{1}{2}(\omega_{max} + \omega_{min})$$ (12.50)

The change in energy becomes

$$(\Delta E)_{max} = I_g(\omega_{max} - \omega_{min})(\omega_{avg})$$

$$(\Delta E)_{max} = I_g \frac{\omega_{max} - \omega_{min}}{\omega_{avg}} \omega_{avg}^2$$ (12.51)

A common term to define the permissible speed variation is the coefficient of fluctuation:

$$C_F = \frac{\dot{\theta}_{max} - \dot{\theta}_{min}}{1/2(\dot{\theta}_{max} + \dot{\theta}_{min})}$$ (12.52)

The maximum and minimum rotational speeds are given as $\dot{\theta}_{max}$ and $\dot{\theta}_{min}$, respectively. The denominator defines the average rotational speed.

$$C_F = \frac{\omega_{max} - \omega_{min}}{\omega_{avg}}$$ (12.53)

The coefficient of fluctuation is used as a measure of the performance of a flywheel. Actual values will vary depending on the application. Values range from around 0.2 to 0.002, and values for several applications can be found in machine design handbooks. The smaller values of the coefficient occur when the allowable range of rotational speeds is small.

Substituting the coefficient of fluctuation into Equation 12.51 gives

$$(\Delta E)_{max} = I_g C_F \omega_{avg}^2$$ (12.54)

Therefore, the required moment of inertia of the flywheel is

$$I_g = \frac{(\Delta E)_{max}}{C_F \omega_{avg}^2}$$ (12.55)

The numerator is determined by integration of the turning moment diagram between locations of maximum and minimum energies. The average speed is the desired operational speed for the system. The coefficient of fluctuation is determined based on the permissible fluctuation of speed. For systems with large allowable fluctuation, such as crushers or punching presses, the coefficient of fluctuation is generally greater than 0.2. Systems that must remain very uniform in speed may have coefficients less than 0.002.

Commonly, the turning moment diagrams encountered in practical applications are too complicated to integrate by hand. Computer methods for flywheel analysis can be developed by expanding the modules developed for force analysis. The function of input torque versus crank angle can be developed by performing the force analysis at multiple locations within the cycle. Using numerical integration methods, the areas can be determined that will give changes in energy.

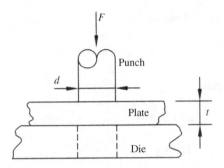

Figure 12.32   Punch and die for a punching press

12.7.6 Flywheel Analysis for a Punching Press

Now we will examine a very common application of flywheels for a punching press or shearing machine. A punching press is a machine used to punch holes in material such as metal plates. To design the flywheel for a punching press, we need to determine the work required for a punching process. Figure 12.32 shows a round punch used to punch a circular plug from the plate. Shear stress is equal to force divided by area.

Therefore, the force required for shearing will equal the shearing stress times the area. Considering a circular hole that is being punched, the area will be the circumference of the hole times the thickness of the plate $A = (\pi d)(t)$. The peak punch force is determined from

$$F = 0.7(t)(L)(\text{UTS}) \tag{12.56}$$

where t is the sheet thickness, L is the length of the shear surface (perimeter of the hole), and UTS is the ultimate tensile strength (UTS) of the material.

Work required for the punching process can be determined by analysis of a plot of force during the punch. Work and energy analysis can then be used to determine the change in kinetic energy of the system, which can be used to design the flywheel. The following example illustrates the process.

Example Problem 12.5

A punching press is used to punch 1.5 inch diameter holes in 0.5 inch thick plate with an ultimate tensile strength of 58 ksi. Figure 12.33 shows the force–displacement curve for the punching of one hole. The flywheel rotates at 80 rpm. It is required that the speed of the flywheel after a punching operation maintains 92% of the original speed. Determine the required moment of inertia of the flywheel.

Solution: The rotation speed must be converted to rad/s:

$$\omega_1 = 80\,\frac{\text{rev}}{\text{min}}\left(2\pi\,\frac{\text{rad}}{\text{rev}}\right)\left(\frac{1\,\text{min}}{60\,\text{s}}\right) = 8.38\,\frac{\text{rad}}{\text{s}}$$

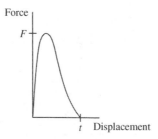

Figure 12.33

The peak punching force required is

$$F = 0.7(t)(L)(\text{UTS})$$

$$F = 0.7(0.5 \text{ in.})(\pi \times 1.5 \text{ in.})\left(58\,000\,\frac{\text{lb}}{\text{in.}^2}\right) = 95\,661 \text{ lb}$$

Referring to the force–displacement curve shown in Figure 12.33, the area under the curve can be approximated by a triangle. Therefore, the work required for a punching operation is approximately equal to

$$W = \frac{Ft}{2} = \frac{95\,661 \text{ lb}(0.5 \text{ in.})}{2} = 23915 \text{ in. lb}$$

The rotational speed after a punching operation will be

$$\omega_2 = 0.92\omega_1 = 0.92\left(8.38\,\frac{\text{rad}}{\text{s}}\right) = 7.71\,\frac{\text{rad}}{\text{s}}$$

The required moment of inertia can be determined using the work and energy equation:

$$KE_1 - W = KE_2$$

$$\frac{1}{2}I\omega_1^2 - W = \frac{1}{2}I\omega_2^2$$

$$I(\omega_1^2 - \omega_2^2) = 2W$$

$$I = \frac{2W}{\omega_1^2 - \omega_2^2} = \frac{2(23915 \text{ in. lb})}{(8.38 \text{ rad/s})^2 - (7.71 \text{ rad/s})^2}\left(\frac{1 \text{ ft}}{12 \text{ in.}}\right)$$

Answer: $I = 370$ ft lb s^2

12.8 Measurement Devices

Equipment with a rotating unbalance will vibrate, so it is useful to measure the vibration in order to determine what is required to improve performance. Vibration measurement devices

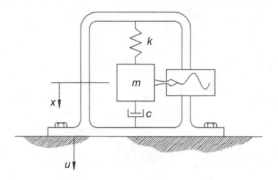

Figure 12.34   Seismic instrument

produce an output that replicates an input (amplitude, velocity, or acceleration) as closely as possible. In general, the measurement is recorded with a seismic instrument as shown in its basic form (spring, mass, and damper) in Figure 12.34.

Newton's second law gives the following equation for the seismic mass m:

$$-k(x - u) - c(\dot{x} - \dot{u}) = m\ddot{x} \tag{12.57}$$

In Equation 12.57, u represents the motion of the frame and x represents the absolute motion of the seismic mass. We will assume the motion of the frame to be harmonic:

$$u(t) = U \sin \omega t \tag{12.58}$$

where U is the magnitude of motion.

From Equation 12.57, we can see that the spring constant is proportional to the relative displacement between the mass and the frame. Likewise, the damping coefficient is proportional to the relative velocity between the mass and the frame. The seismic instrument will therefore measure the relative motion not the true motion of the frame, which is the motion of interest. The absolute motion x is not of real importance. We will define the relative motion as

$$z = x - u \tag{12.59}$$

Substituting into Equation 12.57 gives

$$m(\ddot{z} + \ddot{u}) + c\dot{z} + kz = 0$$

$$m\ddot{z} + c\dot{z} + kz = m\omega^2 U \sin \omega t \tag{12.60}$$

Assume that

$$z = Z \sin (\omega t - \phi) \tag{12.61}$$

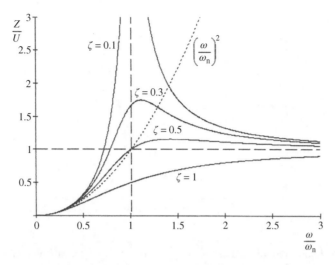

Figure 12.35 Response of seismic instrument

From methods for support motion discussed in Chapter 10, a dimensionless ratio can be developed:

$$\frac{Z}{U} = \frac{(\omega/\omega_n)^2}{\sqrt{\left(1 - (\omega/\omega_n)^2\right)^2 + (2\varsigma(\omega/\omega_n))^2}} \qquad (12.62)$$

Figure 12.35 shows a plot of the acceleration ratio versus frequency ratio for several damping values.

Example Problem 12.6

A vibration measuring instrument has a damping coefficient $\varsigma = 0.4$. Determine the error in the reading if the frequency of the measured motion is 3.5 times the natural frequency of the instrument.

Solution: From Equation 12.62,

$$\frac{Z}{U} = \frac{(\omega/\omega_n)^2}{\sqrt{\left(1 - (\omega/\omega_n)^2\right)^2 + (2\varsigma(\omega/\omega_n))^2}}$$

using $\varsigma = 0.4$ and $\omega = 3.5\omega_n$

$$\frac{Z}{U} = \frac{3.5^2}{\sqrt{\left(1 - 3.5^2\right)^2 + (2(0.4)(3.5))^2}} = 1.057$$

Answer: 5.7% high

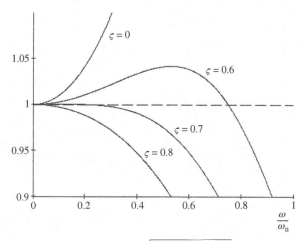

Figure 12.36 Plot of the function $1/\sqrt{(1-r^2)^2 + (2\varsigma r)^2}$ for different values of ς

A vibration-measuring device with a high frequency in comparison to the frequency to be measured is known as an accelerometer, and the instrument shows acceleration. Accelerometers are used in a wide variety of technologies. Smartphones use accelerometers for user interface and automatic switching between portrait and landscape view. Accelerometers are used in vehicles for automatic collision notification. Cameras use accelerometer for image stabilization, and video game systems use accelerometers for motion input.

Examination of Figure 12.35 shows that Z/U approaches zero for low values of ω/ω_n (say less than 0.25). Therefore, the denominator approaches a value of 1. If $\sqrt{\left(1 - (\omega/\omega_n)^2\right)^2 + (2\varsigma(\omega/\omega_n))^2} = 1$, then from Equation 12.62

$$\frac{Z}{U} = \frac{\omega^2}{\omega_n^2}$$

$$Z = \frac{\omega^2 U}{\omega_n^2} \tag{12.63}$$

The term $\omega^2 U$ is an acceleration term. Therefore, Z becomes proportional to acceleration with a ratio of $1/\omega_n^2$.

If we plot the function $1/\sqrt{(1-r^2)^2 + (2\varsigma r)^2}$ (where r is the frequency ratio) for different values of damping, we get the plot shown in Figure 12.36. It is desired to have the function equal to 1 for as wide a range of frequency as possible. A damping ratio of 0.7 gives a useful frequency range from 0 to approximately 0.2 with error less than 0.01%.

12.9 Computer Methods

12.9.1 Balancing

Problems in balancing allow for several possible solutions. The designer has some control over the location and size of balancing masses. It is often desirable to try several possible solutions to determine the best solution for a particular application. Making adjustments to solutions to attempt to optimize would be very time consuming without utilizing computer methods. Software, such as spreadsheets, could be used to aid in optimizing dynamic balancing. Balancing planes can be modified to determine the effects on balancing masses required.

Graphical methods for both static and dynamic balancing of masses in pure rotation can be done easily with the aid of CAD software. The use of CAD software can allow for quick balancing as well as some level of optimization. Vector polygons of the inertial forces can be drawn to scale to determine the direction and magnitude of the balancing inertial force required.

12.9.2 Flywheels

Computer code can be developed to calculate the required input torque for a complete cycle and generate a torque curve. Numerical integration can then be used to calculate changes in energy and aid in flywheel design. The input torque function could be broken into a piecewise function and the areas could be summed to approximate the integral.

12.10   Review and Summary

Unbalance in rotating machinery is common and often unwanted. Unbalanced systems cause vibration, which was demonstrated in Section 12.4. This chapter focused principally on methods of balancing simple rotating masses, but also introduced concepts of balancing more complex linkage systems.

It was determined that complete linkage balancing can be accomplished, but it is very complicated. Many additional methods of linkage balancing exist, and readers interested in a more complete development of methods to balance linkage mechanisms should refer to Further Reading section to determine other sources. It should be noted that the attempt of balancing linkage mechanisms focused on a particular operating speed. Therefore, it has been balanced to have the highest efficiency at that operating speed.

Problems

P12.1 The system shown in Figure 12.37 has three masses equally spaced at 120° intervals and rotates at a constant speed. Each mass is located at the same radial distance. Use a force polygon to show that the system is not statically balanced when $m_1 = 0.9$ lb, $m_2 = 1.1$ lb, and $m_3 = 1$ lb. Would it be possible to statically balance the system by only adding or reducing the weight of mass 3?

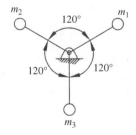

Figure 12.37

P12.2 It is desired to add a third mass to statically balance the system shown in Figure 12.38. What mass is required if the mass is to be located at the same radial distance r? What will be the angular location of the balancing mass? Solve the problem using (a) the graphical and (b) the analytical method.

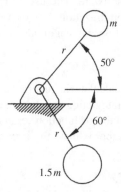

Figure 12.38

P12.3 It is desired to add a third mass equal to $1.5m$ to statically balance the system shown in Figure 12.39. What will be the angular and radial location of the balancing mass? Solve the problem using (a) the graphical and (b) the analytical method.

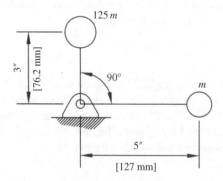

Figure 12.39

P12.4 It is desired to add a fourth mass to statically balance the system shown in Figure 12.40. Determine the location of the balancing mass if it is to have a mass m. Solve the problem using (a) the graphical and (b) the analytical method.

Answer: $\theta_B = -163.91°$
$r_B = 1.59 \text{ mm}$

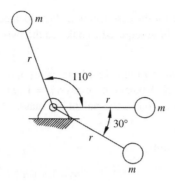

Figure 12.40

P12.5 It is desired to add a third mass to statically balance the system shown in Figure 12.41. Determine the location of the balancing mass if it is to have a mass 1.5*m*. Solve the problem using (a) the graphical and (b) the analytical method.

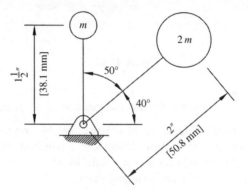

Figure 12.41

P12.6 Solve Problem P12.5 for a case when the large mass is reduced from 2*m* to 1.2*m*.

P12.7 A 10 inch diameter circular plate has three 1 inch diameter holes located, as shown in Figure 12.42. It is desired to drill one additional 1 inch diameter hole to cause the plate to be statically balanced. Where would that hole be located on the plate?

Answer: $\theta_B = 21.59°$
$r_B = 1.6$ in.

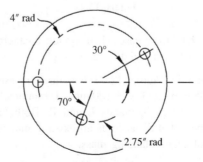

Figure 12.42

P12.8 Solve Problem P12.7 for the case when the hole located at a radius of 2.75 inches was removed. Locate the hole required to balance the two remaining holes on the 4 inch radius.

P12.9 For the system shown in Figure 12.43, $m_2 = m_3$ and $m_1 = 2m_2$. The radial distance for each mass is given. A balancing mass $m_b = 1.5m_2$ is to be added. Determine the angular and radial location of the balancing mass.

Answer:
$$\theta_B = 176.1°$$
$$r_B = 62.26 \text{ mm}$$

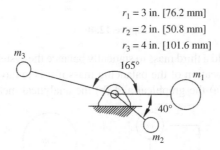

$r_1 = 3$ in. [76.2 mm]
$r_2 = 2$ in. [50.8 mm]
$r_3 = 4$ in. [101.6 mm]

Figure 12.43

P12.10 Solve Problem P12.9 when the angular position of m_2 is moved from 40° to 90°.

P12.11 The system shown in Figure 12.44 has two weights located as shown. Balancing weights are to be placed in planes L and R to dynamically balance the system. Both balancing weights are to be located at a radial location of 90 mm. Determine the mass and angular position of each balancing mass.

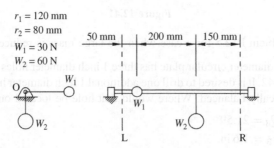

$r_1 = 120$ mm
$r_2 = 80$ mm
$W_1 = 30$ N
$W_2 = 60$ N

Figure 12.44

P12.12 Solve Problem P12.11 for a case when the left balancing plane is relocated to the midpoint position between weight 1 and weight 2.

P12.13 The system shown in Figure 12.45 has three masses located as shown with $m = 10$ kg, $r = 35$ mm, and $d = 100$ mm. Dynamically balance the system by placing balancing masses in planes L and R. Both balancing masses are to be located at the same radial distance as all existing masses. Determine the mass and angular position of each balancing mass.

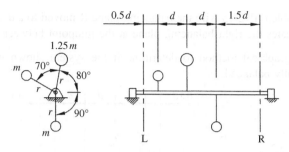

Figure 12.45

P12.14 The system shown in Figure 12.46 has two weights located as shown. Balancing weights are to be placed in planes L and R to dynamically balance the system. Both balancing weights are to be located at a radial location of 6 inches. Determine the mass and angular position of each balancing mass.

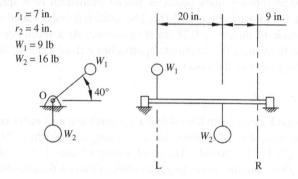

Figure 12.46

P12.15 The system of masses shown in Figure 12.47 has four masses $m_1 = m_2 = m_3 = m_4 = 6$ kg and $r = 20$ mm. Each mass is located at 90° intervals. The masses are located along the shaft as shown with $d = 140$ mm. Dynamically balance the system by placing balancing masses in planes L and R. Determine the radial and angular location of each balancing mass if each balancing mass is 4 kg.

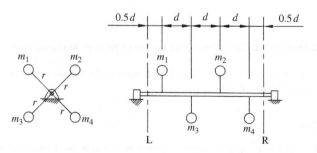

Figure 12.47

P12.16   Solve Problem P12.15 with the balancing plane R moved to a distance d to the left, which locates the right balancing plane at the midpoint between masses 2 and 4.

P12.17   Use the graphical method to determine if the system shown in Figure 12.48 is dynamically balanced.

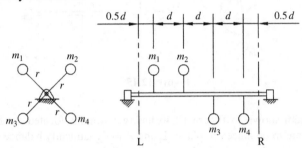

Figure 12.48

P12.18   An eccentric rotating mass produces forced oscillation of a spring–mass system similar to that shown in Figure 12.18a. The rotation speed is varied and it is noted that the amplitude of motion is 0.78 cm at resonance. At a speed considerably beyond resonance frequency, the amplitude approaches a fixed value of 0.1 cm. Determine the damping ratio of the system.

Answer: $\varsigma = 0.64$

P12.19   A slider–crank mechanism has a 4 inch long crank and a 17 inch long connecting rod. The crank weight is 9.2 pounds, the connecting rod weight is 24 pounds, and the slider weight is 10.8 pounds. The crank center of mass is located 1.2 inches from the fixed pivot, and the connecting rod center of mass is 6 inches from the crank end. The crank rotates at a constant rate of 1200 rpm clockwise. Perform a complete kinematic analysis and kinetic analysis on the system. Design a balancing mass to balance the mechanism.

P12.20   A flywheel is required for a mechanism running at 300 rad/s to give a coefficient of fluctuation of 0.4, which resulted in a required flywheel moment of inertia of 0.47 in lb s^2. What is the speed range ($\omega_{max} - \omega_{min}$) for the flywheel in rad/s? What would be the required moment of inertia of the flywheel if the coefficient of fluctuation was changed to 0.3?

References

Arakelian, A. and Makhsudyan, N. (2010) Generalized Lanchester balancer. *Mechanics Research Communications*, **37**, 647–649.

Berkof, R.S. and Lowen, G.G. (1969) A new method for completely force balancing simple linkages. *Journal of Engineering for Industry*, **91**(1), 21–26.

Further Reading

Qi, N.M. and Pennestri, E. (1991) Optimum balancing of four-bar linkages. *Mechanism and Machine Theory*, **26**, 337–348.

13

Applications of Machine Dynamics

13.1 Introduction

This chapter will present some common applications of machine dynamics. The first section will continue on the work done with cam kinematics presented in Chapter 9 and will develop dynamic forces within the cam mechanism. Essential vibration concepts covered in Chapter 10 will now allow us to better examine the system response of cam mechanisms. Though several cam functions were developed in Chapter 9, we will develop some more advanced cam displacement functions in this chapter. Methods will be discussed on combining some basic cam functions into a new function. These combinations are commonly done to reduce peak acceleration values and improve performance. The basic functions developed in Chapter 9 each had certain advantages. Combination functions allow us to utilize the individual advantages. This chapter will also examine forces developed in cam mechanisms. These forces would be required for further design considerations such as stress and wear. Knowledge of the forces within the system will also allow for the calculation of the torque required to drive the cam mechanism.

The second section of this chapter will explore the dynamics of multicylinder engines. Section 12.5 discussed ideas of balancing a single slider–crank mechanism, which is very difficult to complete balance. It is common in machinery, such as engines and compressors, to utilize pairs of slider–crank mechanisms to balance the machine and provide smoother operation. This chapter will discuss the basic concepts of the use of multicylinder configurations in common mechanisms.

Section 1: Cam Dynamics

13.2 Cam Response for Simple Harmonic Functions

13.2.1 Background

In Chapter 9 we developed functions used for follower motion, and it was assumed that the follower's motion would be exactly as the function describes. In reality, the follower will not move exactly with the displacement function due to elasticity in the system. Chapter 9 assumed rigid systems, but all systems will have vibration and deflection of the elements within the system.

Machine Analysis with Computer Applications for Mechanical Engineers, First Edition. James Doane.
© 2016 John Wiley & Sons, Ltd. Published 2016 by John Wiley & Sons, Ltd.
Companion Website: www.wiley.com/go/doane0215

Cam follower mechanisms will always have some form of vibration. A major contributor to vibration, as mentioned in Chapter 9, would be discontinuities in the acceleration function. Careful design of the displacement function can greatly reduce vibrations, but additional sources of vibration are still present. Other sources of vibration are cam unbalance, irregularities in the cam surface (e.g., those due to manufacturing), and large impulsive forces from external loads. Vibration also becomes more significant at higher speeds.

13.2.2 General Equation of Motion

First, we will develop a method for determining cam response when the displacement function is in a convenient mathematical form. Consider a plate cam with a roller follower such as the one shown in Figure 13.1a. The system is force closed, so contact is maintained with a compression spring.

Figure 13.1b shows a simplified model of the system. The stiffness k will be an effective stiffness taking into account the spring and elasticity of other components within the system. The actual spring will, in general, have a very low stiffness compared with other components and will be in series with the other components. Therefore, the spring stiffness can be considered alone. Likewise, the damping coefficient is the total combination of sources of resistance.

Figure 13.1c shows the free body diagram. From Newton's law,

$$m\ddot{y} + c(\dot{y} - \dot{y}_c) + k(y - y_c) = 0 \tag{13.1}$$

Next, we want to separate our inputs (values for y_c) and outputs (values of y). Keeping the inputs on the left-hand side of the equation and moving the inputs to the right-hand side gives

$$m\ddot{y} + c\dot{y} + ky = c\dot{y}_c + ky_c \tag{13.2}$$

The term y_c is the input motion determined from the cam profile and will serve as the forcing function. Equation 13.2 differs from those developed in Chapter 10 in two terms on the right-hand side of the equation. The differential equation can quickly become complicated if the cam profile function y_c is complex, which are mostly complex to provide smooth motion.

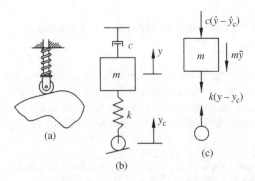

Figure 13.1 (a) Force closed follower. (b) Single DOF model. (c) Free body diagram

13.2.3 Response for Simple Harmonic Cam Function

Consider a cam used for a dwell mechanism using a simple harmonic cam function for the rise. We will examine the response of the system to a rise from $s_1 = 0$ to $s_2 = h$ for a cam rotation from $\theta_1 = 0$ to $\theta_2 = \lambda$. From Chapter 9, the input motion will be

$$y_c = \frac{h}{2}\left[1 - \cos\left(\frac{\pi}{\lambda}\theta\right)\right] \tag{13.3}$$

We know that $\theta = \omega t$ giving

$$y_c(t) = \frac{h}{2}\left[1 - \cos\left(\frac{\pi\omega}{\lambda}t\right)\right] \tag{13.4}$$

The first derivative with respect to time will be

$$\dot{y}_c(t) = \frac{h\pi\omega}{2\lambda}\sin\left(\frac{\pi\omega}{\lambda}t\right) \tag{13.5}$$

Substituting Equations 13.4 and 13.5 into Equation 13.2 gives the equation of motion:

$$m\ddot{y} + c\dot{y} + ky = c\left(\frac{h\pi\omega}{2\lambda}\right)\sin\left(\frac{\pi\omega}{\lambda}t\right) + k\left(\frac{h}{2}\right)\left[1 - \cos\left(\frac{\pi\omega}{\lambda}t\right)\right] \tag{13.6}$$

Dividing by the mass and recalling from Chapter 10 that $\omega_n = \sqrt{k/m}$ and $\varsigma = c/2m\omega_n$ give

$$\ddot{y} + 2\varsigma\omega_n\dot{y} + \omega_n^2 y = \frac{\varsigma h\pi\omega\omega_n}{\lambda}\sin\left(\frac{\pi\omega}{\lambda}t\right) + \frac{\omega_n^2 h}{2}\left[1 - \cos\left(\frac{\pi\omega}{\lambda}t\right)\right]$$

$$\ddot{y} + 2\varsigma\omega_n\dot{y} + \omega_n^2 y = \frac{\omega_n^2 h}{2}\left[\frac{2\varsigma\pi\omega}{\lambda}\sin\left(\frac{\pi\omega}{\lambda}t\right) - \cos\left(\frac{\pi\omega}{\lambda}t\right) + 1\right] \tag{13.7}$$

In order to develop an initial understanding of system response, we will simplify the system by assuming negligible damping. If damping is assumed to be very small ($\varsigma \approx 0$), Equation 13.7 can be simplified to give

$$\ddot{y} + \omega_n^2 y = \frac{\omega_n^2 h}{2}\left[1 - \cos\left(\frac{\pi\omega}{\lambda}t\right)\right] \tag{13.8}$$

From Chapter 10, we know that the homogeneous solution to Equation 13.8 will be

$$y_h(t) = A_1 \cos \omega_n t + A_2 \sin \omega_n t \tag{13.9}$$

The particular solution will be in the form

$$y_p(t) = A + B\sin\left(\frac{\pi\omega}{\lambda}t\right) + C\cos\left(\frac{\pi\omega}{\lambda}t\right) \tag{13.10}$$

The second derivative of the particular solution will be

$$\ddot{y}_p(t) = -B\left(\frac{\pi\omega}{\lambda}\right)^2\sin\left(\frac{\pi\omega}{\lambda}t\right) - C\left(\frac{\pi\omega}{\lambda}\right)^2\cos\left(\frac{\pi\omega}{\lambda}t\right) \tag{13.11}$$

Substituting Equations 13.10 and 13.11 into Equation 13.8 gives

$$-B\left(\frac{\pi\omega}{\lambda}\right)^2\sin\left(\frac{\pi\omega}{\lambda}t\right) - C\left(\frac{\pi\omega}{\lambda}\right)^2\cos\left(\frac{\pi\omega}{\lambda}t\right) + \omega_n^2\left[A + B\sin\left(\frac{\pi\omega}{\lambda}t\right) + C\cos\left(\frac{\pi\omega}{\lambda}t\right)\right]$$
$$= \frac{\omega_n^2 h}{2} - \frac{\omega_n^2 h}{2}\cos\left(\frac{\pi\omega}{\lambda}t\right) \tag{13.12}$$

$$\left[B\omega_n^2 - B\left(\frac{\pi\omega}{\lambda}\right)^2\right]\sin\left(\frac{\pi\omega}{\lambda}t\right) + \left[C\omega_n^2 - C\left(\frac{\pi\omega}{\lambda}\right)^2\right]\cos\left(\frac{\pi\omega}{\lambda}t\right) + A\omega_n^2$$
$$= \frac{\omega_n^2 h}{2} - \frac{\omega_n^2 h}{2}\cos\left(\frac{\pi\omega}{\lambda}t\right) \tag{13.13}$$

Equating terms gives

$$\left[\omega_n^2 - \left(\frac{\pi\omega}{\lambda}\right)^2\right]B = 0 \Rightarrow B = 0 \tag{13.14a}$$

$$\left[\omega_n^2 - \left(\frac{\pi\omega}{\lambda}\right)^2\right]C = \frac{\omega_n^2 h}{2} \Rightarrow C = \frac{h}{2}\frac{\omega_n^2}{\omega_n^2 - (\pi\omega/\lambda)^2} \tag{13.14b}$$

$$\omega_n^2 A = \frac{\omega_n^2 h}{2} \Rightarrow A = \frac{h}{2} \tag{13.14c}$$

Substituting Equation 13.14 into Equation 13.10 gives the final particular solution:

$$y_p(t) = \frac{h}{2} + \frac{h}{2}\frac{\omega_n^2}{\omega_n^2 - (\pi\omega/\lambda)^2}\cos\left(\frac{\pi\omega}{\lambda}t\right) \tag{13.15}$$

The total solution is the summation of the homogeneous and particular solution, but we will make some substitutions. Making the substitution of $\theta = \omega t$ and the frequency ratio $r = \omega_n/\omega$ along with the initial conditions $y = \dot{y} = 0$ at $t = 0$, the total solution becomes

$$y(\theta) = \frac{h}{2}\frac{1}{(r\lambda/\pi)^2 - 1}\left[\cos(r\theta) - \cos\left(\frac{\pi\theta}{\lambda}\right)\right] \tag{13.16}$$

   Figure 13.2 shows a general example of system response on a rise function. The solid line in Figure 13.2a represents the rise based on Equation 13.3 assuming the follower motion to be exactly equal to the displacement function. The dashed line is the plot of Equation 13.16 with $r = 2.5$. The plots in Figure 13.2b and c are the velocity and acceleration, respectively.

   The response can be continued for the system going into a dwell period. During the high dwell, $y_c = h$ and $\dot{y}_c = 0$. From Equation 13.2,

$$m\ddot{y} + c\dot{y} + ky = kh \tag{13.17}$$

Dividing by mass gives

$$\ddot{y} + 2\varsigma\omega_n\dot{y} + \omega_n^2 y = \omega_n^2 h \tag{13.18}$$

We can combine the rise and high dwell to examine the response of the system going into the high dwell. The right-hand side of Equation 13.7 is only present as the cam rotates from

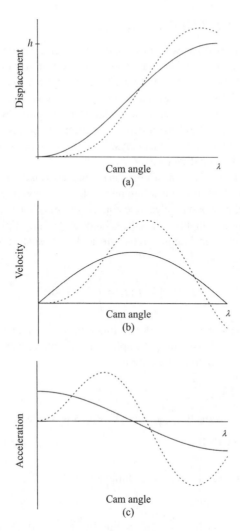

Figure 13.2   Mass response for simple harmonic function. (a) Displacement. (b) Velocity. (c) Acceleration

zero to λ. The right-hand side of Equation 13.18 will be present for cam angles larger than λ. Combining the right-hand side of both equations can be accomplished using step functions, which are discussed in Section 13.3.3.

13.3   General Response Using Laplace Transform Method

13.3.1   Introduction

Section 13.2 discussed the basic response for a system using a simple harmonic function for the cam rise. If a full cycle is to be analyzed, as for a double-dwell mechanism, the process becomes fairly tedious. If the cam function becomes more complex, the analysis can become too complex to be solved. Another approach is to use the Laplace transform method to develop the response of the system.

13.3.2 Basic Concepts of Laplace Transform

The Laplace transform method was developed by Pierre-Simon, marquis de Laplace (1749–1827), a French mathematician and a significant contributor to the field of differential equations. Laplace spent majority of his time working on applications of astronomy (motion of planets and stability of the solar system) and probability (Bayesian interpretation).

The Laplace transform is a powerful tool that can simplify the solution of ordinary differential equations. The method has many applications in physics and engineering such as electrical circuits, vibrations, and beam analysis. The Laplace transform is the most common transform method for vibrational analysis.

The basic principle of the Laplace transformation method is to take a function $f(t)$ and transform it into a different function $F(s)$. If $f(t)$ is a forcing function in a complex differential equation of motion, there will be a corresponding algebraic equation in $F(s)$. Once the algebraic equation is solved, an inverse transform is applied to get the solution $x(t)$ of the original differential equation.

The Laplace transform is an integral transform and the transformation is defined by the following integration:

$$L\{f(t)\} = F(s) = \int_0^\infty e^{-st}f(t)dt \tag{13.19}$$

Provided the integral converges, the result is a function of s. Although it is expected that the reader has a fundamental understanding of Laplace transform, a review of the basic principles will be provided through a couple of examples.

Example Problem 13.1

Evaluate the Laplace transform for $f(t) = c$, where c is a constant.

Solution: From Equation 13.19,

$$F(s) = \int_0^\infty e^{-st}(c)dt = \lim_{b\to\infty} \int_0^b e^{-st}(c)dt$$

$$= \lim_{b\to\infty} \frac{-ce^{-st}}{s}\Big|_0^b = \lim_{b\to\infty}\left(\frac{-c}{s}e^{-sb} + \frac{c}{s}\right)$$

Answer: $F(s) = \dfrac{c}{s}$

The previous example shows that for $f(t)$ equal to a constant, the Laplace transform is equal to c/s. However, this is true only when $s > 0$. This provision is due to the fact that if $s > 0$, the exponent e^{-st} is negative causing it to approach zero as b approaches infinity. If $s < 0$, the exponent would be positive and the integral diverges.

Example Problem 13.2

Evaluate the Laplace transform for $f(t) = t$.

Solution: From Equation 13.19,

$$F(s) = \int_0^\infty e^{-st}tdt$$

This can be solved by integration by parts. Let

$$u = t \qquad du = dt$$
$$dv = e^{-st}dt \qquad v = -\frac{e^{-st}}{s}$$

The result would be

$$F(s) = -\frac{te^{-st}}{s}\bigg|_0^\infty + \frac{1}{s}\int_0^\infty e^{-st}dt$$

The first term goes to zero leaving

$$F(s) = \frac{1}{s}\int_0^\infty e^{-st}dt$$

Notice that the integral is for $f(t) = c$, where c is equal to 1. Using the results from the previous example, we get

$$F(s) = \frac{1}{s}\left(\frac{1}{s}\right)$$

Answer: $F(s) = \frac{1}{s^2}$

In a similar fashion, the Laplace transform of other elementary functions can be computed. Such transforms have been computed and can be found in any differential equations text. An abbreviated list of transforms is given in Table 13.1.

13.3.3 Step Functions

In vibrations, it is very common to have forcing functions that are discontinuous. In mathematics, discontinuities are described using step functions. The most basic step function is the unit step function, which is also called the Heaviside function (named after the English

Table 13.1 Transforms of basic functions

$f(t)$	$F(s)$	$f(t)$	$F(s)$
c	$\dfrac{c}{s}$	$\sin(at)$	$\dfrac{a}{s^2+a^2}$
t	$\dfrac{1}{s^2}$	$\cos(at)$	$\dfrac{s}{s^2+a^2}$
t^2	$\dfrac{2}{s^3}$	$e^{at}\sin(bt)$	$\dfrac{b}{(s-a)^2+b^2}$
t^n	$\dfrac{n!}{s^{n+1}}$	$e^{at}\cos(bt)$	$\dfrac{s-a}{(s-a)^2+b^2}$
e^{at}	$\dfrac{1}{s-a}$	$t\sin(at)$	$\dfrac{2as}{(s^2+a^2)^2}$
$t^n e^{at}$	$\dfrac{n!}{(s-a)^{n+1}}$	$t\cos(at)$	$\dfrac{s^2-a^2}{(s^2+a^2)^2}$

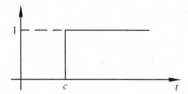

Figure 13.3 Unit step function (Heaviside function)

mathematician Oliver Heaviside). The Heaviside function is typically denoted by

$$u_c(t) = \begin{cases} 0 & t < c \\ 1 & t \geq c \end{cases} \qquad c \geq 0 \tag{13.20}$$

Figure 13.3 illustrates a unit step function.

More complicated step functions can be expressed as linear combinations of unit step functions. For example, the function

$$f(t) = \begin{cases} 3 & 0 < t \leq 3 \\ 1 & 3 < t \leq 8 \\ 2.5 & t > 8 \end{cases}$$

can be expressed as

$$f(t) = 3 - 2u_3(t) + 1.5u_8(t)$$

The graph of the function is shown in Figure 13.4.

Heaviside function can also be used to generate pulse functions. For example, the function

$$f(t) = (u_{1.5}(t) - u_{3.5}(t))(\sin(\pi t) + 1)$$

will be zero for all values except $1.5 < t < 3.5$ and will be $\sin(\pi t) + 1$ between $1.5 < t < 3.5$. Figure 13.5 shows the graph of the function. The complete plot of $\sin(\pi t) + 1$ is shown as a dotted line for reference.

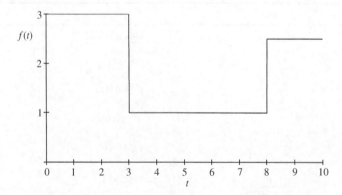

Figure 13.4 Graph of the function $f(t) = 3 - 2u_3(t) + 1.5u_8(t)$

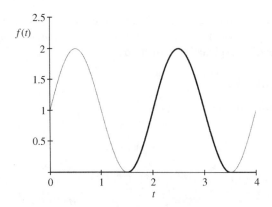

Figure 13.5 Graph of the function $f(t) = [u_{1.5}(t) - u_{3.5}(t)](\sin(\pi t) + 1)$

The Laplace transform of a Heaviside function is

$$L\{u_c(t)\} = U_c(s) = \frac{e^{-cs}}{s} \quad \text{for } s > 0 \tag{13.21}$$

13.3.4 Transforms of Derivatives

The use of the Laplace transform method for solving differential equation requires the Laplace transform of first and second derivative terms. The following equations give the Laplace transforms without proof:

$$L\{f'(t)\} = sF(s) - f(0) \tag{13.22}$$

$$L\{f''(t)\} = s^2 F(s) - sf(0) - f'(0) \tag{13.23}$$

13.3.5 Inverse Transforms

If $F(s)$ is the Laplace transform of the function $f(t)$, then $f(t)$ is the inverse Laplace transform of $F(s)$.

$$L^{-1}\{F(s)\} = f(t) \tag{13.24}$$

In other words, the inverse Laplace transform is a reverse process to find $f(t)$ when $F(s)$ is given. For some simple cases, the inverse transform can be determined from tables of transforms, as shown in Table 13.2.

Table 13.2 Examples of inverse transforms

$L^{-1}\left\{\dfrac{1}{s}\right\} = 1$	$L^{-1}\left\{\dfrac{n!}{s^{n+1}}\right\} = t^n \quad n = 1, 2, \ldots$	$L^{-1}\left\{\dfrac{1}{s-a}\right\} = e^{at}$
$L^{-1}\left\{\dfrac{k}{s^2 - k^2}\right\} = \sin kt$	$L^{-1}\left\{\dfrac{k}{s^2 + k^2}\right\} = \cos kt$	

Inverse Laplace transforms of more complicated cases can be accomplished by using partial fractions. The basic process of partial fraction decomposition is best described with an example. A more detailed explanation will follow.

Example Problem 13.3

Find the partial fractions decomposition of

$$\frac{3x - 5}{(x - 1)(x - 3)}$$

Solution: We can express in terms of partial fractions

$$\frac{3x - 5}{(x - 1)(x - 3)} = \frac{A}{x - 1} + \frac{B}{x - 3}$$

Multiply both sides of the expression by $(x - 1)(x - 3)$ to get

$$3x - 5 = A(x - 3) + B(x - 1)$$
$$3x - 5 = (A + B)x - (3A + B)$$

Equate coefficients of like powers of x to get

$$\begin{cases} 3 = A + B \\ 5 = 3A + B \end{cases}$$

Solving the system of equation gives $A = 1$ and $B = 2$.

Answer: $\dfrac{1}{x - 1} + \dfrac{2}{x - 3}$

In general, the transform will take the form of the quotient of two polynomials:

$$F(s) = \frac{A(s)}{B(s)} \tag{13.25}$$

We will consider a case when $B(s)$ is of higher order than $A(s)$ and has n distinct roots.

$$B(s) = (s - a_1)(s - a_2)(s - a_3) \cdots (s - a_n) \tag{13.26}$$

Partial fraction expansion gives

$$F(s) = \frac{A(s)}{B(s)} = \frac{C_1}{s - a_1} + \frac{C_2}{s - a_2} + \frac{C_3}{s - a_3} + \cdots + \frac{C_n}{s - a_n} \tag{13.27}$$

The constants are determined from

$$C_k = \lim_{s \to a_k} \frac{(s - a_k)A(s)}{B(s)} \tag{13.28}$$

and the inverse transform becomes

$$f(t) = \sum_{k=1}^{n} C_k e^{a_k t} \tag{13.29}$$

Example Problem 13.4

Find the inverse transform

$$L^{-1}\left\{\frac{s}{(s-2)(s-3)(s-6)}\right\}$$

Solution:

$$F(s) = \frac{A(s)}{B(s)} = \frac{s}{(s-2)(s-3)(s-6)}$$

From Equation 13.26, we determine that $a_1 = 2$, $a_2 = 3$, and $a_3 = 6$.

$$F(s) = \frac{C_1}{s-2} + \frac{C_2}{s-3} + \frac{C_3}{s-6}$$

The constants are determined using Equation 13.28.

$$k = 1 \Rightarrow C_1 = \lim_{s \to 2} \frac{(s-2)s}{(s-2)(s-3)(s-6)} = \frac{2}{(-1)(-4)} = \frac{1}{2}$$

$$k = 2 \Rightarrow C_2 = \lim_{s \to 3} \frac{(s-3)s}{(s-2)(s-3)(s-6)} = \frac{3}{(1)(-3)} = -1$$

$$k = 3 \Rightarrow C_3 = \lim_{s \to 6} \frac{(s-6)s}{(s-2)(s-3)(s-6)} = \frac{6}{(4)(3)} = \frac{1}{2}$$

The inverse transform is then determined using Equation 13.29.

$$f(t) = \sum_{k=1}^{n} C_k e^{a_k t} = C_1 e^{a_1 t} + C_2 e^{a_2 t} + C_3 e^{a_3 t}$$

Answer: $f(t) = \frac{1}{2}e^{2t} - e^{3t} + \frac{1}{2}e^{6t}$

The previous example was a case with simple poles. We will now consider a case with higher order poles. As before, we will begin with the transform as a quotient of two polynomials with $B(s)$ of higher order than $A(s)$.

$$F(s) = \frac{A(s)}{B(s)} \tag{13.30}$$

Now, $B(s)$ will have a kth order pole at a_1.

$$B(s) = (s-a_1)^k (s-a_2)(s-a_3)\cdots \tag{13.31}$$

Partial fraction expansion gives

$$F(s) = \frac{C_{11}}{(s-a_1)^k} + \frac{C_{12}}{(s-a_1)^{k-1}} + \cdots + \frac{C_{1k}}{(s-a_1)} + \frac{C_2}{s-a_2} + \frac{C_3}{s-a_3} + \cdots \tag{13.32}$$

The constants C_k are determined as before using Equation 13.28, and the remaining constants are determined from

$$C_{1n} = \frac{1}{(n-1)!}\left[\frac{d^{n-1}}{ds^{n-1}}(s-a_1)^k F(s)\right]_{s=a_1} \tag{13.33}$$

The inverse transform is

$$f(t) = \left[C_{11}\frac{t^{k-1}}{(k-1)!} + C_{12}\frac{t^{k-2}}{(k-2)!} + \cdots\right]e^{a_1 t} + C_2 e^{a_2 t} + C_3 e^{a_3 t} + \cdots \tag{13.34}$$

Example Problem 13.5

Find the inverse transform

$$L^{-1}\left\{\frac{2s-1}{s^2(s+1)^3}\right\}$$

Solution: For this problem, $B(s) = (s-0)^2(s+1)^3$ and therefore $a_1 = 0$ and $a_2 = -1$. From Equation 13.32,

$$F(s) = \frac{C_{11}}{(s-0)^2} + \frac{C_{12}}{(s-0)} + \frac{C_{21}}{(s+1)^3} + \frac{C_{22}}{(s+1)^2} + \frac{C_{23}}{(s+1)}$$

Constants are determined from Equation 13.33. For the first two constants,

$$C_{1n} = \frac{1}{(n-1)!}\left[\frac{d^{n-1}}{ds^{n-1}}(s-a_1)^k F(s)\right]_{s=a_1}$$

$$n = 1 \Rightarrow C_{11} = \frac{1}{0!}\left[s^2\frac{2s-1}{s^2(s+1)^3}\right]_{s=0} = -1$$

$$n = 2 \Rightarrow C_{12} = \frac{1}{1!}\left[\frac{d}{ds}s^2\frac{2s-1}{s^2(s+1)^3}\right]_{s=0} = \left[\frac{2}{(s+1)^3} - \frac{3(2s-1)}{(s+1)^4}\right]_{s=0} = 5$$

Similarly, for the last three constants,

$$C_{2n} = \frac{1}{(n-1)!}\left[\frac{d^{n-1}}{ds^{n-1}}(s-a_2)^k F(s)\right]_{s=a_2}$$

$$n = 1 \Rightarrow C_{21} = \frac{1}{0!}\left[(s+1)^3\frac{2s-1}{s^2(s+1)^3}\right]_{s=-1} = -3$$

$$n = 2 \Rightarrow C_{22} = \frac{1}{1!}\left[\frac{d}{ds}(s+1)^3\frac{2s-1}{s^2(s+1)^3}\right]_{s=-1} = \left[\frac{2}{s^2} - \frac{2(2s-1)}{s^3}\right]_{s=-1} = -4$$

$$n = 3 \Rightarrow C_{23} = \frac{1}{2!}\left[\frac{d^2}{ds^2}(s+1)^3\frac{2s-1}{s^2(s+1)^3}\right]_{s=-1} = \frac{1}{2}\left[\frac{-8}{s^3} + \frac{6(2s-1)}{s^4}\right]_{s=-1} = -5$$

The final inverse transform is determined using Equation 13.34:

$$f(t) = \left[C_{11}\frac{t}{1!} + C_{12}\frac{1}{0!}\right]e^{0t} + \left[C_{21}\frac{t^2}{2!} + C_{22}\frac{t}{1!} + C_{23}\frac{1}{0!}\right]e^{-t}$$

$$f(t) = \left[-1\frac{t}{1} + 5\frac{1}{1}\right](1) + \left[-3\frac{t^2}{2} - 4\frac{t}{1} - 5\frac{1}{1}\right]e^{-t}$$

Answer: $f(t) = -t + 5 + \left[\frac{-3t^2}{2} - 4t - 5\right]e^{-t}$

13.3.6 Vibration Analysis with Laplace Transforms

We can use Laplace transforms to solve our equation of motion $m\ddot{x} + c\dot{x} + kx = f(t)$, where $f(t)$ is an arbitrary excitation. Transforming both sides gives

$$m\left(s^2X(s) - sx(0) - \dot{x}(0)\right) + c(sX(s) - x(0)) + kX(s) = F(s) \tag{13.35}$$

We can rearrange to get

$$\left(ms^2 + cs + k\right)X(s) = F(s) + msx(0) + m\dot{x}(0) + cx(0)$$

$$X(s) = \frac{F(s)}{ms^2 + cs + k} + \frac{(ms + c)x(0) + m\dot{x}(0)}{ms^2 + cs + k} \tag{13.36}$$

Equation 13.36 is the subsidiary equation of the differential equation of motion. The term on the left-hand side of the equation is the response transform. The first term on the right-hand side of the equation is the transform of the particular solution, and the second term on the right-hand side is the transform of the homogeneous solution. $F(s)$ is the driving function transform, and $ms^2 + cs + k$ is the characteristic function. Equation 13.36 can be expressed using $\omega_n = \sqrt{k/m}$ and $\varsigma = c/2m\omega_n$.

$$X(s) = \frac{(1/m)F(s)}{s^2 + 2\varsigma\omega_n s + \omega_n^2} + \frac{(s + 2\varsigma\omega_n)x(0) + \dot{x}(0)}{s^2 + 2\varsigma\omega_n s + \omega_n^2} \tag{13.37}$$

The system response $x(t)$ is found from the inverse Laplace transform using methods discussed in Section 13.3.2.

The general process for solution is fairly basic. First, perform the Laplace transform to both sides of the differential equation. This process will result in an algebraic equation. The second step is to input the initial condition into the algebraic equation and solve the equation. The last step is to perform the inverse Laplace transform to obtain the final solution to the original differential equation. Although the process is straightforward, the inverse Laplace transform of the resulting algebraic equation can be difficult. Alternatively, software with built-in Laplace transform functions such as MathCAD can be used. This process is best illustrated in an example.

Example Problem 13.6

A 1 kg mass is suspended from a spring with a stiffness of 16 N/m and subjected to the force shown in Figure 13.6. Perform Laplace transform in MathCAD and plot the response of the system for the first 6 seconds of motion.

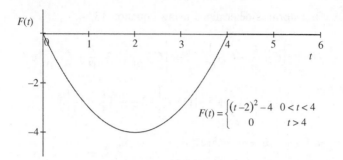

$$F(t) = \begin{cases} (t-2)^2 - 4 & 0 < t < 4 \\ 0 & t > 4 \end{cases}$$

Figure 13.6

Solution: The system properties are entered in MathCAD:

$$k:=16 \quad m:=1$$
$$v0:=0 \quad x0:=0$$

and the natural frequency is calculated:

$$\omega := \sqrt{\frac{k}{m}} \qquad \omega = 4$$

The function can be presented using Heaviside functions:

$$f(t) = \left((t-2)^2 - 4\right)u(t) - \left((t-2)^2 - 4\right)u_4(t)$$

The force function is inserted in MathCAD as

$$f(t):=(\Phi(t-0) - \Phi(t-4)) \cdot [(t-2)^2 - 4]$$

The Laplace transform is then

$$F(s) := \frac{f(t)}{m} \, \text{laplace}, t \to \frac{2}{s^3} - \frac{4}{s^2} - \frac{\exp(-4 \cdot s)}{s^2} - 2 \cdot \frac{\exp(-4 \cdot s)}{s^3}$$

and

$$X(s) := \frac{F(s) + s \cdot x0 + v0}{s^2 + \dfrac{k}{m}}$$

The inverse Laplace transform is then

$$x(t):=X(s) \text{invlaplace}, s \to \frac{1}{16} \cdot t^2 - \frac{1}{128} + \frac{1}{128} \cdot \cos(4 \cdot t) - \frac{1}{4} \cdot t + \frac{1}{16} \cdot \sin(4 \cdot t) + \frac{1}{4} \cdot \Phi(t-4) \cdot t +$$
$$\frac{1}{128} \cdot \Phi(t-4) + \frac{1}{16} \cdot \Phi(t-4) \cdot \sin(4 \cdot t - 16) - \frac{1}{16} \cdot \Phi(t-4) \cdot t^2 - \frac{1}{128} \cdot \Phi(t-4) \cdot \cos(4 \cdot t - 16)$$

The plot of the response is shown in Figure 13.7.

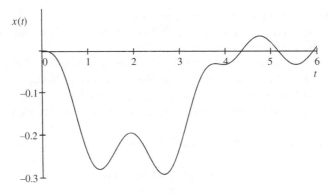

Figure 13.7

13.4 System Response Using Numerical Methods

The process described in the previous sections can be used for any cam input function y_c. However, the process becomes far more complex for more advanced cam functions. Numerical methods that are easier to use can be used to determine the system response for any cam function. Numerical methods allow for analysis of very mathematically complex cam functions or even cam profiles made from combinations of different cam profiles (e.g., those presented in Section 13.5). Another advantage of numerical methods is that the actual mathematical function of the cam profile is not required. System response can be calculated using discrete data points of the cam profile. Numerical methods can also be used when the cam profile is only given graphically.

Timoshenko and Young (1948) demonstrate a numerical method, which is also thoroughly described by Rothbart (1956). The numerical method requires breaking the acceleration function into small intervals with constant time steps. Follower displacement and velocity at the end of interval i can be determined using

$$y_i = y_{i-1} \cos(\omega_n \Delta t) + \frac{dy_{i-1}/dt}{\omega_n} \sin(\omega_n \Delta t) + \frac{a_i}{\omega_n^2}(1 - \cos(\omega_n \Delta t)) \qquad (13.38)$$

$$\frac{dy_i}{dt} \frac{1}{\omega_n} = -y_{i-1} \sin(\omega_n \Delta t) + \frac{dy_{i-1}/dt}{\omega_n} \cos(\omega_n \Delta t) + \frac{a_i}{\omega_n^2} \sin(\omega_n \Delta t) \qquad (13.39)$$

The values of y_i and dy_i/dt are then used as the initial displacement and velocity for the following interval. The total displacement of the follower will be the cam displacement function plus the displacement term defined in Equation 13.38.

Although the numerical method is rather straightforward, use of a simple table will help organize the calculations. The calculations will use acceleration data for the particular cam profile being analyzed. The acceleration data may not be available directly for a particular cam system. The table developed will depend on what data are available. The accuracy will improve for smaller time steps (Δt). Also, note that the equations do not account for damping, but it can be included in the equations. The numerical method process is best understood by an example.

As an illustrative example, let us consider a case of a rise of 4 inches using a 3-4-5 polynomial function, and the rise occurs in 120° of cam rotation. The cam rotation speed is 900 rpm. The natural frequency of the system is 55 Hertz.

Based on the rotation speed of 900 rpm (94.5 rad/s), the rise of 120° (2.09 radians) will take place in 0.022 seconds. We will discretize the acceleration data into 20 steps, giving a time step of $\Delta t = 0.0011$ s. This time step was arbitrarily chosen, and a smaller time step would provide a higher level of accuracy. The system natural frequency is 55 Hertz, which gives $\omega_n = 345.575$ rad/s.

We know from Chapter 9 that a 3-4-5 polynomial displacement is determined using

$$y(\beta) = 10\beta^3 - 15\beta^4 + 6\beta^5 \tag{13.40}$$

and the acceleration is found using

$$\frac{d^2 y}{d\beta^2} = 60\beta - 180\beta^2 + 120\beta^3 \tag{13.41}$$

Using the procedures described in Chapter 9, the displacement function for the rise will be

$$\frac{s - s_1}{s_2 - s_1} = 10\left(\frac{\theta - \theta_1}{\theta_2 - \theta_1}\right)^3 - 15\left(\frac{\theta - \theta_1}{\theta_2 - \theta_1}\right)^4 + 6\left(\frac{\theta - \theta_1}{\theta_2 - \theta_1}\right)^5$$

$$\frac{s}{4} = 10\left(\frac{\theta}{120}\right)^3 - 15\left(\frac{\theta}{120}\right)^4 + 6\left(\frac{\theta}{120}\right)^5$$

$$s = 40\left(\frac{\theta}{120}\right)^3 - 60\left(\frac{\theta}{120}\right)^4 + 24\left(\frac{\theta}{120}\right)^5 \tag{13.42}$$

The acceleration function will be

$$\frac{d^2 s}{d\theta^2} = \left(\frac{s_2 - s_1}{(\theta_2 - \theta_1)^2}\right)\left(60\left(\frac{\theta - \theta_1}{\theta_2 - \theta_1}\right) - 180\left(\frac{\theta - \theta_1}{\theta_2 - \theta_1}\right)^2 + 120\left(\frac{\theta - \theta_1}{\theta_2 - \theta_1}\right)^3\right)$$

$$\frac{d^2 s}{d\theta^2} = \left(\frac{4}{120^2}\right)\left(60\left(\frac{\theta}{120}\right) - 180\left(\frac{\theta}{120}\right)^2 + 120\left(\frac{\theta}{120}\right)^3\right)$$

$$\frac{d^2 s}{d\theta^2} = a = 0.017\left(\frac{\theta}{120}\right) - 0.05\left(\frac{\theta}{120}\right)^2 + 0.033\left(\frac{\theta}{120}\right)^3 \tag{13.43}$$

The values obtained from Equation 13.43 will be in inches per degree squared. To convert into inches per seconds squared, the values must be multiplied by the rotation speed of 5400°/s squared. Table 13.3 gives the displacement and acceleration information for the full rise portion of the cycle.

Figure 13.8 shows the plot of the cam acceleration for the full rise portion. The phantom line represents Equation 13.43. The plot is then broken into intervals based on data from Table 13.3.

Table 13.3   Displacement and acceleration data for illustrative example

Interval	Time (s)	θ (°)	s (in.)	a (in./s^2)
1	0.0000	0	0.000	0
2	0.0029	6	0.005	20 776.5
3	0.0057	12	0.034	34 992
4	0.0086	18	0.106	43 375.5
5	0.0114	24	0.232	46 656
6	0.0143	30	0.414	45 562.5
7	0.0171	36	0.652	40 824
8	0.0200	42	0.941	33 169.5
9	0.0229	48	1.270	23 328
10	0.0257	54	1.627	12 028.5
11	0.0286	60	2.000	0
12	0.0314	66	2.373	−12 028.5
13	0.0343	72	2.730	−23 328
14	0.0371	78	3.059	−33 169.5
15	0.0400	84	3.348	−40 824
16	0.0429	90	3.586	−45 562.5
17	0.0457	96	3.768	−46 656
18	0.0486	102	3.894	−43 375.5
19	0.0514	108	3.966	−34 992
20	0.0543	114	3.995	−20 776.5
21	0.0571	120	4.000	0

Now, Equations 13.38 and 13.39 are used for each interval, and the results are shown in Table 13.4. The displacement from Equation 13.42 is also provided as a reference. The actual system response is the sum of y and s as shown.

Figure 13.9 shows the system response in the solid line. The dashed line is the displacement from Equation 13.42.

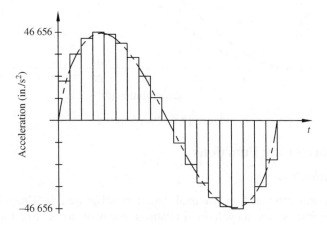

Figure 13.8   Cam acceleration

Table 13.4 System response for illustrative example

Interval	y (in.) (from Equation 13.38)	$\dfrac{dy}{dt}\dfrac{1}{\omega_n}$ (from Equation 13.39)	s (in.) (from Equation 13.42)	Response (in.) $(y + s)$
1	0.0000	0.0000	0.000	0.000
2	0.0127	0.0652	0.005	0.017
3	0.0575	0.1654	0.034	0.092
4	0.1417	0.2679	0.106	0.248
5	0.2602	0.3417	0.232	0.492
6	0.3970	0.3622	0.414	0.811
7	0.5287	0.3152	0.652	1.181
8	0.6285	0.1982	0.941	1.569
9	0.6712	0.0215	1.270	1.941
10	0.6377	−0.1938	1.627	2.265
11	0.5187	−0.4186	2.000	2.519
12	0.3168	−0.6201	2.373	2.689
13	0.0472	−0.7668	2.730	2.777
14	−0.2637	−0.8327	3.059	2.796
15	−0.5813	−0.8014	3.348	2.766
16	−0.8670	−0.6682	3.586	2.719
17	−1.0826	−0.4411	3.768	2.686
18	−1.1954	−0.1395	3.894	2.698
19	−1.1820	0.2087	3.966	2.784
20	−1.0304	0.5712	3.995	2.965
21	−0.7414	0.9155	4.000	3.259

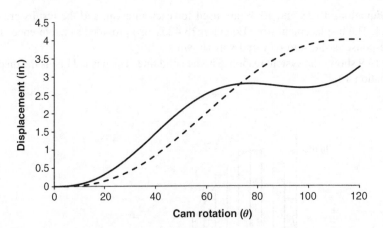

Figure 13.9 Cam response

13.5 Advanced Cam Functions

13.5.1 Introduction

The basic trigonometric and polynomial functions discussed in Chapter 9 are great introductory functions, but they have limitations and will not satisfy the requirements for all applications. Other function types have been developed to improve overall performance, especially for high-speed applications. This section will discuss a few different advanced cam functions.

13.5.2 Combination of Basic Cam Functions

Each basic cam function discussed in Chapter 9 has its own unique advantages. Basic curves can be combined in such a way to utilize those advantages. This section will discuss some of the more common combination of curves. Regardless of the curves to be combined, the boundary conditions are based on continuous displacement and velocity curves. At any coupling point between curves, the pressure angle must be the same for both curves. Pressure angle is a function of velocity, so a continuous velocity curve satisfies the pressure angle requirement. All the basic curves discussed have zero acceleration at their midpoint location. A simple way to ensure continuous acceleration is to use the midpoint location as the coupling point, although that will not always be a convenient coupling location.

A typical goal in cam design is to limit forces by limiting peak acceleration. The simplest approach to minimizing peak acceleration amplitude is to use a square wave constant acceleration function such as that shown in Figure 13.10a. Obviously, the constant acceleration will have the problem of infinite jerk at the discontinuities. An improvement would be a trapezoidal acceleration function, such as the one shown in Figure 13.10b. Although the trapezoidal acceleration improves performance, the overall area under the curves must remain the same. The peak amplitude of acceleration for a trapezoidal acceleration function will be a little larger than a continuous acceleration function. Differentiation of the trapezoidal acceleration function will yield a finite square wave jerk function.

The trapezoidal acceleration function can be modified by combining cycloidal functions and constant acceleration. This modified trapezoid function will have lower peak acceleration than the cycloidal function. Figure 13.11 shows the basic development of a rise function using the modified trapezoidal acceleration curve. The final rise function will provide a displacement h over an angular displacement β of the cam. Development will begin with a cycloidal acceleration curve, shown in Figure 13.11a, which completes one cycle in a rotation of $\beta/2$. The first step to create the modified trapezoidal acceleration is to divide the cycloidal acceleration curve into four sections equal to $\beta/8$. The next step is to add the constant acceleration portions, as shown in Figure 13.11b, which completes the modified trapezoidal acceleration curve.

The peak acceleration value for the curve shown in Figure 13.11b is unknown and must be determined based on the desired total rise distance. Section AB of the modified trapezoidal curve is simply the first one-fourth of the cycloidal acceleration curve, as shown in Figure 13.11a, which is the second derivative of a function s' having a rise of h'. Using the equations developed in Chapter 9, the displacement shown in

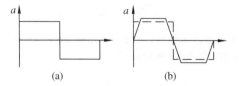

(a)   (b)

Figure 13.10   (a) Constant acceleration function. (b) Trapezoidal acceleration function

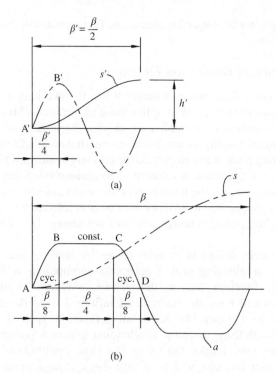

Figure 13.11 (a) Cycloidal function. (b) Modified trapezoidal acceleration curve (phantom line) and displacement curve (solid line)

Figure 13.11a is

$$s'(\theta) = 2h' \frac{\theta}{\beta} - \frac{h'}{2\pi} \sin\left(4\pi \frac{\theta}{\beta}\right) \tag{13.44}$$

The final cam displacement curve will follow s' for $0 < \theta < \beta/8$. Therefore, the final displacement curve's value at $\beta/8$ (point B on the displacement curve shown in Figure 13.11b) will be

$$s_B = 2h' \left(\frac{1}{8}\right) - \frac{h'}{2\pi} \sin\left(4\pi \frac{1}{8}\right) = 0.09085h' \tag{13.45}$$

The velocity and acceleration magnitudes at point B can be found in a similar manner:

$$v_B = \frac{2h'\omega}{\beta}$$

$$a_B = \frac{8\pi h' \omega^2}{\beta^2} \tag{13.46}$$

Section BC on the final acceleration curve will have constant acceleration. Therefore, the displacement will be the parabolic function:

$$s_{B-C} = v_B t + \frac{1}{2} a_B t^2 \tag{13.47}$$

The time from B to C will be

$$\omega t = \frac{\beta}{4} \Rightarrow t = \frac{\beta}{4\omega} \tag{13.48}$$

Substituting Equations 13.46 and 13.48 into Equation 13.47 gives

$$s_{B-C} = \left(\frac{2h'\omega}{\beta}\right)\left(\frac{\beta}{4\omega}\right) + \frac{1}{2}\left(\frac{8\pi h'\omega^2}{\beta^2}\right)\left(\frac{\beta}{4\omega}\right)^2 = 1.2854h' \tag{13.49}$$

The velocity at point C is determined based on the constant acceleration:

$$v_C = v_B + a_B t \tag{13.50}$$

Substituting Equations 13.46 and 13.48 into Equation 13.50 gives

$$v_C = \frac{2h'\omega}{\beta} + \frac{8\pi h'\omega^2}{\beta^2}\left(\frac{\beta}{4\omega}\right)$$

$$v_C = \frac{8.283h'\omega}{\beta} \tag{13.51}$$

The acceleration at point C is the same as the acceleration at point B given in Equation 13.46.

To move from point C to point D, we again consider the cycloidal function. It can be determined that the cycloidal function is constructed based on the displacement being the sum of a constant velocity displacement and a harmonic displacement:

$$s_{C-D} = v_C t + \frac{h'}{2\pi} \tag{13.52}$$

The time from C to D is

$$\omega t = \frac{\beta}{8} \Rightarrow t = \frac{\beta}{8\omega} \tag{13.53}$$

Substituting Equations 13.51 and 13.53 into Equation 13.52 gives

$$s_{C-D} = \frac{8.283h'\omega}{\beta}\left(\frac{\beta}{8\omega}\right) + \frac{h'}{2\pi} = 1.19455h' \tag{13.54}$$

Everything up to point D is half of the total displacement. The total displacement h is determined from

$$2(s_A + s_{B-C} + s_{C-D}) = h$$

$$2(0.09085h' + 1.2854h' + 1.19455h') = h$$

$$h' = 0.1945h \tag{13.55}$$

Equation 13.55 can be substituted back into the previous equations to give the complete rise function, which will be defined as piecewise function. Equation 13.56 gives a normalized displacement function for a rise distance h.

$$\frac{s}{h} = \begin{cases} 0.38898\dfrac{\theta}{\beta} - 0.03095\sin\left(4\pi\dfrac{\theta}{\beta}\right) & 0 < \theta < \dfrac{\beta}{8} \\[2ex] 2.44406\left(\dfrac{\theta}{\beta}\right)^2 - 0.22203\dfrac{\theta}{\beta} + 0.00723 & \dfrac{\beta}{8} < \theta < \dfrac{3\beta}{8} \\[2ex] 1.61102\dfrac{\theta}{\beta} - 0.03095\sin\left(4\pi\dfrac{\theta}{\beta} - \pi\right) - 0.30551 & \dfrac{3\beta}{8} < \theta < \dfrac{\beta}{2} \\[2ex] 1.61102\dfrac{\theta}{\beta} + 0.03095\sin\left(4\pi\dfrac{\theta}{\beta} - 2\pi\right) - 0.30551 & \dfrac{\beta}{2} < \theta < \dfrac{5\beta}{8} \\[2ex] -2.44406\left(\dfrac{\theta}{\beta}\right)^2 + 4.66609\dfrac{\theta}{\beta} - 1.22927 & \dfrac{5\beta}{8} < \theta < \dfrac{7\beta}{8} \\[2ex] 0.38898\dfrac{\theta}{\beta} + 0.03095\sin\left(4\pi\dfrac{\theta}{\beta} - 3\pi\right) + 0.61102 & \dfrac{7\beta}{8} < \theta < \beta \end{cases} \tag{13.56}$$

The normalized displacement can be plotted against a normalized cam rotation, which is shown in Figure 13.12.

Velocity and acceleration equations can be developed as piecewise functions. The follower velocity in inches per second can be determined from

$$v = C_v h \frac{6N}{\beta} \tag{13.57}$$

where N is the cam rotation speed in revolutions per minute, β is the angular displacement for the total rise, and C_v is a velocity coefficient determined from

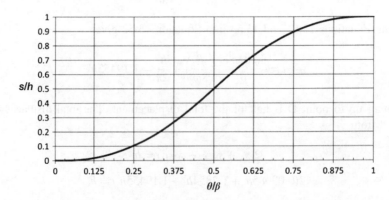

Figure 13.12 Normalized displacement diagram for modified trapezoid

$$
C_v = \begin{cases}
0.38898 - 0.38898 \cos\left(4\pi\dfrac{\theta}{\beta}\right) & 0 < \theta < \dfrac{\beta}{8} \\[2ex]
-0.22203 + 4.88812\dfrac{\theta}{\beta} & \dfrac{\beta}{8} < \theta < \dfrac{3\beta}{8} \\[2ex]
1.61102 - 0.38898 \cos\left(4\pi\dfrac{\theta}{\beta} - \pi\right) & \dfrac{3\beta}{8} < \theta < \dfrac{\beta}{2} \\[2ex]
1.61102 + 0.38898 \cos\left(4\pi\dfrac{\theta}{\beta} - 2\pi\right) & \dfrac{\beta}{2} < \theta < \dfrac{5\beta}{8} \\[2ex]
4.66609 - 4.88812\dfrac{\theta}{\beta} & \dfrac{5\beta}{8} < \theta < \dfrac{7\beta}{8} \\[2ex]
0.38898 + 0.38898 \cos\left(4\pi\dfrac{\theta}{\beta} - 3\pi\right) & \dfrac{7\beta}{8} < \theta < \beta
\end{cases}
\tag{13.58}
$$

The follower acceleration in inches per second squared is

$$
a = C_a h \left(\frac{6N}{\beta}\right)^2
\tag{13.59}
$$

where N is the cam rotation speed in revolutions per minute and β is the angular displacement for the total rise. The acceleration coefficient is determined from

$$
C_a = \begin{cases}
4.88812 \sin\left(4\pi\dfrac{\theta}{\beta}\right) & 0 < \theta < \dfrac{\beta}{8} \\[2ex]
4.88812 & \dfrac{\beta}{8} < \theta < \dfrac{3\beta}{8} \\[2ex]
4.8812 \sin\left(4\pi\dfrac{\theta}{\beta} - \pi\right) & \dfrac{3\beta}{8} < \theta < \dfrac{\beta}{2} \\[2ex]
-4.8812 \sin\left(4\pi\dfrac{\theta}{\beta} - 2\pi\right) & \dfrac{\beta}{2} < \theta < \dfrac{5\beta}{8} \\[2ex]
-4.88812 & \dfrac{5\beta}{8} < \theta < \dfrac{7\beta}{8} \\[2ex]
-4.8812 \sin\left(4\pi\dfrac{\theta}{\beta} - 3\pi\right) & \dfrac{7\beta}{8} < \theta < \beta
\end{cases}
\tag{13.60}
$$

Figure 13.13 shows the plots for both coefficients.

Example Problem 13.7

A cam function is required to produce a rise from 0 to 2.5 inches for a cam rotation from 0 to 80° using a modified trapezoid function. The cam rotates at a speed of 120 rpm. Determine and plot the displacement, velocity, and acceleration.

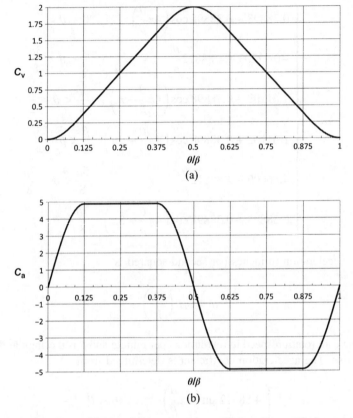

Figure 13.13 (a) Velocity coefficient. (b) Acceleration coefficient for modified trapezoidal function

Solution: For our problem, $h = 2.5$ and $\beta = 80°$. Substituting these values into Equation 13.56 gives

$$
s = \begin{cases}
0.9725\dfrac{\theta}{\beta} - 0.0774 \sin\left(4\pi\dfrac{\theta}{80}\right) & 0 < \theta < 10 \\[2ex]
6.1015\left(\dfrac{\theta}{80}\right)^2 - 0.5551\dfrac{\theta}{80} + 0.0181 & 10 < \theta < 30 \\[2ex]
4.0276\dfrac{\theta}{80} - 0.0774 \sin\left(4\pi\dfrac{\theta}{80} - \pi\right) - 0.7638 & 30 < \theta < 40 \\[2ex]
4.0276\dfrac{\theta}{80} + 0.0774 \sin\left(4\pi\dfrac{\theta}{80} - 2\pi\right) - 0.7638 & 40 < \theta < 50 \\[2ex]
-6.1102\left(\dfrac{\theta}{80}\right)^2 + 11.6652\dfrac{\theta}{80} - 3.0732 & 50 < \theta < 70 \\[2ex]
0.9725\dfrac{\theta}{80} + 0.0774 \sin\left(4\pi\dfrac{\theta}{80} - 3\pi\right) + 1.5276 & 70 < \theta < 80
\end{cases}
$$

The complete rise function is shown in Figure 13.14.

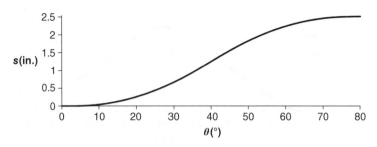

Figure 13.14

The velocity is determined using Equations 13.57 and 13.58. Using $N = 120\,\text{rpm}$, the velocity function is

$$
v = \begin{cases}
8.7521 - 8.7521 \cos\left(4\pi\dfrac{\theta}{80}\right) & 0 < \theta < 10 \\[2ex]
-4.9957 + 109.8270\dfrac{\theta}{80} & 10 < \theta < 30 \\[2ex]
36.2480 - 8.7521 \cos\left(4\pi\dfrac{\theta}{80} - \pi\right) & 30 < \theta < 40 \\[2ex]
36.2480 + 8.7521 \cos\left(4\pi\dfrac{\theta}{80} - 2\pi\right) & 40 < \theta < 50 \\[2ex]
104.9870 - 109.8270\dfrac{\theta}{80} & 50 < \theta < 70 \\[2ex]
8.7521 + 8.7521 \cos\left(4\pi\dfrac{\theta}{80} - 3\pi\right) & 70 < \theta < 80
\end{cases}
$$

From Equations 13.59 and 13.60, the acceleration is

$$
a = \begin{cases}
989.8443 \sin\left(4\pi\dfrac{\theta}{80}\right) & 0 < \theta < 10 \\[2ex]
989.8443 & 10 < \theta < 30 \\[2ex]
989.8443 \sin\left(4\pi\dfrac{\theta}{80} - \pi\right) & 30 < \theta < 40 \\[2ex]
-989.8443 \sin\left(4\pi\dfrac{\theta}{80} - 2\pi\right) & 40 < \theta < 50 \\[2ex]
-989.8443 & 50 < \theta < 70 \\[2ex]
-989.8443 \sin\left(4\pi\dfrac{\theta}{80} - 3\pi\right) & 70 < \theta < 80
\end{cases}
$$

The velocity and acceleration plots are shown in Figure 13.15.

Another common class of combination functions would be the modified sine function, which combines cycloidal and harmonic curves. Development of the equations is done similar

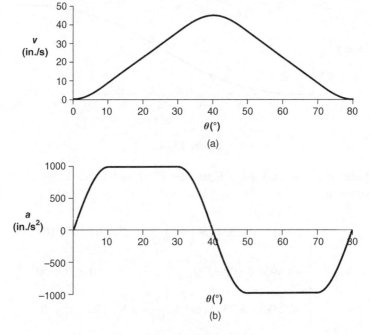

Figure 13.15

to those for the modified trapezoidal function. The displacement plot for the modified sine function is shown in Figure 13.16.

The cam displacement can be determined from the following piecewise function:

$$\frac{s}{h} = \begin{cases} 0.4399\dfrac{\theta}{\beta} - 0.03501 \sin\left(4\pi\dfrac{\theta}{\beta}\right) & 0 < \theta < \dfrac{\beta}{8} \\[3mm] 0.4399\dfrac{\theta}{\beta} - 0.31505 \cos\left(4\pi\dfrac{\theta}{\beta} - \dfrac{\pi}{6}\right) + 0.28005 & \dfrac{\beta}{8} < \theta < \dfrac{7\beta}{8} \\[3mm] 0.4399\dfrac{\theta}{\beta} - 0.03501 \sin\left(2\pi\left(2\dfrac{\theta}{\beta} - 1\right)\right) + 0.5601 & \dfrac{7\beta}{8} < \theta < \beta \end{cases} \qquad (13.61)$$

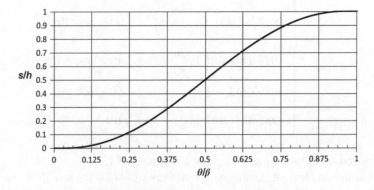

Figure 13.16   Modified sine function

Follower velocity is determined from Equation 13.57 where the velocity coefficient is

$$C_v = \begin{cases} 0.4399\left(1 - \cos\left(4\pi\dfrac{\theta}{\beta}\right)\right) & 0 < \theta < \dfrac{\beta}{8} \\[3mm] 0.4399 + 1.31967\sin\left(\dfrac{4\pi}{3}\dfrac{\theta}{\beta} - \dfrac{\pi}{6}\right) & \dfrac{\beta}{8} < \theta < \dfrac{7\beta}{8} \\[3mm] 0.4399\left(1 - \cos\left(4\pi\dfrac{\theta}{\beta} - 2\pi\right)\right) & \dfrac{7\beta}{8} < \theta < \beta \end{cases} \qquad (13.62)$$

Acceleration is determined from Equation 13.59 using the following acceleration coefficient:

$$C_a = \begin{cases} 5.528\sin\left(4\pi\dfrac{\theta}{\beta}\right) & 0 < \theta < \dfrac{\beta}{8} \\[3mm] 5.528\cos\left(\dfrac{4\pi}{3}\dfrac{\theta}{\beta} - \dfrac{\pi}{6}\right) & \dfrac{\beta}{8} < \theta < \dfrac{7\beta}{8} \\[3mm] 5.528\sin\left(4\pi\dfrac{\theta}{\beta} - 2\pi\right) & \dfrac{7\beta}{8} < \theta < \beta \end{cases} \qquad (13.63)$$

The velocity and acceleration coefficients for the modified sine function are shown in Figure 13.17.

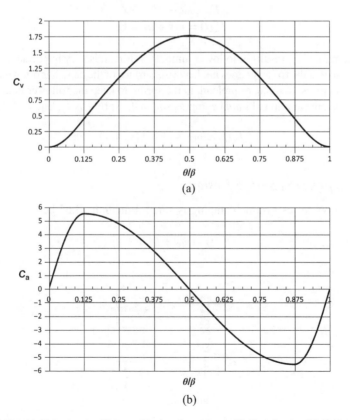

Figure 13.17   (a) Velocity coefficient. (b) Acceleration coefficient for modified sine function

13.5.3 Higher Order Polynomial Functions

Chapter 9 introduced two common polynomial function types: the 3-4-5 polynomial and the 4-5-6-7 polynomial. Higher order polynomial functions can be developed to fit specific requirements. Methods of curve fitting can be used to adjust displacement curves to match specific operating requirements. Locations of maximum velocity, acceleration, or jerk can be shifted. Though higher order polynomials can provide very precise desired motion, the solution becomes more complex. Also, the higher order polynomials will have very slow initial and final displacements. The very small changes require very accurate manufacturing. Rothbart (1956) provides great details on higher order polynomial functions and curve adjusting methods.

13.6 Forces Acting on the Follower

13.6.1 Basic Concepts

Once the overall cam geometry has been determined, the next step is to consider forces acting on the system. Forces need to be determined at peak acceleration points. Force analysis is critical for sizing the spring and determining contact stress values. Consider the basic cam follower system shown in Figure 13.18a. The compression spring keeps the follower in contact with the cam.

Figure 13.18b shows the free body diagram of the lumped mass model. Using summation of forces,

$$F_c - kx - c\dot{x} = m\ddot{x}$$
$$F_c(t) = m\ddot{x} + c\dot{x} + kx \tag{13.64}$$

A negative cam force F_c results in follower jump; therefore, it is very important to keep the cam force positive so that the follower does not leave the surface of the cam. It is often required to have some initial compressive force, or preload, in the spring to keep the follower in contact with the cam. If we define the amount of initial compression in the spring as x_p, Equation 13.64 becomes

$$F_c(t) = m\ddot{x} + c\dot{x} + k(x_p + x) \tag{13.65}$$

13.6.2 Compression Spring Design

For simplification, let us assume damping is small. Equation 13.65 becomes

$$F_c(t) = m\ddot{x} + k(x_p + x) \tag{13.66}$$

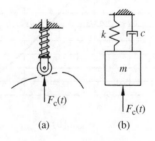

(a)   (b)

Figure 13.18   (a) Cam follower system. (b) Lumped mass model

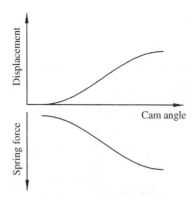

Figure 13.19 Cam displacement and spring force

which contains an inertial force $m\ddot{x}$ and a spring force $k(x_p + x)$. Cam mechanisms have moving parts in contact, which will result in frictional forces opposing motion. This section will briefly discuss the different forces involved in cam mechanisms and the concepts of designing the compression spring.

Forced closed cam mechanisms require a compression spring to maintain contact between the follower and the cam. If the spring force is too low, the follower will lose contact with the cam, which is known as follower jump. Follower jump needs to be avoided for smooth operation of the cam mechanism. Consider the general rise function shown in Figure 13.19. Assuming a linear spring with spring constant k, the spring force will be proportional to displacement as shown. The spring force at zero displacement will be the preload.

From Equation 13.66, we can see that the spring force required to maintain a positive cam force (negative force will indicate follower jump) is such that the following is satisfied:

$$m\ddot{x} + k(x_p + x) > 0 \tag{13.67}$$

Therefore, we must look at the acceleration plot to design the spring. Consider the general acceleration plot shown in Figure 13.20. Multiplying by the mass will give the inertial force. The phantom line shown in Figure 13.20 is the spring force. Notice that the spring force needs

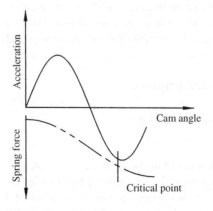

Figure 13.20 Spring design critical point

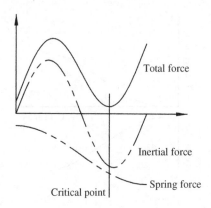

Figure 13.21   Total force

to be designed based on the critical point. The critical point will be the location when the difference between inertial force and spring force is minimum, resulting in the smallest contact force.

Figure 13.21 is the combination of the inertial force plot and the restoring spring force plot. For Equation 13.67 to be satisfied, the plot of the total force in Figure 13.21 must be completely positive. Other forces, such as frictional forces and external loads, can be included in the process illustrated in Figure 13.21.

13.7   Computer Applications of Cam Response

System response to standard cam functions is very suited for computer work. Many software packages, such as MathCAD, will solve Laplace transforms making the solution process very simple and fast. Examples on using the Laplace transform functions in MathCAD to solve system response were presented in Section 13.3. One major advantage of using software is the ability to examine the effects of system parameters on the response. Simple trial and error can be used to get an initial idea of required spring stiffness to get optimal response.

The numerical method discussed in Section 13.4 is easily automated with the use of spreadsheets or MATLAB®. The calculations presented in Tables 13.3 and 13.4 can be automatically performed in spreadsheets. System response can then easily be plotted.

Some software packages have built-in cam analysis options. Autodesk Inventor, for example, has cam design options that will aid in the analysis of cams. Users can select the desired function and size parameters, and the software will display kinematic plots for the complete cycle along with force plots. Once the design is complete, the software will generate the cam profile.

Section 2: Multicylinder Engines

13.8   Internal Combustion Engines

13.8.1   Introduction

Before we examine the dynamics of multicylinder engines, we will explore the basic principles of internal combustion engines and some general terminology. Figure 13.22 shows a cross-sectional view of a cylinder in an engine. The valves at the top of the bore allow for intake and exhaust, which will both be explained in the following paragraph.

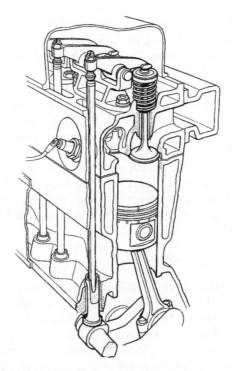

Figure 13.22   Engine cross-sectional view showing piston and valves

Figure 13.23a shows some important terms for the reciprocating piston. The extreme positions of the piston are called top dead center (TDC) and bottom dead center (BDC). The distance between TDC and BDC is the stroke.

This section will discuss two types of engine cycles: the four-stroke engine and the two-stroke engine. Figure 13.23b illustrates a typical four-stroke engine cycle. The four-stroke

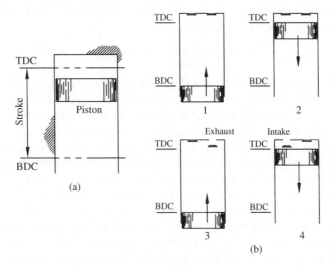

Figure 13.23   (a) Piston stroke and terminology (b) Four-stroke combustion cycle

combustion cycle consists of the compression stroke, combustion stroke, exhaust stroke, and the intake stroke. Step 1, which is the compression stroke, begins with both valves closed and a gas–air mixture contained in the cylinder. The piston moves up from BDC to TDC to compress the mixture. The combustion stroke, which is shown as step 2, occurs when the spark plug fires to force the piston down. The combustion stroke is also called the power stroke or the expansion stroke. As the piston moves back up, shown in step 3, the gases are exhausted out the open exhaust valve. During the final step, step 4, the piston returns to BDC while the intake valve is open to allow a new air–gas mixture to enter the cylinder. The motion continues smoothly with the aid of a flywheel and the power stroke of piston in another cylinder.

A two-stroke engine completes the process in only two steps, which are the power stroke and compression stroke. Two-stroke engines have fewer moving parts compared with a four-stroke engine, and are therefore lighter. Due to their lightweight and high power-to-weight ratio, they are commonly used in handheld equipment, such as chain saws, and small machines, such as scooters and lawn mowers.

13.8.2 Engine Force Analysis

Chapter 11 focused on dynamic force analysis of linkage mechanisms, which included force analysis of slider–crank mechanisms. This section will focus on the application of engine force analysis and will use the illustrative example for the slider–crank mechanism shown in Figure 13.24. Mabie (1987) provides a complete analysis of this system. The mechanism is for a single-cylinder four-stroke internal combustion engine. As indicated, the crankshaft is rotating at a constant rate of 3000 rpm. For this illustrative example, we wish to determine the loads on the mechanism when the crank is at an angle of 60°. The required forces are the forces transmitted through the wrist pin bearing, the crank pin bearing, the main bearings, and the required crankshaft torque.

We must first determine the force on the piston due to the gas pressure. In the upper right portion of Figure 13.24 is a plot of gas pressure versus crank angle, which would be determined from a thermodynamic analysis or from experimental measurements. At the required crank angle of 60°, the system is in the expansion phase. The dashed line represents the crank angle of 60° and shows that the gas pressure is 200 psig. To determine the force exerted on the piston, we must multiply the pressure by the piston area, which is listed in Figure 13.24 as 7.05 square inches. The force acting on the piston is

$$P = p(A_p) = 200 \frac{\text{lb}}{\text{in.}^2} (7.05 \text{ in.}^2) = 1410 \text{ lb} \tag{13.68}$$

In addition to the force due to the gas pressure, there will also be an inertial force acting on the piston. Figure 13.24 shows the acceleration polygon and lists the necessary acceleration values. Note that the counterweight on the crankshaft will balance the crank, which will cause A_{g_2} to be zero. The inertial forces are all calculated in Figure 13.24.

The force analysis presented here will be done graphically using superposition (see Section 11.2 for details on the superposition method), although an analytical solution procedure would also be valid. Figure 13.25 shows the system with the forces $\vec{P}$ and $\vec{F}_{O_4}$ acting on the piston. In the figure, the force $\vec{F}_4$ is the resultant force. The direction of the connecting rod force $\vec{F}'_{34}$ is known due to the fact that the connecting rod will be a two-force member. Assuming a frictionless case, there will be a horizontal force on the piston from the wall. All three forces are shown concurrent at B in the second figure, and a force polygon can be created to determine the force magnitudes. The two parallel but noncollinear forces acting on the crank will create a moment couple, which must be offset by crankshaft torque. All force magnitudes, as

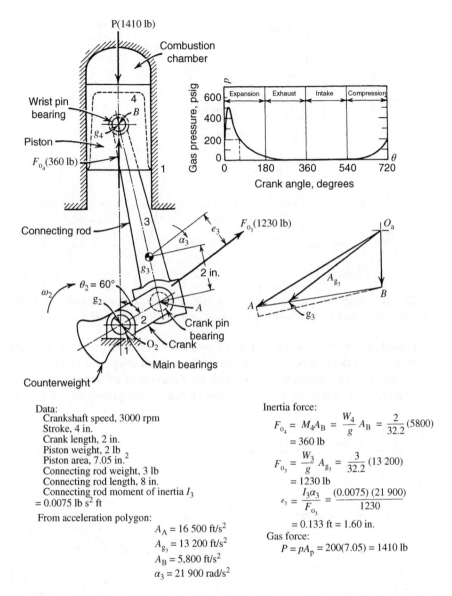

P(1410 lb)

Combustion chamber

Wrist pin bearing

Piston

F_{o_4}(360 lb)

Connecting rod

$\theta_2 = 60°$

ω_2

Crank pin bearing

Crank

Main bearings

Counterweight

F_{o_3}(1230 lb)

Gas pressure, psig

600 | Expansion | Exhaust | Intake | Compression
400
200
0
0   180   360   540   720
Crank angle, degrees

Figure 13.24 Force analysis example. Reproduced from Mabie and Reinholtz, Mechanisms and Dynamics of Machinery, 4th edition, John Wiley & Sons, © 1987

Data:
Crankshaft speed, 3000 rpm
Stroke, 4 in.
Crank length, 2 in.
Piston weight, 2 lb
Piston area, 7.05 in.^{2}
Connecting rod weight, 3 lb
Connecting rod length, 8 in.
Connecting rod moment of inertia I_3
= 0.0075 lb s^2 ft

From acceleration polygon:
A_A = 16 500 ft/s^2
A_{g_3} = 13 200 ft/s^2
A_B = 5,800 ft/s^2
α_3 = 21 900 rad/s^2

Inertia force:
$$F_{o_4} = M_4 A_B = \frac{W_4}{g} A_B = \frac{2}{32.2}(5800)$$
$$= 360 \text{ lb}$$
$$F_{o_3} = \frac{W_3}{g} A_{g_3} = \frac{3}{32.2}(13\ 200)$$
$$= 1230 \text{ lb}$$
$$e_3 = \frac{I_3 \alpha_3}{F_{o_3}} = \frac{(0.0075)(21\ 900)}{1230}$$
$$= 0.133 \text{ ft} = 1.60 \text{ in.}$$
Gas force:
$$P = pA_p = 200(7.05) = 1410 \text{ lb}$$

determined graphically from the force polygon, are given in Figure 13.25. The crank torque is calculated from the moment couple, and the final value is also given in Figure 13.25.

Next, as shown in Figure 13.26, we will perform a force analysis based on the equivalent offset force $\vec{F}_{o_3}$ (see Section 11.2.2 for details on determining the equivalent offset force and offset distance). Calculations for the equivalent offset force and its offset distance are provided in Figure 13.24. The direction of the force vector $\vec{F}_{o_3}$ was determined from the acceleration analysis shown in Figure 13.24. The direction of the piston force $\vec{F}''_{43}$ is horizontal because it

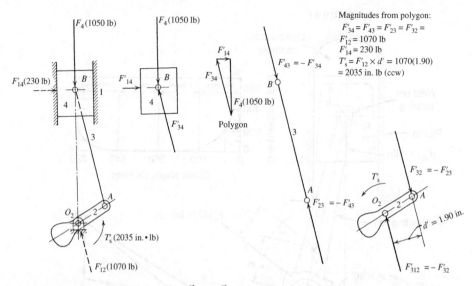

Figure 13.25 Force analysis for forces $\vec{P}$ and $\vec{F}_{O_4}$ acting on the piston. Reproduced from Mabie and Reinholtz, Mechanisms and Dynamics of Machinery, 4th edition, John Wiley & Sons, © 1987

must act perpendicular to the cylinder wall. The coupler is a three-force member; therefore, all three forces must pass through point k as shown on the isolated coupler in Figure 13.26. The crank force must also pass through point k. With the direction of all forces known, a force polygon is constructed, and force magnitudes can be determined graphically. The resulting

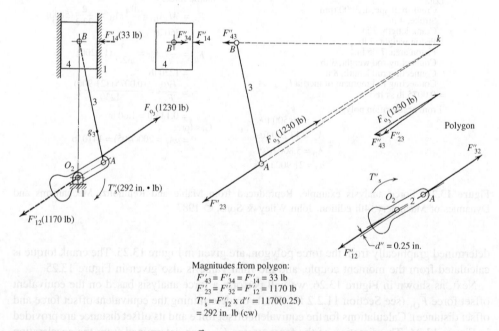

Figure 13.26 Force analysis for force $\vec{F}_{O_3}$. Reproduced from Mabie and Reinholtz, Mechanisms and Dynamics of Machinery, 4th edition, John Wiley & Sons, © 1987

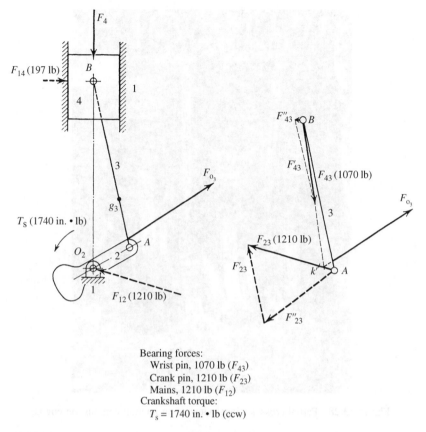

Bearing forces:
  Wrist pin, 1070 lb (F_{43})
  Crank pin, 1210 lb (F_{23})
  Mains, 1210 lb (F_{12})
Crankshaft torque:
  $T_s = 1740$ in. • lb (ccw)

Figure 13.27 Superposition of forces. Reproduced from Mabie and Reinholtz, Mechanisms and Dynamics of Machinery, 4th edition, John Wiley & Sons, © 1987

force magnitudes are given in Figure 13.26. The crank torque is calculated by isolating the crank and placing the crank force on the crank pin. The main bearing will have the same force in the opposite direction. The two forces will create a moment couple, which must be counteracted by the crank torque. Calculations for the crank torque are shown in Figure 13.26.

The last step, which is illustrated in Figure 13.27, is the superposition of the previous results. Forces obtained from Figures 13.25 and 13.26 are added together to get the final force and torque values.

13.9 Common Arrangements of Multicylinder Engines

13.9.1 Introduction

Balancing methods for single slider–crank mechanisms were presented in Chapter 12. Counterbalancing weight was added to the crank to offset shaking forces. It was seen that complete balancing was not possible by the addition of the counterbalance due to the introduction of new vertical shaking forces. Also recall that the secondary shaking forces were not balanced because the secondary shaking force is twice the frequency of the primary. Therefore, the approach was a partial balance. Lanchester balancing was introduced in Section 12.5.5 and it

Figure 13.28 Partial cross-sectional view of an internal combustion engine

offers a more complete balancing solution and allows for balancing primary and secondary shaking forces. The method required the addition of balancing gears, which requires space and additional expense to balance the mechanism.

Another approach, which is often more convenient and cost-effective, to balance slider–crank mechanisms is to utilize pairs of identical sliders. A very common example would be multicylinder engines where the goal is to have a complete balance of the total engine. Figure 13.28 shows a cross-sectional view of a multicylinder engine. Because the individual slider cranks are each identical, strategic arrangement of each pair of slider–crank mechanisms will cause inertial forces (or shaking forces) to essentially cancel one another, leaving a balanced machine. Different arrangements of multicylinder machines exist, and this section will focus on some common arrangements. The arrangements will be shown as examples of automotive engine arrangements, although other applications, such as compressors, exist. Though the purpose of using pairs of slider–cranks is balancing, another advantage is the increased power output of the machine.

Each cylinder will operate in phase relationships, and the phase relationship will depend on the number of cylinders. A crank phase diagram is typically used to illustrate the phase angles. The phase relationships can be chosen such that the shaking forces completely balance. Cylinder firing order is also important to consider for engine balancing. Changing the firing order while keeping all other parameters constant will cause changes in the dynamics of the engine.

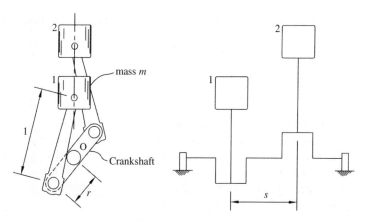

Figure 13.29   In-line engine arrangement with two cylinders

13.9.2 In-Line Engines

The first arrangement we will consider is the in-line engine, which is also called a straight engine. All cylinders will lie in one line, which allows for easier fabrication of the engine. Figure 13.29 shows the general arrangement of an in-line engine with two cylinders. The crank rotates about point O and cylinders 1 and 2 operate 180° out of phase. The cylinders are numbered sequentially from one end of the crankshaft to the other. Each cylinder will have the same mass m, connecting rod length l, and crank length r. Each cylinder will drive the common crankshaft.

Each cylinder will have the shaking force determined in Chapter 12 and given in Equation 12.25. The equation is repeated here using the crank and connecting rod lengths r and l along with a mass m for the piston:

$$F = mr\omega^2 \cos{(\theta)} + m\omega^2 \frac{r^2}{l} \cos{(2\theta)} \tag{13.69}$$

As defined in Section 12.5.2, the first term on the right-hand side is the primary force and the second term is the secondary force. In a multicylinder engine, the resultant shaking force is the summation of all inertia forces.

$$F = \sum_{n=1}^{N} \left(F_{\text{p}} + F_{\text{s}} \right) \tag{13.70}$$

In Equation 13.70, N is the number of cylinders, F_{p} is each individual cylinder's primary force, and F_{s} is each individual cylinder's secondary force. The vibration characteristics of the engine will differ based on the number of cylinders. In some cases, the primary forces may be balanced where the secondary forces are not balanced. Other cases may have balanced secondary forces and unbalanced primary forces.

To fully understand the forces, we will look at the summation of primary forces and the summation of secondary forces separately. Let us first consider the summation of primary forces. Because all cylinders have the same sliding mass, crank length, and connecting rod length, these terms can be pulled outside of the summation. The final summation for primary forces would be

$$\sum_{n=1}^{N} F_p = mr\omega^2 \sum_{n=1}^{N} \cos(\theta_n) \tag{13.71}$$

The angular location of each individual cylinder θ_n can be expressed based on the angular relationships. The angle θ_1 is the clockwise angular position for cylinder 1, but all remaining cylinder locations will be known based on the desired phase angles. The angular position for cylinder n can be defined by ψ_n, which is the angular location measured clockwise with respect to cylinder 1. Making the substitution of $\theta_n = \theta_1 + \psi_n$, Equation 13.71 becomes

$$\sum_{n=1}^{N} F_p = mr\omega^2 \sum_{n=1}^{N} \cos(\theta_1 + \psi_n)$$

$$\sum_{n=1}^{N} F_p = mr\omega^2 \sum_{n=1}^{N} \left[\cos(\theta_1)\cos(\psi_n) - \sin(\theta_1)\sin(\psi_n)\right] \tag{13.72}$$

Noting that $\cos(\theta_1)$ and $\sin(\theta_1)$ will be constant, we can pull these terms outside the summation to give

$$\sum_{n=1}^{N} F_p = mr\omega^2 \cos(\theta_1) \sum_{n=1}^{N} \cos(\psi_n) - mr\omega^2 \sin(\theta_1) \sum_{n=1}^{N} \sin(\psi_n) \tag{13.73}$$

A similar process can be used for the summation of secondary forces to give

$$\sum_{n=1}^{N} F_s = m\omega^2 \frac{r^2}{l} \cos(2\theta_1) \sum_{n=1}^{N} \cos(2\psi_n) - m\omega^2 \frac{r^2}{l} \sin(2\theta_1) \sum_{n=1}^{N} \sin(2\psi_n) \tag{13.74}$$

It can be seen from Equation 13.73 that both $\sum \cos\psi$ and $\sum \sin\psi$ must be zero to balance primary forces. Similarly, from Equation 13.74, both $\sum \cos 2\psi$ and $\sum \sin 2\psi$ must be zero to balance secondary forces. For complete balance, all four summations must equal zero. Because the cylinders are not located in the same plane in the z-axis, we must also consider the effects of shaking moments caused by each cylinder. The concept of balancing the shaking moments is much like dynamic balancing of rotating masses compared with static balancing of rotating masses.

13.9.3 Opposed Engines

Next, we will examine an opposed engine arrangement. Figure 13.30 shows an example of a two-cylinder opposed engine arrangement. The cylinders are on a common line, as with the in-line arrangement, but now the cylinders are on opposite sides of the crank axis. Opposed engines tend to be better balanced compared with in-line engines.

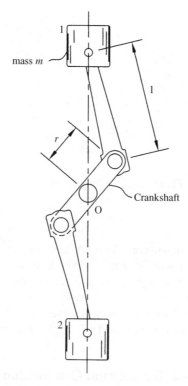

Figure 13.30   Opposed engine arrangement with two cylinders

If the arrangement consists of the cylinders offset a distance d apart, as shown in Figure 13.31a, the net shaking force will be zero but a shaking couple will exist. An arrangement using a double connecting rod, as shown in Figure 13.31b, will eliminate the shaking couple.

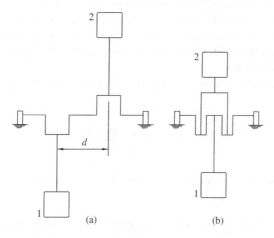

Figure 13.31   (a) Opposed engine with two cylinders at a distance d. (b) Opposed engine with two cylinders on double connecting rod

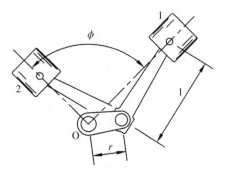

Figure 13.32 V engine arrangement

13.9.4 V Engines

The last engine arrangement we will consider is a V arrangement. Figure 13.32 shows a general arrangement of a V engine. Viewing the engine along the crankshaft will show that the pistons form a V shape, where the actual angle ϕ between pistons will differ for different engines. In fact, the opposed engine is simply a case of a V engine with an angle of 180° between pistons. The V engine arrangement is compact allowing for many common applications. The calculations for determining balanced behavior of the V engine are more complex and beyond the scope of this text.

13.10 Flywheel Analysis for Internal Combustion Engines

Flywheels are commonly used in internal combustion engines to smooth torque. Consider the single-cylinder engine illustrated in Figure 13.33a. The system is different from most of the slider–crank systems discussed in previous sections. It has been generally treated that the crank was the input and the slider was the output. The system for an internal combustion engine is not driven by the crank, but it is driven by forces on the piston due to gas pressure. Therefore, the crank becomes the output link for the system.

Using the concepts discussed for lumped mass approximation, we know that the total reciprocating mass is equal to the mass of the piston plus the connecting rod lumped mass at point B. The gas pressure will produce a force equal to the pressure times the cross-sectional area of the piston. This force will be delivered to the crank through the massless connecting rod to produce a turning moment.

From the free body diagram of the reciprocating mass shown in Figure 13.33b,

$$F_{gas} + m_{rec}\ddot{x} = F_{cr} \cos \phi$$

$$F_{cr} = \frac{F_{gas} + m_{rec}\ddot{x}}{\cos \phi} \tag{13.75}$$

Using the approximation of the acceleration of the reciprocating mass determined in Equation 11.40,

$$F_{cr} = \frac{F_{gas} - m_{rec}(O_2A)\,\omega_2^2\left[\cos\left(\theta\right) + (O_2A/AB)\cos\left(2\theta\right)\right]}{\cos \phi} \tag{13.76}$$

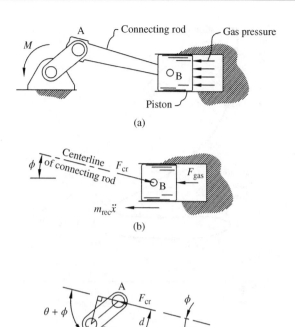

Figure 13.33 (a) Single-cylinder engine. (b) Free body diagram for piston. (c) Free body diagram for the crank

The connecting rod force will be transmitted to the crank as shown in Figure 13.33c. The perpendicular distance d is equal to

$$d = O_2A \sin(\theta + \phi) \tag{13.77}$$

The turning moment is equal to

$$M = F_{cr}d$$

$$M = F_{gas} - m_{rec}(O_2A)\,\omega_2^2 \left[\cos(\theta) + \frac{O_2A}{AB}\cos(2\theta)\right] \frac{O_2A\,\sin(\theta + \phi)}{\cos\phi} \tag{13.78}$$

The gas force will vary with θ because the pressure will not be constant. The change in pressure was illustrated in Figure 13.2, which showed the change in gas pressure versus crank angle. Figure 13.34 shows how the pressure changes during one complete cycle of the four-stroke engine versus volume.

Also, the angle ϕ will vary with θ and is determined by the geometry of the linkage mechanism.

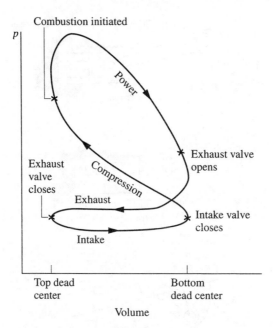

Figure 13.34 Pressure–volume diagram for a four-stroke combustion engine. Reproduced from Moran and Shapiro, Fundamentals of Engineering Thermodynamics, 6th edition, John Wiley & Sons, © 2007

13.11 Review and Summary

This chapter focused on two primary topic areas that relate to machine dynamics. The first section covered the general concepts of cam dynamics. Methods for determining cam response were presented. Also, advanced cam functions were introduced to allow for improved performance. Internal combustion engines were introduced as the second application. Coverage included engine forces as well as common multicylinder arrangements for engine balancing. The content in this chapter only served as an introduction. Texts included in References and Further Reading sections should be consulted for more detailed analysis.

Problems

P13.1   Derive the general differential equation of motion for a cycloidal cam function for a rise from $s_1 = 0$ to $s_2 = h$ for a cam rotation from $\theta_1 = 0$ to $\theta_2 = \lambda$.

P13.2   Derive the general differential equation of motion for a 3-4-5 polynomial cam function for a rise from $s_1 = 0$ to $s_2 = h$ for a cam rotation from $\theta_1 = 0$ to $\theta_2 = \lambda$.

P13.3   Derive the general differential equation of motion for a 4-5-6-7 polynomial cam function for a rise from $s_1 = 0$ to $s_2 = h$ for a cam rotation from $\theta_1 = 0$ to $\theta_2 = \lambda$.

P13.4   Consider a double-dwell cam that will cause a rise of 1.4 inches in 40° of cam rotation using a 3-4-5 polynomial cam function. The cam rotates at a constant speed of 50 rpm. Develop a computer code to analyze the system response during the rise and dwell periods using Laplace transforms. Examine the effects of system parameters (mass, spring stiffness, and damping ratio) on the response and describe your conclusions.

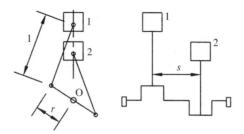

Figure 13.35

P13.5 Solve Example Problem 10.14 using Laplace transforms in MathCAD.

P13.6 Solve Problem P10.22 using Laplace transforms in MathCAD.

P13.7 Perform a force analysis and spring design for a double-dwell cam mechanism. The cam will rise to 1.5 inches in a cam rotation of 100° using a modified sine function. The follower will remain in a high dwell for 40° and then return to 0 in 100° using a modified sine function. The follower will be in a low dwell for the remainder of the cycle. The cam rotation speed is 1300 rpm.

P13.8 Determine the equations of unbalance for the two-cylinder in-line arrangement shown in Figure 13.35.

References

Mabie, H.H. and Reinholtz, C.F. (1987) *Mechanisms and Dynamics of Machinery*, John Wiley & Sons, Inc., New York.

Rothbart, H.A. (1956) *Cams: Design, Dynamics, and Accuracy*, John Wiley & Sons, Inc., New York.

Timoshenko, S. and Young, D.H. (1948) *Advanced Dynamics*, McGraw-Hill, New York.

Further Reading

Wilson, C.E. and Sadler, J.P. (1993) *Kinematics and Dynamics of Machinery*, Harper Collins, New York.

Appendix A

Center of Mass

If a body is balanced at its center of mass, or centroid, the net torque will be zero and the body will remain at rest. The centroid of a rigid body is the unique fixed point representing the mean location of the weight of the body. For structural members in bending, it is necessary to find the section's neutral axis (NA). The neutral axis is a line perpendicular to the applied load and represents the location of zero stress. It is important to note that the neutral axis is not necessarily located at the section's centroid.

Calculating the centroid can be thought of as a weighted average in which each element is weighted by its area (or mass). Consider n elements each with area A_i and centroid location (x_i, y_i). The centroid of all elements would be determined from the weighted average.

$$\bar{x} = \frac{\sum_{i=1}^{n} A_i x_i}{\sum_{i=1}^{n} A_i} \qquad \bar{y} = \frac{\sum_{i=1}^{n} A_i y_i}{\sum_{i=1}^{n} A_i} \tag{A.1}$$

To extend the concept to finding centroids of complex shapes, consider the area bound by a function as shown in Figure A.1. We can take a small rectangular differential area of height y and width dx. The differential element has an area of $y dx$ and a centroid located at $\bar{x}_{el} = x$ and $\bar{y}_{el} = y/2$.

The summation process is now replaced by integration, and the centroid of the total area can be determined from

$$\bar{x} = \frac{\int \bar{x}_{el} dA}{A} \qquad \bar{y} = \frac{\int \bar{y}_{el} dA}{A} \tag{A.2}$$

Example Problem A.1

Determine the centroid of the area bound by the X-axis and the function $y(x) = 5x^3$ in the region $2 < x < 4$, as shown in Figure A.2.

Machine Analysis with Computer Applications for Mechanical Engineers, First Edition. James Doane.
© 2016 John Wiley & Sons, Ltd. Published 2016 by John Wiley & Sons, Ltd.
Companion Website: www.wiley.com/go/doane0215

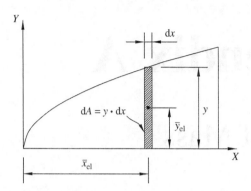

Figure A.1 Centroid of an area by integration

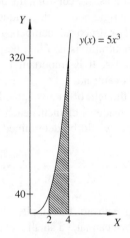

Figure A.2

Solution: First we need to have the total area under the curve in the region of interest. The area of the differential element shown in Figure A.3 is

$$dA = y\,dx = 5x^3\,dx$$

The total area is found by integrating the element area over the range of x-values:

$$A = \int y\,dx = \int_2^4 5x^3\,dx = 5\left[\frac{x^4}{4}\right]_2^4 = 5\left(\frac{4^4 - 2^4}{4}\right) = 300$$

The centroid is then found using Equation A.2:

$$\bar{x} = \frac{\int \bar{x}_{el}\,dA}{A} = \frac{\int x\left(5x^3\,dx\right)}{300}$$

$$\bar{x} = \frac{\int_2^4 5x^4\,dx}{300} = \frac{5\left[x^5/5\right]_2^4}{300} = \frac{4^5 - 2^5}{300}$$

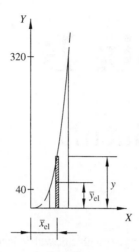

Figure A.3

Answer: $\bar{x} = 3.31$

$$\bar{y} = \frac{\int \bar{y}_{el}\,dA}{A} = \frac{\int (5x^3/2)\,(5x^3\,dx)}{300}$$

$$\bar{y} = \frac{\int_{2}^{4} (25/2)x^6\,dx}{300} = \frac{(25/2)[x^7/7]_2^4}{300} = \frac{25((4^7 - 2^7)/7)}{600}$$

Answer: $\bar{y} = 96.76$

Appendix B

Moments of Inertia

B.1 Basic Concepts

If we consider a mass m moving along a straight line with velocity v, the kinetic energy is given as

$$\text{KE} = \frac{1}{2}mv^2 \tag{B.1}$$

Next consider the same particle moving along a circular path. The velocity magnitude, now defined by $v = r\omega$, is substituted into Equation B.1 to give

$$\text{KE} = \frac{1}{2}\left(r^2 m\right)\omega^2 \tag{B.2}$$

The term in parenthesis is defined as the moment of inertia

$$I = r^2 m \tag{B.3}$$

giving the more familiar expression for rotational kinetic energy:

$$\text{KE} = \frac{1}{2}I\omega^2 \tag{B.4}$$

By simple comparison of Equations B.1 and B.4, it can be concluded that the moment of inertia for rotation is similar to mass in translation. Moment of inertia is a measure of rigidity or resistance to rotation and is determined from the section geometry.

For a rigid body, moment of inertia is determined by integration. Consider a disk rotating about its centroid, as shown in Figure B.1.

For a small element of mass m_i located at a distance r_i from the axis of rotation, the kinetic energy would be $\text{KE} = 1/2\left(r_i^2 m_i\right)\omega^2$. The total kinetic energy for the rotating disk would be determined by the summation of the kinetic energy of all the elements:

$$\text{KE} = \frac{1}{2}\omega^2 \sum_{i=1}^{n} r_i^2 m_i \tag{B.5}$$

Machine Analysis with Computer Applications for Mechanical Engineers, First Edition. James Doane.
© 2016 John Wiley & Sons, Ltd. Published 2016 by John Wiley & Sons, Ltd.
Companion Website: www.wiley.com/go/doane0215

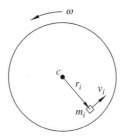

Figure B.1 Kinetic energy for rotating disk

The summation term is defined as the moment of inertia. For continuous systems, however, the summation must be replaced with an integral:

$$KE = \frac{1}{2}\omega^2 \int r^2 dm$$

$$I = \int r^2 dm$$

(B.6)

B.2 Rectangular Moment of Inertia by Integration

Consider the general shape shown in Figure B.2a and the small square area dA. Calculation of the moment of inertia of the general shape would require the use of a double integral.

Commonly thin rectangular elements can be used, as shown in Figure B.2b and c, in place of small square elements. Moments of inertia can then be calculated with respect to the X-axis or Y-axis using the following single integrations:

$$I_X = \int y^2 dA$$

$$I_Y = \int x^2 dA$$

(B.7)

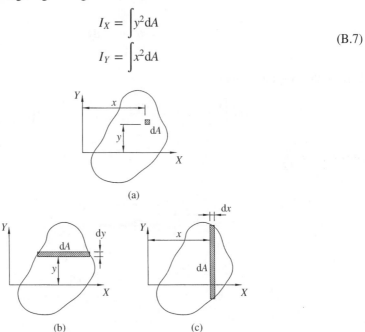

Figure B.2 Rectangular moments of inertia by integration

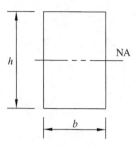

Figure B.3

Example Problem B.1

Calculate the moment of inertia, with respect to the neutral axis (NA), for the rectangle shown in Figure B.3.

Solution: Because the width is constant, the easiest method of solving this problem is to pick a differential area as a rectangular strip, as shown in Figure B.4. The differential area is then $dA = b\,dy$.

The moment of inertia is then calculated from Equation B.7:

$$I = \int y^2\, dA = \int_{-h/2}^{h/2} y^2 b\, dy$$

The width b is constant, and can be pulled out of the integral:

$$I = b \int_{-h/2}^{h/2} y^2\, dy$$

Solving the integration gives

$$I = b \left[\frac{y^3}{3}\right]_{-h/2}^{h/2} = b \left[\frac{(h/2)^3 - (-h/2)^3}{3}\right] = b\left(\frac{h^3/4}{3}\right)$$

Answer: $I = \dfrac{bh^3}{12}$

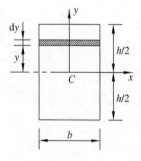

Figure B.4

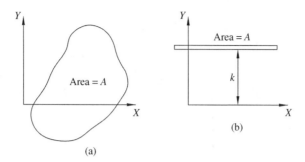

Figure B.5   Radius of gyration

B.3   Radius of Gyration

Consider the general shape shown in Figure B.5a with area A and moment of inertia I_X with respect to the X-axis. Assume that shape is flattened into a thin strip, as shown in Figure B.5b, so that the thin strip has the same area as the original shape. If that thin strip was located at a distance k from the X-axis, the moment of inertia would be the area times the distance squared:

$$I = k^2 A \tag{B.8}$$

If we desire the moment of inertia of the thin strip to be equal to the moment of inertia of the original shape, the distance k must be a specific value. The distance k for which the moment of inertia of the thin strip is equal to the moment of inertia of the original area is known as the *radius of gyration*:

$$k = \sqrt{\frac{I}{A}} \tag{B.9}$$

B.4   Parallel Axis Theorem

The moment of inertia depends on the shape of the body and the axis of rotation. Moments of inertia of common geometric shapes are available; however, the equations typically only give moment of inertia about the centroid of the shape. It is not uncommon to require the moment of inertia with respect to an axis not passing through the shape's centroid. Though the moment of inertia about the new axis could be determined by integration, the calculations can be greatly simplified by use of the parallel axis theorem.

Consider the thin circular disk shown in Figure B.6. Moment of inertia with respect to axis X–X, which passes through its centroid, is easily calculated. Now assume we require the moment of inertia about axis X'–X' (parallel to the original axis X–X) passing through point C'.

Calculating the new moment of inertia by integration would be complicated. However, if the new axis is located at a distance d from the original axis, we can calculate the new moment of inertia by using the parallel axis theorem:

$$I = I_C + Ad^2 \tag{B.10}$$

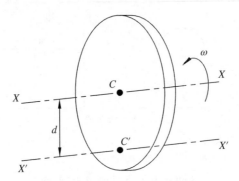

Figure B.6 Parallel axis theorem

Example Problem B.2

Using the parallel axis theorem calculate the moment of inertia, with respect to the base, for the rectangle in Example Problem B.1.

Solution: The moment of inertia with respect to the neutral axis was determined to be

$$I_C = \frac{bh^3}{12}$$

The moment of inertia about the base can be determined using the parallel axis theorem:

$$I = I_C + Ad^2$$

Referring to Figure B.7, the distance d will be equal to $h/2$.

$$I = \frac{bh^3}{12} + (bh)\left(\frac{h}{2}\right)^2$$

$$I = \frac{bh^3}{12} + \frac{bh^3}{4}$$

Answer: $I = \dfrac{bh^3}{3}$

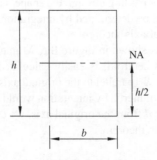

Figure B.7

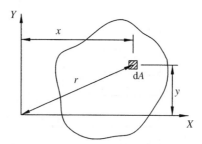

Figure B.8   Polar moment of inertia

B.5   Polar Moment of Inertia

Consider the area shown in Figure B.8. Moments of inertia can then be calculated with respect to the X-axis or Y-axis using the small differential area dA, as discussed in the section on rectangular moments of inertia by integration. Next, let us determine the moment of inertia of the area about the Z-axis, which is perpendicular to the plane of the area. The *polar moment of inertia* is a moment of inertia with respect to an axis perpendicular to the plane of the area.

The polar moment of inertia for the area shown in Figure B.8 about the Z-axis would be

$$J_Z = \int r^2 dA \qquad\qquad (B.11)$$

From Pythagorean theorem we know that $r^2 = x^2 + y^2$, which gives

$$J_Z = \int (x^2 + y^2) dA$$
$$J_Z = \int x^2 dA + \int y^2 dA \qquad\qquad (B.12)$$
$$J_Z = I_X + I_Y$$

Example Problem B.3

Calculate the moment of inertia, with respect to the X-axis, for the quarter circle shown in Figure B.9.

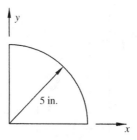

Figure B.9

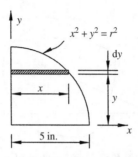

Figure B.10

Solution: Setting up the differential area as a rectangle as shown in Figure B.10, we need to use the equation of a circle to get an expression for x and dA:

$$x = \sqrt{r^2 - y^2}$$
$$dA = x\,dy = \sqrt{r^2 - y^2}\,dy$$

Now using Equation B.7

$$I = \int y^2 dA = \int_0^r y^2 \sqrt{r^2 - y^2}\,dy$$

which is a complex integral. An easier approach is to utilize polar moment of inertia. The differential area is now set up as an arc, as shown in Figure B.11.

$$dA = \frac{2\pi r}{4}\,dr = \frac{\pi r}{2}\,dr$$

$$J = \int r^2 dA = \frac{\pi}{2}\int_0^5 r^3\,dr = \frac{\pi}{2}\left[\frac{r^4}{4}\right]_0^5 = \frac{625\pi}{8}$$

We will now use $J = I_X + I_Y$. Noting that $I_X = I_Y = I$ due to symmetry,

$$J = 2I = \frac{625\pi}{8}$$

Answer: $I = \dfrac{625\pi}{16}$

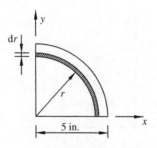

Figure B.11

B.6 Determining Moment of Inertia by Composite Areas

Previous examples involved calculating moment of inertia of relatively simple geometric shapes. Often, however, it is necessary to calculate moment of inertia of more complex shapes. Though integration could still be used, it is often very tedious. A more straightforward approach involves representing the complex shape as a composite of simple geometric shapes. The section is first broken into simple geometric shapes, each of which has a moment of inertia about its own centroid (see Figure B.12). Using the parallel axis theorem, the moments of inertia of each geometric shape can be transformed to the entire section's centroid (or other desired location) and then summed to give the moment of inertia of the entire section.

As an example, let us say we need to calculate the moment of inertia of the gusset plate shown in Figure B.13a. The complex geometry of the plate can be easily represented as simple geometric shapes, as shown in Figure B.13b. Moments of inertia for each individual simple geometric shape can be easily calculated. Calculating the moment of inertia of the initial plate, however, is not a simple summation of individual moments of inertia of the simple shapes.

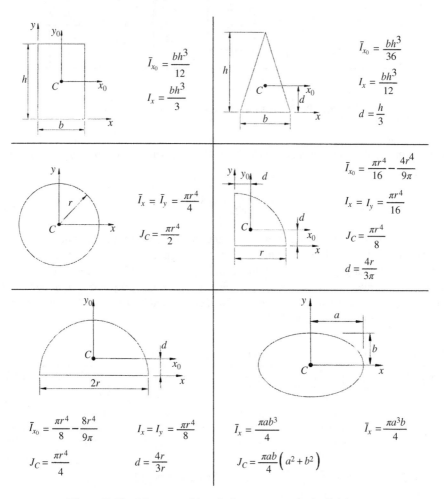

Figure B.12 Moments of inertia for common geometric shapes

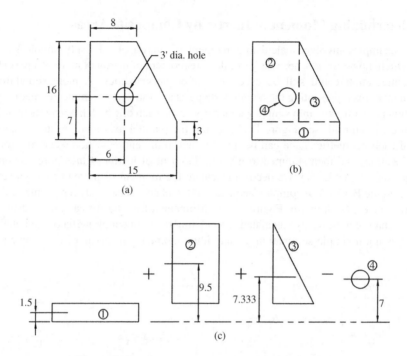

Figure B.13 Illustrative example for composite body

Individual moments of inertia are calculated, and then the parallel axis theorem is used for each simple shape giving

$$I_X = \sum \bar{I} + \sum Ad^2 \qquad\qquad (B.13)$$

Figure B.13c illustrates the process of calculating the total moment of inertia. The individual moments of inertia can all be calculated about their respective centroids using equations in Figure B.12. Summary of results for moments of inertia are given in Table B.1. The moment of inertia for the circle is negative because the circular shape is a hole and is removed from the final gusset plate. The product of area times distance squared is computed for each shape. The results are again summarized in Table B.1.

Table B.1 Calculation summary for illustrative example of composite body

Shape – No.	$\bar{I}$ (in.^{4})	A (in.^{2})	d (in.)	d^2 (in.^{2})	Ad^2 (in.^{4})
Rectangle – 1	33.750	45.000	1.500	2.250	101.250
Rectangle – 2	1464.667	104.000	9.500	90.250	9386.00
Triangle – 3	427.194	45.500	7.333	53.778	2446.889
Circle – 4	−63.617	−28.274	7.000	49.000	−1385.426
Total	1861.994				10 548.713

$$I_X = \sum \bar{I} + \sum Ad^2 = 1861.994 + 10\,548.713$$
$$I_X = 12\,410.7 \text{ in.}^4$$

Appendix C

Fourier Series

C.1 Frequency Content and Combination of Harmonic Functions

Fourier analysis is fundamentally the study of the effects of the summation of sine and cosine functions. Therefore, a prerequisite to the study of Fourier analysis is the understanding of concepts of linear combinations of harmonic functions. The summation of harmonic function will, in general, create a periodic function.

To first illustrate combinations of harmonic functions, let us consider the combination of sine waves. Figure C.1a shows the functions $2 \sin (8\pi t)$ and $3 \sin (2\pi t)$ plotted separately. The combination of the two functions can be seen in Figure C.1b. Note that the combination gives a periodic function but not a harmonic function.

Next let us consider a more general illustrative example consisting of a linear combination of both sine and cosine functions using the summation

$$f(t) = \sum_{k=1}^{2} A_k \sin (2\pi\omega_k t) + B_k \cos (2\pi\omega_k t)$$

and letting

$$A_1 = 1 \quad B_1 = 0.5 \quad \omega_1 = 3$$
$$A_2 = 1.5 \quad B_2 = 3 \quad \omega_2 = 1$$

The individual sine and cosine functions can be seen in Figure C.2a and b. The combination is shown in Figure C.2c. Once again the combination is a periodic function. Unlike the combination shown in Figure C.1b, due to the inclusion of cosine functions, the combination function shown in Figure C.2c is not equal to zero at time equal to zero.

From these two illustrative examples, you can begin to see that combination of harmonic functions allows for the development of very complex periodic functions. In these examples the coefficients $(A_k \ B_k \ \omega_k)$ were chosen somewhat arbitrarily to generate a periodic function. It will soon be seen that Fourier analysis is essentially the opposite procedure. In Fourier analysis, a known periodic function is given and the coefficients of the individual harmonic functions are determined from equations.

Machine Analysis with Computer Applications for Mechanical Engineers, First Edition. James Doane.
© 2016 John Wiley & Sons, Ltd. Published 2016 by John Wiley & Sons, Ltd.
Companion Website: www.wiley.com/go/doane0215

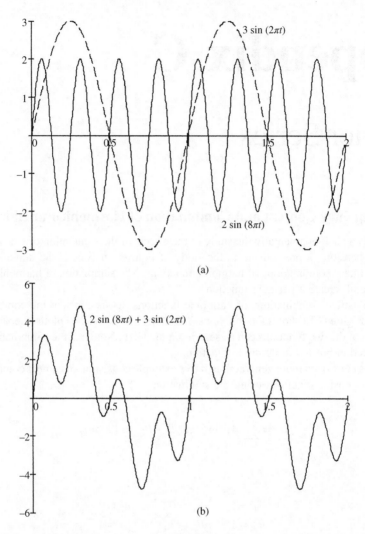

Figure C.1 Linear combination of sine functions. (a) Two sine functions plotted individually. (b) Combination of the two functions

C.2 Periodic Functions

C.2.1 Introduction

If a given periodic function $f(x)$ has a period of 2π, then the summation of sine and cosine functions is

$$f(x) = \frac{a_0}{2} + \sum_{n=1}^{\infty} (a_n \cos nx + b_n \sin nx) \qquad (C.1)$$

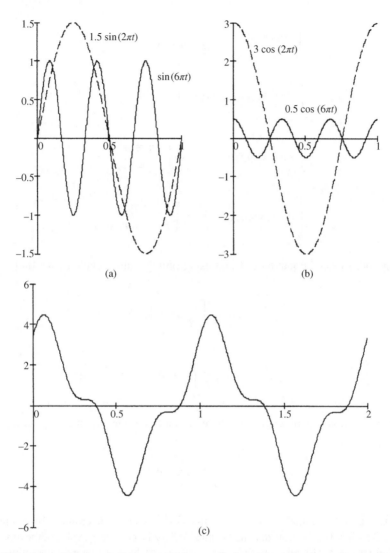

Figure C.2 General linear combination of harmonic functions. (a) Two sine functions plotted individually. (b) Two cosine functions plotted individually. (c) Combination of the four harmonic functions

Because the sine and cosine terms within the series all have a period of 2π, the summation will yield a function that also has a period of 2π.

If the function $f(x)$ is known, we can solve for the constants a_0, a_n, and b_n. The values can be found by using the orthogonality relations:

$$\int_{-\pi}^{\pi} \sin(nx)\mathrm{d}x = 0$$

$$\int_{-\pi}^{\pi} \cos (nx) dx = \begin{cases} 0 & n \neq 0 \\ 2\pi & n = 0 \end{cases}$$

$$\int_{-\pi}^{\pi} \sin (nx) \cos (mx) dx = 0$$

$$\int_{-\pi}^{\pi} \sin (nx) \sin (mx) dx = \begin{cases} 0 & n \neq m \\ \pi & n = m \end{cases}$$

$$\int_{-\pi}^{\pi} \cos (nx) \cos (mx) dx = \begin{cases} 0 & n \neq m \\ \pi & n = m \neq 0 \end{cases}$$

Integrating both sides of Equation C.1 and using orthogonality relations, we find

$$\int_{-\pi}^{\pi} f(x) dx = \int_{-\pi}^{\pi} \frac{a_0}{2} dx = \pi(a_0)$$

$$a_0 = \frac{1}{\pi} \int_{-\pi}^{\pi} f(x) dx \tag{C.2}$$

Multiplying Equation C.1 by $\cos (mx)$, where m is a fixed integer, and integrating, we find

$$a_n = \frac{1}{\pi} \int_{-\pi}^{\pi} f(x) \cos nx dx \tag{C.3}$$

Equation C.2 is not essential because the value for a_0 can be determined by simply using Equation C.3 for $n = 0$. Also note that the magnitude $a_0/2$ is the average value of the function $f(x)$ over the full period 2π. Finally, multiplying Equation C.1 by $\sin(mx)$ and integrating, we find

$$b_n = \frac{1}{\pi} \int_{-\pi}^{\pi} f(x) \sin nx \, dx \tag{C.4}$$

C.2.2 Arbitrary Period

The previous equations were developed for a special case when the period was equal to 2π. The Fourier series can be broadened to include functions of any period. For a periodic function $f(x)$ having period $2L$, the Fourier series is defined as

$$f(x) = \frac{a_0}{2} + \sum_{n=1}^{\infty} \left(a_n \cos \frac{n\pi x}{L} + b_n \sin \frac{n\pi x}{L} \right) \tag{C.5}$$

where the coefficients are

$$a_0 = \frac{1}{L} \int_{-L}^{L} f(x) dx \tag{C.6}$$

$$a_n = \frac{1}{L} \int_{-L}^{L} f(x) \cos \frac{n\pi x}{L} dx \tag{C.7}$$

$$b_n = \frac{1}{L} \int_{-L}^{L} f(x) \sin \frac{n\pi x}{L} dx \tag{C.8}$$

C.2.3 Even and Odd Functions

Sine and cosine functions have symmetry properties that lead to simplifications in determining coefficients in a Fourier series. A cosine function is symmetrical about the vertical axis, where a sine function is antisymmetrical. If $f(x)$ is an even periodic function, the coefficients are determined from

$$a_n = \frac{1}{L} \int_{-L}^{L} f(x) \cos \frac{n\pi x}{L} dx \qquad n = 0, 1, 2, 3, \ldots \tag{C.9a}$$

$$b_n = 0 \qquad n = 1, 2, 3, \ldots \tag{C.9b}$$

and if $f(x)$ is an even periodic function, the coefficients are determined from

$$a_n = 0 \qquad n = 0, 1, 2, 3, \ldots \tag{C.10a}$$

$$b_n = \frac{1}{L} \int_{-L}^{L} f(x) \sin \frac{n\pi x}{L} dx \qquad n = 1, 2, 3, \ldots \tag{C.10b}$$

As an illustrative example, consider the function

$$f(x) = \begin{cases} 1.5x & 0 < x < 1 \\ -1.5x + 3 & 1 < x < 2 \end{cases}$$

which is plotted in Figure C.3. If the function is periodic, we can continue plotting the function similar to that represented by the dashed line. Because of the symmetry about the vertical line, the function is even. An even function cannot be expressed by a summation of odd functions. Therefore, for the function shown in Figure C.3, $b_n = 0$ for all values of n.

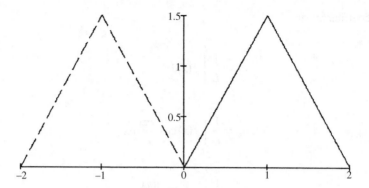

Figure C.3 Illustrative example

The first two coefficients are

$$a_0 = \frac{1}{L}\int_0^{2L} f(x)dx = \frac{1}{1}\left(\int_0^1 1.5xdx + \int_1^2 -1.5x + 3dx \right) = 1.5$$

$$a_1 = \frac{1}{L}\int_0^{2L} f(x)\cos\frac{n\pi x}{L}dx$$

$$= \frac{1}{1}\left(\int_0^1 1.5x \cos(\pi x)dx + \int_1^2 (-1.5x + 3) \cos(\pi x)dx \right) = -0.608$$

The partial series representation $f(x) = a_0/2 + a_1 \cos(\pi x/L)$ using the first two coefficients is shown in Figure C.4.

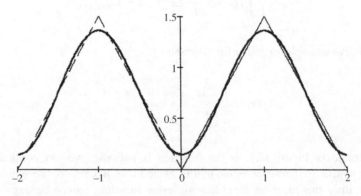

Figure C.4 Fourier series approximation $f(x) = 0.75 - 0.608 \cos(\pi x)$

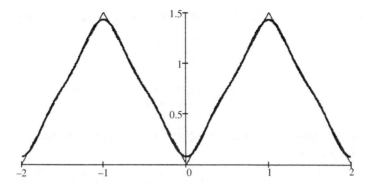

Figure C.5 Fourier approximation $f(x) = 0.75 - 0.608 \cos(\pi x) - 0.061 \cos(3\pi x)$

The approximation can be improved by adding more terms to the series. The next two terms in the series are

$$a_2 = \frac{1}{L}\int_0^{2L} f(x) \cos\frac{n\pi x}{L}\,dx$$

$$= \frac{1}{1}\left(\int_0^1 1.5x \cos(2\pi x)dx + \int_1^2 (-1.5x + 3)\cos(2\pi x)dx\right) = 0$$

$$a_3 = \frac{1}{L}\int_0^{2L} f(x) \cos\frac{n\pi x}{L}\,dx$$

$$= \frac{1}{1}\left(\int_0^1 1.5x \cos(3\pi x)dx + \int_1^2 (-1.5x + 3)\cos(3\pi x)dx\right) = -0.061$$

The partial series representation $f(x) = a_0/2 + a_1 \cos(\pi x/L) + a_2 \cos(2\pi x/L) + a_3 \cos(3\pi x/L)$ is shown in Figure C.5.

C.3 Vibration Response to General Periodic Loading

C.3.1 Undamped Systems

It is common to have a forcing function that is periodic but not harmonic. Some specific examples were discussed in Chapter 10, but Fourier series will allow for a more general solution. Consider a periodic forcing function represented by the Fourier series:

$$F(t) = \frac{f_0}{2} + \sum_{n=1}^{\infty} f_n \sin(n\omega t + \phi_n) \tag{C.11}$$

The constants are now f_n and ϕ_n, which can be determined using the Fourier constants:

$$f_n = \sqrt{a_n^2 + b_n^2} \qquad \text{(C.12)}$$

$$\phi_n = \tan^{-1}\left(\frac{a_n}{b_n}\right) \qquad \text{(C.13)}$$

For an undamped system, the steady-state response to the individual force excitation $f_n \sin(n\omega t + \phi_n)$ will be

$$x_n = \frac{f_n/k}{1 - (n\omega/\omega_n)^2} \sin(n\omega t + \phi_n) \qquad \text{(C.14)}$$

The total system response will be the sum of all individual responses:

$$x = \sum_{n=1}^{\infty} \frac{f_n/k}{1 - (n\omega/\omega_n)^2} \sin(n\omega t + \phi_n) \qquad \text{(C.15)}$$

It is very important to note that the subscript in ω_n denotes natural frequency and does not change with the summation.

Index

Machine Analysis with Computer Applications for Mechanical Engineers, First Edition. James Doane.
© 2015 John Wiley & Sons, Ltd. Published 2015 by John Wiley & Sons, Ltd.